Reconstructing Urban Ambiance in Smart Public Places

Hisham Abusaada
Housing and Building National Research Center (HBRC), Egypt

Ashraf M. Salama
University of Strathclyde, Glasgow, UK

Abeer Elshater
Ain Shams University, Egypt

IGI Global
PUBLISHER of TIMELY KNOWLEDGE

A volume in the Advances in Civil and Industrial Engineering (ACIE) Book Series

Published in the United States of America by
IGI Global
Engineering Science Reference (an imprint of IGI Global)
701 E. Chocolate Avenue
Hershey PA, USA 17033
Tel: 717-533-8845
Fax: 717-533-8661
E-mail: cust@igi-global.com
Web site: http://www.igi-global.com

Library of Congress Cataloging-in-Publication Data

Names: Abusaada, Hisham, editor. | Salama, Ashraf M. A., editor. | Elshater, Abeer, editor.
Title: Reconstructing urban ambiance in smart public places / Hisham Abusaada, Ashraf M. Salama, and Abeer Elshater, editors.
Description: Hershey, PA : Engineering Science Reference, an imprint of IGI Global, [2020] | Includes bibliographical references and index. | Summary: "This book provides contemporary insights into a comprehensive interpretation of urban ambiances in smart places as it relates to the development of cities or to various levels of intervention in extant urban environments. The book also examines the impact of architectural design on the creation of urban ambience in the artworks and how to reflect this technique in the fields of professional architectural practice"-- Provided by publisher.
Identifiers: LCCN 2019059521 (print) | LCCN 2019059522 (ebook) | ISBN 9781799838562 (hardcover) | ISBN 9781799851462 (paperback) | ISBN 9781799838579 (ebook)
Subjects: LCSH: Urban renewal. | City planning. | Smart cities.
Classification: LCC HT170 .R427 2020 (print) | LCC HT170 (ebook) | DDC 711--dc23
LC record available at https://lccn.loc.gov/2019059521
LC ebook record available at https://lccn.loc.gov/2019059522

This book is published in the IGI Global book series Advances in Civil and Industrial Engineering (ACIE) (ISSN: 2326-6139; eISSN: 2326-6155)

British Cataloguing in Publication Data
A Cataloguing in Publication record for this book is available from the British Library.

For electronic access to this publication, please contact: eresources@igi-global.com.

Advances in Civil and Industrial Engineering (ACIE) Book Series

Ioan Constantin Dima
University Valahia of Târgovişte, Romania

ISSN:2326-6139
EISSN:2326-6155

Mission

Private and public sector infrastructures begin to age, or require change in the face of developing technologies, the fields of civil and industrial engineering have become increasingly important as a method to mitigate and manage these changes. As governments and the public at large begin to grapple with climate change and growing populations, civil engineering has become more interdisciplinary and the need for publications that discuss the rapid changes and advancements in the field have become more in-demand. Additionally, private corporations and companies are facing similar changes and challenges, with the pressure for new and innovative methods being placed on those involved in industrial engineering.

The **Advances in Civil and Industrial Engineering (ACIE) Book Series** aims to present research and methodology that will provide solutions and discussions to meet such needs. The latest methodologies, applications, tools, and analysis will be published through the books included in **ACIE** in order to keep the available research in civil and industrial engineering as current and timely as possible.

Coverage

- Materials Management
- Coastal Engineering
- Earthquake engineering
- Production Planning and Control
- Optimization Techniques
- Structural Engineering
- Hydraulic Engineering
- Urban Engineering
- Quality Engineering
- Operations Research

IGI Global is currently accepting manuscripts for publication within this series. To submit a proposal for a volume in this series, please contact our Acquisition Editors at Acquisitions@igi-global.com or visit: http://www.igi-global.com/publish/.

The Advances in Civil and Industrial Engineering (ACIE) Book Series (ISSN 2326-6139) is published by IGI Global, 701 E. Chocolate Avenue, Hershey, PA 17033-1240, USA, www.igi-global.com. This series is composed of titles available for purchase individually; each title is edited to be contextually exclusive from any other title within the series. For pricing and ordering information please visit http://www.igi-global.com/book-series/advances-civil-industrial-engineering/73673. Postmaster: Send all address changes to above address.

Titles in this Series

For a list of additional titles in this series, please visit:
https://www.igi-global.com/book-series/advances-civil-industrial-engineering/73673

Multi-Objective Optimization of Industrial Power Generation Systems Emerging Research and Opportunities
Timothy Ganesan (Royal Bank of Canada, Canada)
Engineering Science Reference • © 2020 • 233pp • H/C (ISBN: 9781799817109) • US $195.00

Impact of Industry 4.0 on Architecture and Cultural Heritage
Cecilia Maria Bolognesi (Politecnico of Milano, Italy) and Cettina Santagati (Università di Catania, Italy)
Engineering Science Reference • © 2020 • 422pp • H/C (ISBN: 9781799812340) • US $225.00

Handbook of Research on Urban-Rural Synergy Development Through Housing, Landscape, and Tourism
Aleksandra Krstić-Furundžić (University of Belgrade, Serbia) and Aleksandra Djukić (University of Belgrade, Serbia)
Engineering Science Reference • © 2020 • 437pp • H/C (ISBN: 9781522599326) • US $275.00

Re-Coding Homes Through Flexible Interiors Emerging Research and Opportunities
Nilüfer Saglar Onay (Politecnico di Torino, Italy) S. Banu Garip (Istanbul Technical University, Turkey) and Ervin Garip (Istanbul Technical University, Turkey)
Engineering Science Reference • © 2020 • 165pp • H/C (ISBN: 9781522589587) • US $175.00

Green Building Management and Smart Automation
Arun Solanki (Gautam Buddha University, India) and Anand Nayyar (Duy Tan University, Vietnam)
Engineering Science Reference • © 2020 • 312pp • H/C (ISBN: 9781522597544) • US $215.00

Handbook of Research on Implementation and Deployment of IoT Projects in Smart Cities
Krishnan Saravanan (Anna University Chennai – Regional Office Tirunelveli, India) Golden Julie (Anna University, India) and Harold Robinson (SCAD College of Engineering and Technology, India)
Engineering Science Reference • © 2019 • 415pp • H/C (ISBN: 9781522591993) • US $295.00

Handbook of Research on Digital Research Methods and Architectural Tools in Urban Planning and Design
Hisham Abusaada (Housing and Building National Research Center, Egypt) Carsten Vellguth (German Academic Exchange Service, Germany) and Abeer Elshater (Ain Shams University, Egypt)
Engineering Science Reference • © 2019 • 445pp • H/C (ISBN: 9781522592389) • US $265.00

Recycled Waste Materials in Concrete Construction Emerging Research and Opportunities
Jahangir Mirza (Research Institute of Hydro-Québec, Canada) Mohd Warid Hussin (Universiti Teknologi Malaysia, Malaysia) and Mohamed A. Ismail (Miami College of Henan University, China)
Engineering Science Reference • © 2019 • 169pp • H/C (ISBN: 9781522583257) • US $165.00

IGI Global
PUBLISHER of TIMELY KNOWLEDGE

701 East Chocolate Avenue, Hershey, PA 17033, USA
Tel: 717-533-8845 x100 • Fax: 717-533-8661
E-Mail: cust@igi-global.com • www.igi-global.com

Editorial Advisory Board

Table of Contents

Foreword xiii

Preface xx

Acknowledgment xxvi

Chapter 1
The Atmospheres of Urban Environments: Ambiguity and Plausibility in Urban Studies 1
Hisham Abusaada, Housing and Building National Research Center (HBRC), Egypt

Chapter 2
The Concept of Atmospheres: From Goethe to Bratton – How Atmosphere Is Key to Creating Smart Cities 23
Diana Soeiro, Centre for Socioeconomic and Territorial Studies (DINÂMIA'CET), ISCTE, University of Lisbon, Portugal

Chapter 3
Affective Atmospheres, Essence of Architecture, and Spirit of Place 41
Hisham Abusaada, Housing and Building National Research Center (HBRC), Egypt
Abeer Elshater, Ain Shams University, Egypt

Chapter 4
Potentials of ICT and IoT in the Design of Smart Public Spaces 60
Rafee Ibrahim Hakky, Al Rasheed University, Damascus, Syria

Chapter 5
Transformation of Urban Space by Smart Technologies: Evidence and Traces From Istanbul Historical Peninsula 82
Burak Belli, Istanbul Technical University, Turkey
Turgay Kerem Koramaz, Istanbul Technical University, Turkey

Chapter 6
Towards a Smart Neighbourhoods Simulated Model 114
Nada M. Alhakkak, Baghdad College of Economic Sciences University, Iraq

Chapter 7
Impact of Urban Ambiance on the Productivity of the Built Environment Using Cluster Analysis: Sustainable Built Environment 126
Saurav Dixit, RICS School of Built Environment, Amity University, Noida, India

Chapter 8
Theorizing the Critical Placemaking: Impact of Creative Ambiance on Social Activities 145
Aya Elgobashi, Arab Academy for Science, Technology, and Maritime Transport, Egypt
Yasmeen Mohamed Elsemary, Arab Academy for Science, Technology, and Maritime Transport, Egypt

Chapter 9
The Gastronomy of the Eye: A Tale of Two Cities in India 175
Vrushali Dhage, ArtForum SF, USA

Chapter 10
Combinatorial Architecture: Urban Causality, Creativity, and Socialisation 199
Caterina Tiazzoldi, Rhode Island School of Design, USA
Abeer Elshater, Ain Shams University, Egypt

Chapter 11
Situational Atmospheres: Filmmaking Approaches for Conveying Urban Design 221
Hisham Abusaada, Housing and Building National Research Center (HBRC), Egypt
Abeer Elshater, Ain Shams University, Egypt

Compilation of References 246

About the Contributors 276

Index 279

Detailed Table of Contents

Foreword xiii

Preface xx

Acknowledgment xxvi

Chapter 1
The Atmospheres of Urban Environments: Ambiguity and Plausibility in Urban Studies 1
Hisham Abusaada, Housing and Building National Research Center (HBRC), Egypt

This chapter investigates the ambiguity of the word "atmospheres" in the fields of urban studies. It examines the justifications (plausibility) beyond its uses, with the terms that are focusing on the perceptual qualities. The author investigated the uses of the word "atmospheres" from the beginning of the 17th century to the year 2020, a period which he divided into four stages. The investigation covered the work of 27 thinkers in the fields of natural sciences and humanities, including 10 in architecture disciplines, in addition to 28 manuscripts that addressed the relationship between atmospheres in the areas of architecture, particularly urban planning and design and urban landscape architecture between 1998 and 2020. The outcomes were developed through a comprehensive literature review by gaps analysis and a deductive online survey with 58 specialized participants, using SurveyMonkey. This chapter contributes to the rationale that an urban designer can use to study people's changing feelings, emotions, and moods according to the understanding of the terms related to atmospheres.

Chapter 2
The Concept of Atmospheres: From Goethe to Bratton – How Atmosphere Is Key to Creating Smart Cities 23
Diana Soeiro, Centre for Socioeconomic and Territorial Studies (DINÂMIA'CET), ISCTE, University of Lisbon, Portugal

Architects and urban planners currently face the challenge of the Fourth Industrial Revolution (4IR) that quickly permeates urban environments. In an effort to understand this event as one that goes beyond technology use and technological innovation, the authors investigate how the concept of atmosphere is key to envisioning the future of smart cities. In order to clarify the understanding of the concept, this chapter presents an original alignment of three clusters of authors (Goethe and Wittgenstein, Böhme and Schmitz, and Bratton). The goal is to demonstrate that atmospheres and urban ambiances are fundamental urban design elements. They have the ability to positively shape technology use in cities as decisive elements to promote sustainable smart cities.

Chapter 3
Affective Atmospheres, Essence of Architecture, and Spirit of Place 41
Hisham Abusaada, Housing and Building National Research Center (HBRC), Egypt
Abeer Elshater, Ain Shams University, Egypt

This chapter reviews the implications of using the words "essences" and "spirit" in urban studies and their link with the concept of affective atmospheres in the realms of architecture. Two assumptions are valid when this matter is addressed. The first is that, despite affective atmospheres being considered as the fifth dimension in architecture, they are not their essence or spirit. The second is that these atmospheres impersonate a crucial role in reconstructing different urban environments, which are based on the perceptual dimension. The purpose of this chapter is to provide an informed view bibliographically and conceptually about distinguishing between essence, spirit, and affective atmospheres. The chapter also provides an analysis of the concept of affective atmospheres to verify the hypothesis. The conclusion is latent in the possibility that the expression "affective atmospheres," instead of "essence" or "spirit," can be used for referring to people's emotional impressions in urban environments as a fifth dimension.

Chapter 4
Potentials of ICT and IoT in the Design of Smart Public Spaces 60
Rafee Ibrahim Hakky, Al Rasheed University, Damascus, Syria

This study aims to show that information and communication technologies (ICT) and the internet of things (IoT) can be used in order to improve the atmosphere of a space. This chapter is based on the assumption that if ICT and IoT are used to better the physical environment of the space, its atmosphere will also be enhanced. A summary of the design of public open spaces and how current ICT and IoT can help in this regard is presented. This theoretical background was the base for a proposal that uses current technology to allow constant changes to the physical layout of a space to accommodate the changing needs of users and improve the space's atmosphere. The proposal advanced here is intentionally simple in terms of its idea and its execution. It is also possible to be tested at any scale a designer or a researcher would prefer. The different and economically conscious experiments will lead to continuous improvements to the proposal.

Chapter 5
Transformation of Urban Space by Smart Technologies: Evidence and Traces From Istanbul Historical Peninsula 82
Burak Belli, Istanbul Technical University, Turkey
Turgay Kerem Koramaz, Istanbul Technical University, Turkey

This chapter evaluates the Istanbul Historical Peninsula, which has been the historical, cultural, and financial core of the city, in its quest for becoming smart(er) with reference to the five principles that are put forth by Mitchell, in his 1999 book e-topia. The chapter aims to define the role of ICTs in urban planning and management of Istanbul Historical Peninsula by depicting its fields of use in redesigning the societal, financial, and cultural aspects of urban life in the area. Additionally, the transformative feature of these technologies, used in urban planning, conservation, and heritage management, will be validated whether they serve for a vision and a strategic framework of transitioning the community into an information society.

Chapter 6
Towards a Smart Neighbourhoods Simulated Model 114
Nada M. Alhakkak, Baghdad College of Economic Sciences University, Iraq

The literature generally describes a conventional neighbourhood as an urban area that consists of houses, schools, shopping, religious buildings, and open spaces. This work focuses on the challenge of establishing smart neighbourhoods. The contribution here is in presenting a new model for a smart neighbourhood that contains two frameworks. The first framework is related to input specific parameters from the user and gets reports for timetable, map, and materials. The second framework uses the virtual reality aspect and compound of multiple layers. The concluding remarks figured out that the importance of using the model's work depends on reusing and ABC clustering algorithm.

Chapter 7
Impact of Urban Ambiance on the Productivity of the Built Environment Using Cluster Analysis: Sustainable Built Environment 126
Saurav Dixit, RICS School of Built Environment, Amity University, Noida, India

Productivity is the key measure of success and efficiency of any process or activity, and it is dependent on a number of factors. The ambiance of cities and the use of latest technology and environment also affect the productivity of the residents working in the city. The objective of this chapter is to identify the attributes of urban ambiance from the literature review and to analyse the impact of satisfactory urban/city ambiance on the productivity of the built environment. The methodology adopted for the study is to collect the data using a structured questionnaire survey on the impact of urban ambiance on the productivity form the different cities of the world. The collected data shall be analysed using reliability analysis, relative importance index, exploratory factor analysis, and cluster analysis. The finding of the study shall provide an understanding of urban ambiance on the productivity of the built environment and the ways and means to improve the situation.

Chapter 8
Theorizing the Critical Placemaking: Impact of Creative Ambiance on Social Activities 145
Aya Elgobashi, Arab Academy for Science, Technology, and Maritime Transport, Egypt
Yasmeen Mohamed Elsemary, Arab Academy for Science, Technology, and Maritime Transport, Egypt

All over the world, lots of challenges are affecting the urban development of spaces in the cities, especially spaces attached to the heritage site. It provides various ambiances that touch the senses of the users. Although the attached public space to heritage sites is full of rich, deep knowledge and plays an active role in the community, it was neglected. Those spaces contribute to the entrenchment of culture and social heritage of the community to let future generations get benefit from it. In this chapter, the researchers theorize the "critical placemaking" with creative ambiance as a tool for successful attached to public spaces. This has been drawn from key concepts in environmental psychology of the users such as physical environment, atmosphere, and services environment. The proposed framework seeks to theorize the processes of placemaking and its key concepts and how it helps in the preservation process. The argument that the placemaking process needs a creative ambiance, which contributes to transforming community narratives and that space, becomes more inclusive.

Chapter 9
The Gastronomy of the Eye: A Tale of Two Cities in India 175
Vrushali Dhage, ArtForum SF, USA

Works of art can be read at various levels: from being objects of simple retinal pleasure to the other extreme of being significant critical statements of their time. This chapter aims to strike a cerebral dialogue through the works of art. The current study shall consider the latter function of art and analyze the methods in which contemporary Indian artists have made attempts to provide a critique of the early initiatives towards developing Delhi and Mumbai as 'smart cities'. The review of works from India concludes the essential role of infrastructural projects and envisioned spaces built in the era of economic liberalization. The study aims at drawing a methodological approach, with an art historical perspective, with the artists analysing and translating the urban experiential phenomenon, into artworks.

Chapter 10
Combinatorial Architecture: Urban Causality, Creativity, and Socialisation 199
Caterina Tiazzoldi, Rhode Island School of Design, USA
Abeer Elshater, Ain Shams University, Egypt

At a time when laptops and Wi-Fi connections proliferate, it is possible to work and socialise anywhere. The pressing issue is how people should perceive their ambiance when released from physical interactions in public. This chapter addresses the possibility of designing a space that combines users' plurality with the coherence of the entire design. Argumentatively, the authors show professional and educational case studies with a mixture of physical and virtual environments, exploring their impact in the reconstruction of urban ambiance. These cases are organised into groups, moving from the most straightforward implementations of technology in order to create a new realm of interactions between the physical and virtual. A review of these cases showed collaborations between different institutions and artists. The concluding remarks call for the identification of a set of transdisciplinary design actions. These actions could respond to the transformations of public space with cooperation between academics and professionals.

Chapter 11
Situational Atmospheres: Filmmaking Approaches for Conveying Urban Design 221
Hisham Abusaada, Housing and Building National Research Center (HBRC), Egypt
Abeer Elshater, Ain Shams University, Egypt

This chapter examines how placemaking and filmmaking can improve teaching urban design-based situations of atmospheres. The qualitative-deductive form of content analysis addresses atmospheres in four Egyptian and American films and thus creates a conceptual framework for design strategy in studio of urban design. In this chapter, the authors discuss this framework with regard to the works of the Egyptian novelist Naguib Mahfouz. They show that filmmaking and placemaking are similar in the intellectual discourse because both produce ideas, themes, and events together with architectural elements and experience of everyday life. The findings highlight the impact of local architecture on the quantitative and nonquantitative factors that create affective atmospheres. The theoretical argument proved that the process of filmmaking could constitute a method for teaching in urban design studios and urban design history. The conclusion gives an action plan to use cinematic works as a tool for enhancing teaching affective atmospheres in urban-design studios.

Compilation of References 246

About the Contributors 276

Index 279

Foreword

Experiencing and Inhabiting Space: Multi-Sensory Atmospheres in Lived Spaces

The Hegemony of Vision

The characteristics of urban and architectural entities, places and spaces, are usually described and analyzed in geometric, formal and material terms. In fact, visual forms and qualities, as well as their relationships, have been the predominant interest in the modern planning and design theories and practices. The vision-centered attitude is exemplified by Le Corbusier's credo: "Architecture is the masterly, correct and magnificent play of the masses brought together in light" (LeCorbusier, 1927, p. 31). This confession concretizes the hegemonic position of vision, which has been supported by the tradition of western philosophy all the way since Aristotle´s hierarchy of the senses (R.Flynn, 1993, p. 274).

Le Corbusier's many other exclamations, such as: "I am and I remain an impenitent visual – everything is in the visual" (LeCorbusier, 1991, p. 7), and "Architecture is a plastic thing. I mean by "plastic" what is seen and measured by the eye" (ibid., 191), underline the vision-centered attitude, which even had ethical connotations, as in his notion of "the eye of truth" [1]. When quoting Le Corbusier as an example of sensory reductionism, I need to emphasize the undeniable artistic and existential power of his architectural masterworks. Declarations by other leading modernist architects reveal that the obsession with vision was not Le Corbusier's alone. "He [the designer] has to adapt knowledge of the scientific facts of optics and thus obtain a theoretical ground that will guide the hand giving shape, and create an objective basis" (Walter Gropius, 1956, pp. 15-25), Walter Gropius, director of the Bauhaus School, claimed. Laszlo Moholy-Nagy, one of the reformers of artistic thinking at the Bauhaus, supported the progressive nature of vision equally enthusiastically: "The hygiene of the optical, the health of the visible is slowly filtering through" [2]. The preference for immaterial whiteness is another example of interpreting visual qualities in symbolic and moral terms.

Whiteness is extremely moral. Suppose there were a decree requiring all rooms in Paris to be given a coat of whitewash. I maintain that that would be a police task of real stature and a manifestation of high morality, the sign of great people. (LeCorbusier, 1925)

Whiteness and the importance of daylight also reflected hygienic ideas of Louis Pasteur from the 1870s. As the modernist interest center on pure form, only focused vision has been considered, while the role of unfocused, peripheral vision, and shapeless atmospheres, feelings and moods has rarely been recognized, not to speak of the essential interaction of all the senses. This conscious limitation gave the Gestalt laws their dominant position in understanding visual and spatial perception, as well as the appreciation of architectural form. This focus on form also regarded spatial experiences as geometric, immaterial and negative forms. The visual and geometric understanding of space has had especially negative consequences in urban planning. As a consequence of this intellectual and perceptual bias, pleasurable modern and contemporary urban spaces, squares and streetscapes are rare anywhere in the world.

Atmospheres[3] and Multi-Sensory Experience

Our real experiences of the world are inherently multi-sensory, peripheral, fragmented, and formally vague. In *The Experience of Place*, Tony Hiss (1991) uses the notion "simultaneous perception" of the perceptual mode that we use in experiencing surroundings. On the other hand, Paul Klee wrote about the significance of dispersed attention, the polyphonic structure of artistic works, and a "multi-dimensional attention", when describing the essential sensory multiplicity in artistic perception [4]. However, this is also the way we normally observe, with all the senses at once, as Maurice Merleau-Ponty (1964) notes:

My perception is . . . not a sum of visual, tactile, and audible givens: I perceive in a total way with my whole being: I grasp a unique structure of the thing, a unique way of being, which speaks to all my senses at once. (p. 48)

In real life experiences, the integrated sensations have a more significant role than vision alone, and our ability of sensing extends beyond the five Aristotelian senses.

Sonic and tactile qualities, as well as smells and even ideated tastes, are significant factors in the experience of reality. Each architectural and urban situation is a condition of sensitized reality, not an artistic fantasy. An especially significant sensory quality is the unconscious or ideated touch concealed in vision. "Through vision we touch the sun and the stars", Merleau-Ponty suggests poetically [5]. Entire landscapes or regions can be sensuously dominated even by smells, as in the case of distinct agricultural practices or polluting industries. Smells are connected with our primal levels of memory and they can thus evoke vivid recollections of places and events, as Marcel Proust's (1996) *In Search of Lost Time* claims repeatedly. The ears, nose and skin (experiences of temperature, moisture, wind) can create atmospheric experiences that are as domineering as any visually defined situations.

Regardless of the fact that already Greek thinking assigned vision its hegemonic role among the senses, vision acquired its dominant role rather late. "The sixteenth century did not see first: it heard and smelled, it sniffed the air and caught sounds. It was only later that it seriously and actively became engaged in geometry [...] It was then that vision was unleashed in the world of beauty as well", Lucien Febvre argues [6]. Robert Mandrou, another historian, makes a parallel argument: "The hierarchy [of the senses] was not the same [as in the twentieth century] because the eye, which rules today, found itself in third place, behind hearing and touch, and far after them. The eye that organises, classifies and orders was not a favoured organ of a time that preferred hearing" [7].

The Significance of Atmosphere

The experiential character of a space or place is not merely a visual perceptual quality. In fact, focused vision makes us outsiders in relation to what we are seeing in focus. Focused vision covers only 4 percent of the entire field of vision, and the experience of enveloping space has to arise from the dynamics of the entire visual field rather than focused and restricted perception. The sense of spatial reality and insideness calls for unfocused, peripheral, enfolding and fused perceptions and interactions of various sense experiences. This is our reality sense. I wish to argue, that the spatial experience of reality is existential rather than visual; we interact with space with our whole being and sense of self.

Atmosphere or ambience is a kind of a virtual, experiential and multi-sensory condition, which usually has shapeless, undefineable and ephemeral boundaries, and unfocused experiential qualities. An atmosphere can also have dynamically changing characteristics and varying durations, as in the cases of weather, natural illumination or musical experiences. Atmospheres could thus be regarded as limit cases of "placeness", or as "quasi-places" [8]. An atmosphere gives rise to the experience of a specific location or place with distinct experiential qualities and emotive attributes. We are drawn to specific places by their enticing character, ambience and feeling, which are evidently more complex, integrated and dynamic entities than form.

The biased orientation towards focused vision and clear form has regrettably guided the modern architect's conscious interest away from the essential experiential realities and qualities of atmosphere, mood and tuning. Architects have regarded atmospheres as something romantic, entertaining or, perhaps, even kitsch, something that has its place in set design, restaurant and shop decoration, but not in serious architecture.

Multi-Sensory Fusion

The judgement of environmental character is a complex, multi-sensory fusion of numerous factors, which are, however, immediately and synthetically grasped as an overall atmosphere or feeling. The immediate sensing of atmosphere and mood takes place mostly unconsciously. John Dewey, the visionary American philosopher (1859-1952), who already eight decades ago understood the immediate, embodied, emotive, and largely subconscious essence of experience, describes the nature of such existential encounters as follows:

The total overwhelming impression comes first, perhaps in a seizure by a sudden glory of the landscape, or by the effect upon us of entrance into a cathedral when dim light, incense, stained glass and majestic proportions fuse in one indistinguishable whole. We say with truth that a painting strikes us. There is an impact that precedes all definite recognition of what it is about.[9]

We can be similarly struck by a setting, place or space, and the presence and the feeling that it exudes can have an overpowering impact on our mental state, emotive tuning and actions. This unconscious emergence of mood is often the most significant effect of a space or place. It is likely that non-architects sense primarily the atmosphere of a place or building, whereas a formal attention implies a distinct intellectual and theoretical position.

Atmospheres are not necessarily consequences of conscious design intention; the pleasurable ambiences of vernacular settings, for instance, are usually unintentional consequences of traditional building practices and the fusion of man-made constructions and their settings. Even the multifarious and diffuse components of weather create distinct emotive states and moods at a glance. Also the experience of weather has a diffuse and peripheral character, instead of consisting of focused images with clear boundaries or shapes. Weather is a complex experience with a strong qualitative tuning, similar to a landscape, urban or architectural space. These are all enfolding and embracing experiences that encircle us, and in which we are insiders and participants, not mere onlookers. In man-made settings, atmosphere is the over-all fused and enfolding inside experience of the place, space or situation.

Sensing Atmospheres

An atmospheric perception involves judgements beyond the five Aristotelian senses, such as sensations of orientation, motion, duration, continuity, scale, density, intimacy, temperature, humidity, air movement, and the dynamics of illumination. Indeed, the judgement of the character of a space calls for our entire embodied and existential sense, our sense of being. It is perceived in a diffuse and peripheral manner rather than through precise, focused and conscious observation. This complex assessment also includes the dimension of time as atmospheres can be dynamically changing, and all experiences imply duration, and they fuse perception, memory and imagination. All live experiences glide along a temporal axis between memorizing, perceiving and imagining. Moreover, every space and place is always an invitation to and suggestion of distinct acts. Spaces and true architectural experiences are inherently verbs and promises; the floor is a suggestion to stand up, move around, the door is an invitation to enter and exit, the window a call to look out and expand the interior space to the world, and the stair a suggestion to ascend or descend to another level.

In addition to environmental atmospheres, there are cultural, social, work place, family, etc. interpersonal atmospheres, which all have similarly enfolding characteristics. We can even speak of specific atmospheres in the scales of cultural, regional or national entities. Even countries project their identifiable atmospheres deriving from the characteristic landscapes, typical weather, architectural traditions, conventions of dressing and behaviour, food culture, tone of language, etc.. Powerful atmospheric experiences can feel like an embracing touch of a veil or a lover, and they address directly our sense of being. Altogether, I suggest that instead of vision, our most important environmental and architectural sense is our existential sense, the sense of being.

Genius loci, the Spirit of Place, is similarly an ephemeral, unfocused and non-material experiential feeling or tuning, that is closely related to atmosphere ; we can, indeed, speak of the atmosphere or "spirit" of a place, which give the place its unique experiential identity and quality. *Genius loci* is usually attached to the permanent material, formal and dimensional aspects of the place. Yet, a place can have dynamically changing atmospheres; a great theater hall, for instance, has its own, more or less permanent ambience, and the changing imaginative atmospheres of the performances, and these two different atmospheres interact. I am here thinking of Peter Brook's Bouffes-du-Nord theatre space in Paris, located in an classical theatre space, parts of which the director intentionally demolished in order to sensitize the ruined space for the varying emotional situations of theatre performance (Todd & Lecat, 2003, p. 25).

Spatial Experience as an Exchange

We tend to think of space and place as something outside of ourselves, as external contexts for our existence and events of life. Yet, space and the self are fused and inseparable, as the notions of space and place are constituted through experience. Similarly, the aesthetic dimension actualizes in the very experience of the phenomenon or artistic work. Martin Heidegger (1997) linked space indivisibly with the human condition:

When we speak of man and space, it sounds as though man stood on one side, space on the other. Yet space is not something that faces man. It is neither an external object nor an inner experience. It is not that there are men, and over and above them space. (p. 334)

Spatial experience is always a kind of an exchange; as I enter a space, the space enters me, and the experience is essentially a fusion of the object and the subject. The American literary scholar, Robert Pogue Harrison, writes poetically:

In the fusion of place and soul, the soul is as much of a container of place as place is a container of soul, both are susceptible to the same forces of destruction. (Harrison, 2008, p. 130)

Atmosphere is similarly an exchange between the material and immaterial properties of the place or situation, on the one hand, and the imaginative realm of human perception and mind, on the other. Yet, an atmosphere does not consist of physical "things" or facts; it is a human experiential entity, or an imaginative "creation". A musical piece is similarly an experiential atmospheric entity without factual physical or formal features. Yet, it has an experiential interior space and specific tuning. Edward Relph's notion "empathic insideness", gives rise to such a sensation of belonging. Relph (1976) uses the notions of "placelessness" and "existential outsideness" of spatial situations that do not project an inviting image, a coherent experiential meaning of a sense of interiority. Transitional spaces, such as airports, exemplify non-places, that do not have the capacity of envoking a sense of rootedness or belonging, or any kind of domesticity. Also, the atmospheres of human interactions seem to have their own dynamics, which are usually beyond conscious control. A minute emotional dissonance can easily shatter a friendly encounter.

In accordance with Dewey's suggestion, we grasp the atmosphere of a place before we identify its details or understand it intellectually. In fact, we may be completely unable to say anything meaningful about the characteristics of a place or a situation, yet we have a firm image, emotive attitude, and recall of it. In the same way, although we do not consciously analyse or understand the interaction of meteorological factors, we grasp the essence of weather at a glance, and it conditions our mood and intentions. As we enter a new city, we grasp its overall character without having consciously identified or analysed any of its countless material, geometric, dimensional and sensory properties. John Dewey (1998) even extends processes that advance from an initial, but temporary grasp of the whole towards details, all the way to the processes of thinking:

All thought in every subject begins with just such an unanalysed whole. When the subject matter is reasonably familiar, relevant distinctions speedily offer themselves, and sheer qualitativeness may not remain long enough to be readily recalled. (Dewey, 1998, p. 197)

Recent research in the neurosciences provides support to the assumption that contrary to our inherited belief, perceptions and understanding advance from entities down to details. "According to the right hemisphere, understanding is derived from the whole, since it is only in light of the whole that one can truly understand the nature of the parts", philosopher and therapist Iain McGilchrist (2009, p. 184) asserts.

"Understanding" The Lived Image

In architectural education we are usually advised to develop our designs from elementary aspects towards larger entities, but as I have suggested, our perceptions and experiential judgements advance in the reverse manner, from the entity down to details. When experiencing a work of art, the whole gives meaning to the parts, not the other way round. It is evident that we need to grasp and conceive complete poetic images and atmospheres instead of singular formal elements, and in fact, there are hardly any "elements" in the world of artistic expression; there are only complete poetic images, intertwined with distinct emotive charges and orientations.

As I argued earlier, we have been taught to conceive, observe and evaluate architectural spaces and settings primarily as collections of "elements", or formal and aesthetic entities. Yet, the diffuse ambience is often more decisive and powerful in determining our attitude to the setting or place. Even buildings and details that hardly possess any noteworthy aesthetic values, can create a sensorially rich and pleasant atmosphere. Vernacular settings and traditional towns are frequently examples of pleasant atmospheres that often arise from aesthetically rather uninteresting components. Such urban atmospheres are most often created by specific materiality, scale, rhythm, colour, or other themes with variations. Materials, colours, rhythms and illuminations are strongly atmospheric, due to their embodied, haptic and enveloping nature. On the contrary, form and formal cohesion seem to have a closing and externalising effect instead of an inviting embracing and emotional impact.

In order to create inviting and accommodating architectural and urban spaces, we as architects need to shift from formal thinking to experiential and empathic thinking. In education, empathic imagination should be trained instead of formal visual imagination.

Juhani Pallasmaa
Independent Researcher, Helsinki, Finland

REFERENCES

Dewey, J. (1998). Qualitative thought. In L. A. Hichman & T. M. Alexander (Eds.), The Essential Dewey: Volume 1: Pragmatism, Education, Democracy (pp. 195-205). Bloomington, IN: Indian University Press.

Ehrenzweig, A. (1970). *The Hidden Order of Art.* Paladin.

Flynn, R. T. (1993). Foucault and the Eclipse of Vision. In D. M. Levin (Ed.), Modernity and the Hegemony of Vision (pp. 273-286). Berkeley, CA: University of California Press.

Griffero, T. (2017). *Quasi-Things, The Paradigm of Atmospheres.* Sunny Press.

Harrison, R. P. (2008). *Gardens: An Essay on the Human Condition.* The University of Chicago Press. doi:10.7208/chicago/9780226317861.001.0001

Heidegger, M. (1997). *Building, Dwelling, Thinking", Basic Writings.* Harper & Row.

Hiss, T. (1991). *The Experience of Place.* Random House Inc.

Jay, M. (1994). *Downcast Eyes: The Denigration of Vision in Twentieth Century French Thought.* University of California Press.

Johnson, M. (2007). *The Meaning of the Body: Aesthetics of Human Understanding*. The University of Chicago Press. doi:10.7208/chicago/9780226026992.001.0001

LeCorbusier. (1925). *L'art decoratif dáujourd'hui*. Edition G.Grès et Cie.

LeCorbusier. (1927). *Towards a New Architecture*. The Architectural Press.

LeCorbusier. (1991). *Precision*. The MIT Press.

McGilchrist, I. (2009). *The Master and HIs Emissary: The Divided Brain and the Making of the Western World*. Ysle University Press.

Merleau-Ponty, M. (1964). *The Film and the New Psychology", Maurice Merleau-Ponty, Sense and Non-Sense*. Northwestern University Press.

Mostafavi, M., & Leatherbarrow, D. (1993). *On Weathering: The Life of Buildings in Time*. The MIT Press.

Proust, M. (1996). *In Search of Lost Time*. The Random House Group Limited.

Relph, E. (1976). *Place and Placelessness*. Pion Limited.

Sontag, S. (1986). *On Photography*. Penguin Books.

Todd, A., & Lecat, J.-G. (2003). *The Open Circle: Peter Brook's Theatre Environments*. Palgrave MacMillan.

Walter Gropius. (1956). *Architektur*. Fischer.

ENDNOTES

1 Le Corbusier as quoted in Mohsen Mostafavi and David Leatherbarrow, *On Weathering* (Cambridge, Massachusetts, and London, England: The MIT Press, 1993), 76.

2 As quoted in Susan Sontag, *On Photography* (New York: Penguin Books, 1986), 96.

3 As most writers on the subject, I am using the notions of *atmosphere* and *ambience* as synonyms. The English and German writers tend to prefer the first and the French the second notion.

4 Paul Klee, *The Thinking Eye* (1964), as quoted in Anton Ehrenzweig, *The Hidden Order of Art* (Frogmore, St.Abans: Paladin, 1970), 39.

5 As quoted in Martin Jay, *Downcast Eyes: The Denigration of Vision in Twentieth Century French Thought* (Berkeley, California: University of California Press, 1994).

6 Lucien Febvre, as quoted in Martin Jay, *Downcast Eyes – The Denigration of Vision in Twentieth-Century French Thought* (Berkeley and Los Angeles, California: University of California Press, 1994), 34.

7 Robert Mandrou, as quoted in Martin Jay, (ibid., 34-35).

8 The notion of "quasi-place" refers to the title of Tonino Griffero's book Quasi-Things, The Paradigm of Atmospheres (New York: Sunny Press, 2017).

9 John Dewey, Art As Experience (1934) as quoted in Mark Johnson, The Meaning of the Body: Aesthetics of Human Understanding (Evanston, Illinois: The University of Chicago Press, 2007), 75.

Preface

In recent years, the notion of 'smartness' as it relates to the urban environment, the incorporation of smart technologies in urban places, and the associated impact on urban forms and civic life has become one of the widely acknowledged and growing challenges in urban design and planning and design. Within the contemporary body of knowledge, which pertains to civic design at different scales from small urban places to smart cities, scholars from different disciplinary backgrounds started to discuss the profound implications of the use of new technologies (Abusaada, Vellguth, & Elshater, 2019). This is coupled with continuous attempts are made to understand the sensory and perceptual experiences of users within the urban environment in various contexts (Abusaada, 2020; Salama & Azzali, 2015; Salama, Remali, & MacLean, 2017). This provides opportunities to develop a contemporary discourse that aims at understanding the overarching idea of ambiance within the public realm, especially when having technologies in users' hands (McKenna, 2019).

This book aims to develop new insights into a holistic interpretation of urban ambiance as it relates to the development of new cities or as part of intervention strategies in extant urban environments. The key objective is to provide a comprehensive theoretical base related to urban ambiance and smart public places, which stems from the various conceptual positions and scholarly understandings of contemporary urban scholars. Premised on empirical and experimental research, the theoretical discourse presented in this book is coupled with analyses of a series of cases, which address various contexts and capture the essence of urban ambiance. In principle, smartness as a concept is operationalized in a manner that is not limited to emerging technologies, but incorporates speaking to the users' senses and cognition. *Reconstructing Urban Ambiance in Smart Public Places* embraces the belief that the "ambiances" or "atmospheres" represent the non-physical dimension of public places, and that they equally have an impact on people's perceptions and the accompanying changes in their attitudes, modes, and feelings (Abusaada & Elshater, 2020; Anderson, 2009; Hsu, 2017; Pallasmaa, 2013; Zumthor, 2006). Like time, the qualities of a meaningful ambiance are invisible and represent a important design dimension that requires more investigation towards an inclusive urban literature.

As part of shaping new urban environments and appropriating existing ones, the value system and attention of architects, urban designers and planners has dramatically departed from an exclusive emphasis on the physical dimension of the environment to an emphasis on the imperatives that respect the notion of ambiance. In essence, ambiance is manifested in the urban environment and is rooted in cultural norms and socio-spatial practices. Distinctively, the public realm in its social, cultural, and architectural sense is the principal focus of this book. Such a focus is pursued across a series of contributions and is undertaken through an exploration of how affective atmospheres or urban ambiances, which have direct relations to users as individuals and in groups, can develop. This is coupled with the postulation that the formation of urban ambiance ensues from various factors that include the way in which buildings and their immediate surroundings have an impact on the public realm. In particular, such an understanding is developed through systematic observations, examining mental images and cognitive processes, and investigating behavioural settings and the associated boundaries and patterns of activities.

Going beyond the approach of discussing smartness through a focus on digital and virtual tools, contributions within this book can be situated within two different lines of focus inquiry and the associated quests. Both place emphasis on examining the urban environment by studying the impact of environmental ambiance on the perception and satisfaction of users. However, one line of inquiry underscores the concepts of urban ambiance or urban atmosphere from various perspectives, including geographical, socio-cultural, perceptual, morphological, and temporal dimensions. Yet, the second line of inquiry utilises the notion of "atmospheres," as integral to the discussion of smart public places.

While debating the preceding two approaches reveals interesting discussions, there appears to be a lack of clarity on using the two terms "ambiance" and "atmospheres." Recent scholarly work has discussed the significance of differentiating and recognizing the similarities and differences, especially when discussing smart places and what these concepts represent for the academic and professional practice communities. Relevant to this, the work within this volume demonstrates the importance of considering atmospheres/ambiance to express people's feelings and that this needs to be adopted in architectural and urban pedagogy. Likewise, ambiguity manifests in the lack of adopting and practicing these concepts in several universities around the world. *Reconstructing Urban Ambiance in Smart Public Places* divulges the dialectics of the two notions "ambiance" and "atmospheres" and the way in which both instigate a discourse on smart public places through several cases from around the world, including Egypt, Italy, India, France, Syria, Turkey, and the United States through the under-mentioned key areas of focus:

- The ambiguity and plausibility of the two terms "ambiance" and "atmosphere" manifest in professional practice.
- The ambiance/atmospheres need to be viewed as critical dimensions in architecture and urbanism.
- There is a reciprocal relationship between the phenomenon and the quality of urban ambiance/atmospheres and the notion of smart public places.
- Adapting planning and design to address urban ambiance/atmospheres is a vital factor to enhance people's satisfaction of the qualities of public places in an era characterised by technological advancements.
- Urban ambiance can be relevant when discussing installation arts and artworks, which are linked to architectural history.

READERSHIP AND TARGET AUDIENCE

The editors believe that developing new debates on urban ambiance, its constituents, attributes, and qualities enable future urban designers and planners to consider new dimensions for rethinking urban places and their users. Primarily, the design of these urban places speaks to human perceptions and feelings while enhancing their effective use. Authors were invited to contribute to a multitude of topics that advance the understanding of urban ambiance while underscoring ideas related to the introduction of smart technologies in public places.

This book is of significance to academic and professional architects and urban designers, as well as researchers, and both graduate and undergraduate students. Whilst there have been very few articles on the key concepts adopted in this book, through various contributions, the book advances the discussion and provides a comprehensive analysis of the notions of urban ambiance, atmospheres, and smartness in public places. The book covers a number of contexts and thus its theoretical interpretations and practical applications unfold geographic and intellectual connection. *Reconstructing Urban Ambiance in Smart Public Places* is of interest to the international college textbook market, aimed at practitioners as well as academics within universities and schools of architecture and urban design and planning that offer professional programs in architecture, urban design and planning, and landscape architecture. Additionally, the book will be useful for a range of readers across all design-based disciplines, and other fields that include education, urban sociology, politics and the broader humanities readership.

ORGANIZATION OF THE BOOK

Reconstructing Urban Ambiance in Smart Public Places includes eleven chapters that engage with the topic in its fullest sense. It starts by establishing the crux; presenting the overarching concept of urban ambiance and its relevant factors through the provision of reliable and comprehensive theoretical tenets derived from the beliefs and arguments made by key thinkers and theorists. This is followed by a series of chapters that move from principles to experiments and from concepts to practices. Hierarchically, the book is structured around three levels of contributions, contributions at the conceptual level (first, second and third chapters), contributions at an implementation level (from chapter four to chapter eight), and contributions relevant to artistic projects (from chapter nine to chapter eleven).

The first three chapters adopt and argue for the concept of urban atmospheres/ambiance as a dimension that is integral to urban planning and design. They represent a form of advocacy that demonstrates how designers could provide an ambiance conducive to human feelings. Progressing into more pragmatic approaches, the subsequent chapters examine the possibility of creating smartness and urban atmospheres/ambiance in the urban environment in a wide spectrum of diverse contexts worldwide. These chapters offer in-depth discussions on the impact of the atmosphere on built environments. Collectively, they place emphasis on the notions of smart cities, smart public spaces, and places within intelligent neighbourhoods. The last three chapters present an attempt to address the way in which feelings change as an outcome of artistic constructions. Within the American, Egyptian, French, Indian and Italian context, these chapters link aspects of beauty of place and its relevance in instigating immediate impressions while addressing the associated technological dimensions. They show the significant role of the quality-related artistry while linking affective atmospheres and urban ambiance to cinematic and artistic works that provide contextual narratives.

Written by Abussada, the first chapter is entitled "The Atmospheres of Urban Environments: Ambiguity and Plausibility in Urban Studies." It explores the ambiguity embedded in the meaning of 'atmospheres' as it relates to urban studies. Qualitatively through investigating 27 manuscripts published over the past decades, the results establish the rationale of using the change in people's feelings, emotions and moods for a better understanding of various design considerations. Stemming from a quantitative investigation, the findings verify the same results generated from the critical analysis of the literature, while establishing future research directions of design-based atmospheres.

At a strategic level, written by Soeiro, the second chapter is entitled "The Concept of Atmospheres: From Goethe to Bratton – How Atmosphere Is Key to Creating Smart Cities." It discusses the consequences of the Fourth Industrial Revolution (4IR) while demonstrating its effects on the built environment. She reviews a cluster of key figures including Goethe and Wittgenstein; Böhme and Schmitz; and Bratton in the writing domain of atmospheres and urban ambiances. The review concludes with a discussion on the prospects of technology in today's cities and its contribution to the creation of sustainable and smart built environments.

On a different but related encounter with the topic, the third chapter is entitled "Affective Atmospheres, Essence of Architecture, and Spirit of Place" in which Abussada and Elshater discuss the differences between the two terms: "essences" and "spirit" in urban studies and their links with the concept of effective atmospheres. A bibliographical and thematic review of the essence of architecture or spirit of place is undertaken to highlight the possibility of using "affective atmospheres" instead of the "essence" or "spirit" which are generic in nature. Interestingly, this alternative in the use of terminology can better describe people's comprehension of the urban environment and enable insights into the understanding of their emotional impressions and reactions in third places.

At a design implementation level, the fourth chapter, entitled "Potentials of ICT and IoT in the Design of Smart Public Spaces" engages with the notion of smart public places. Written by Hakky the chapter argues for the importance of integrating Information and Communication Technology (ICT) and the Internet of Things (IoT). It concludes with an emphasis on the need for the current technological shift, which would enable better utilisation of space. In essence, this work provides a future direction of research that considers diverse economically conscious experiments.

In the fifth chapter, entitled "Transformation of Urban Space by Smart Technologies: Evidences and Traces From Istanbul Historical Peninsula," Bell and Koramaz evaluate Istanbul Historical Peninsula with a focus on the transformation of urban space through smart technologies. Having a city with such a historical, cultural and financial core enables the debate on the challenges of integrating ICT tools in urban planning and management systems of historical places. Addressing the concept of "e-topia," the authors provide a method for transformation based on ICT technologies. It remains to be validated how such an emerging technology serves the development of a vision of smartness and relevant strategic framework for transforming the local community into an information society.

In a similar area of interest, the sixth chapter, entitled "Towards a Smart Neighbourhoods Simulated Model," Alhakkak offers a definition of smartness and its role in shaping residential neighbourhoods. This work generates two frameworks where the first explicitly integrates parameters related users and generates reports for the changes across time illustrated on maps; and the second framework employs virtual reality and a complex of multiple layers of ABC clustering algorithm. Chapter 7 is entitled "Impact of Urban Ambiance on the Productivity of the Built Environment Using Cluster Analysis." Saurav Dixit discusses the effect of technology on the productivity of the users working in the city. By scanning relevant literature, he identifies the attributes of urban ambiance and extends to exploring the impact of satisfactory city ambiance on the productivity of the urban environment. Utilizing culture analyses, the results gained from the structured questionnaire procedure provide ways to improve the constructed urban environment by creating human-driven spaces that can improve our rate of urban wellness experiences.

A closer examination of another significant aspect of implementing a placemaking approach is evident in Chapter 8 by Elgobashi and Elsemary, entitled "Theorizing the Critical Placemaking: Impact of Creative Ambiance on Social Activities." It discusses the development of urban spaces in areas associated to historical sites that might have been overlooked in some cities of Global South. In attempting to construct a creative ambiance, their results highlight the essential need for deploying the process of placemaking among practitioners in a manner that addresses the sensory experience of the users. Through their proposed framework, they attempt to theorise the processes of placemaking and develop critical concepts in the preservation process of historical sites.

Having discussed various theoretical standpoints followed by demonstrating diverse possibilities for introducing the notion of ambiance in the design of smart places, the concluding three chapters place emphasis on the integration of artistic works as an important indicator of successful ambiance. The work presented in Chapter 9 by Dhage, which is entitled "The Gastronomy of the Eye: A Tale of Two Cities in India," highlights the role of art in city atmospheres through the examination of two cases. Addressing the context of two major cities, Delhi and Mumbai, her work provides a solid argument for translating the urban experiential phenomenon to the real. Analysing two artistic works this chapter provides a pathway towards a critique of the early initiatives that explore aspects of smartness in urban environments.

Along the same line of thinking, in Chapter 10, entitled "Combinatorial Architecture: Urban Causality, Creativity, and Socialization," Tiazzoldi and Elshater review several projects that have been already implemented in the contexts in Italy, the United States, and France. These projects have focused on ambiance as produced by technology in historical and contemporary settings and its practical application in designing spaces. The results of their work call for a shift in the way in which design challenges are approached where assessment and transdisciplinary action are envisioned through collaboration between academia and professional practice. Placing emphasis on design configuration, the chapter underscores the importance of conceiving a design process of combinatorial architecture that engages different actors with attention given to an outcome that fully considers urban ambiance and users' feelings. On different level, in Chapter 11, entitled "Situational Atmospheres: Filmmaking Approaches for Conveying Urban Design," Abusaada and Elshater attempt to link placemaking and filmmaking. Based on content analysis of four films and the accompanying Egyptian narratives written by Najib Mahfouz, their work investigates the impact of local architecture on the quantitative and non-quantitative factors that create affective atmospheres. They conclude with a call for integrating the process of filmmaking as pedagogical method for teaching atmospheres-based design in various courses ranging from design studios and history classes.

It is increasingly evident that the creation of distinct architectural and urban forms in smart cities that are based on respecting the influence of urban ambiances/atmospheres is gaining importance. Collectively, the contributions call for relevant criteria that decision-makers should adopt to achieve successful urban ambiances/atmospheres. Principally, the contributions address the importance of realizing or activating a transdisciplinary design approach that crosses the boundaries of the disciplines of architecture, urban design and planning, public art, cognitive studies and environmental psychology. A key message of the book is that it rejects the argument that the loss of the urban character and identity of new cities and settlements is merely due to failure to follow reasons beyond creating urban ambiance prevailing in a culture of the urban realm of the city. Collectively, three key aspects can be inferred from the contributions: The first is presenting clarity with respect to the differences between affective atmospheres and ambiance and that integrating these in public places is a human necessity. The second brings forward a proposal called a 'smart city of a transdisciplinary paradigmatic approach, which includes the assumptions about environmental planning and designing public places in the newly developed smart cities. The third presents a combination of quantitative and qualitative approaches.

Hisam Abusaada
Housing and Building National Research Center (HBRC), Egypt

Ashraf M. Salama
University of Strathclyde, Glasgow, UK

Abeer Elshater
Ain Shams University, Egypt

REFERENCES

Abusaada, H. (2020). Strengthening the affectivity of atmospheres in urban environments: The toolkit of multi-sensory experience. *Archnet-IJAR: International Journal of Architectural Research*. doi:10.1108/ARCH-03-2020-0039

Abusaada, H., & Elshater, A. (2020). Effects of urban atmospheres on changing attitudes of crowded public places: An action plan. *International Journal of Community Well-Being*, *3*(2), 109–159. doi:10.100742413-019-00042-w

Abusaada, H., Vellguth, C., & Elshater, A. (Eds.). (2019). *Handbook of research on digital research methods and architectural tools in urban planning and design*. IGI Global. doi:10.4018/978-1-5225-9238-9

Anderson, B. (2009). Affective atmospheres. *Emotion, Space and Society*, *2*(2), 77–81. doi:10.1016/j.emospa.2009.08.005

Hsu, H. L. (2017). Literary Atmospherics. *Literary Geographies*, *3*(1), 1–5.

McKenna, H. P. (2019). *Ambient Urbanities as the Intersection Between the IoT and the IoP in Smart Cities*. IGI Global. doi:10.4018/978-1-5225-7882-6

Pallasmaa, J. (2013). Towards a neuroscience of architecture: Embodied mind and imagination. In J. Pallasmaa, H. F. Mallgrave, M. Arbib, & P. Tidwell (Eds.), *Architecture and neuroscience* (pp. 5–22). Tapio Wirkkala—Rut Bryk Foundation.

Salama, A. M., & Azzali, S. (2015). Examining Attributes of Urban Open Spaces in Doha. *Urban Design and Planning*, *168*(2), 75–87. doi:10.1680/udap.14.00011

Salama, A. M., Muhammad, A., & MacLean, L. (2017). Deciphering urban life: A multi-layered investigation of St. Enoch Square, Glasgow City Centr. *Archnet-IJAR*, *11*(2), 137–156. doi:10.26687/archnet-ijar.v11i2.1278

Zumthor, P. (2006). *Atmospheres: Architectural Environments - Surrounding Objects*. Birkhauser.

Acknowledgment

The editors should like to express their gratitude to everyone who participated in the progress of this publication. Special thanks to the chapters' authors of the current book for their enthusiastic contribution this to project and to the editorial advisory board for their constructive recommendations in the reviewing process. Their conscientiousness is very much appreciated. We extend our special gratitude to Juhani Pallasmaa who, in his foreword, put our publication into the wider context of current discourses.

The editors would also like to thank Josh Christ, Assistant Development Editor and Nicole Gonser, Customer Service Manager, at IGI Global, for his untiring support throughout the development of this book. They are indebted to the Publisher, IGI Global, for giving them a platform to inform our professional audience of the innovative views and interesting arguments held and developed by the contributors of this publication.

Chapter 1
The Atmospheres of Urban Environments:
Ambiguity and Plausibility in Urban Studies

Hisham Abusaada
https://orcid.org/0000-0001-6530-7714
Housing and Building National Research Center (HBRC), Egypt

ABSTRACT

This chapter investigates the ambiguity of the word "atmospheres" in the fields of urban studies. It examines the justifications (plausibility) beyond its uses, with the terms that are focusing on the perceptual qualities. The author investigated the uses of the word "atmospheres" from the beginning of the 17th century to the year 2020, a period which he divided into four stages. The investigation covered the work of 27 thinkers in the fields of natural sciences and humanities, including 10 in architecture disciplines, in addition to 28 manuscripts that addressed the relationship between atmospheres in the areas of architecture, particularly urban planning and design and urban landscape architecture between 1998 and 2020. The outcomes were developed through a comprehensive literature review by gaps analysis and a deductive online survey with 58 specialized participants, using SurveyMonkey. This chapter contributes to the rationale that an urban designer can use to study people's changing feelings, emotions, and moods according to the understanding of the terms related to atmospheres.

INTRODUCTION

Many contemporary researchers believe that the concept of atmospheres is still unclear. It is ambiguous in itself, in the level of its usage or even in the location to which it refers (Adams, Smith, Larkham, & Abidin, 2019; Algeo, 2010; Anderson, 2009; Böhme, 1993; Hasse, 2014). Other experts believe that the expression "atmospheres" gives a feeling of what cannot be easily explicated in words (Hasse, 2014); it is "a cultural and environmental product" (Hsu, 2017, p. 2). Besides, "affective atmosphere" is used as "a form of placed assemblage" (Shaw, 2014, p. 87).

DOI: 10.4018/978-1-7998-3856-2.ch001

In this research, the author used the term "ambiguity" to explore the extent of appropriateness of the word "atmospheres" in urban studies, particularly when "atmospheres" aims to identify the reasons for people's changing emotions and moods towards places in the cities. This ambiguity comes for two reasons:

1. People's change of emotions and moods in city design is related to stamped terms, such as the sense of place (Guthey, Whiteman, & Elmes, 2014; Jackson, 1994), place attachments (Jorgensen & Stedman, 2001; Scannell & Gifford, 2010), and place belonging (Kemmis, 1995; Sandercock & Attili, 2009).
2. The term "atmospheres" focuses on the impacts of the physical aspects of urban environments on the perceptual quality (Pallasmaa, 2014; Zumthor, 1998). In addition, it has been recently linked to the natural and physical sciences, such as climatic changes, air quality, climatology and meteorology, and biometeorology.

In the fields of urban studies, the main problem is that some architects have used the word "atmospheres" to refer to the product of architecture (Pallasmaa, 2014; Wigley, 1998; Zumthor, 2006, 1998), while others to measure the change of people's feeling, emotions, and moods in the outdoor spaces of the city (Abusaada, 2019; Abusaada & Elshater, 2020; Adams et al., 2019; Backer, 2019; Bissell, 2010). Urban designers have preferred to use the term "sense of place," which includes the perceptual dimension (Guthey et al., 2014; Jackson, 1994; Jorgensen & Stedman, 2001). Afterwards, the terms associated with the word "atmosphere" varied into "urban ambiance" (Thibaud, 2002), "urban atmospheres" (Böhme, 2014; Gandy, 2017), and "affective atmospheres" (Anderson, 2009; Buser, 2014; Michels & Steyaert, 2017). This confused use worsened when the experts in the fields of natural and physical sciences applied the word "atmospheres" in aerobiology, atmosphere ecosystems, biometeorology, climatology, ecology, meteorology, and microbiology. These fields include atmospheric chemistry and physics, climatic changes, air quality, bioclimatic well-being, and environmental odour (Beggs, Šikoparija, & Smith, 2017; Reichler, 2009).

Moreover, the word "atmosphere" still denotes power in the different fields of the humanities, such as social sciences, pedagogy, anthropology, artworks, economy, geography, political sciences, psychopathology, sociology, and literature (Egan, 2014; Hsu, 2017; Jenner, 2013). These fields consider atmospheres as a human, social, and architectural experience, particularly around human perception and aesthetics (Böhme, 1993, 2000, 2013, 2014; Labuhn, 2016; Pallasmaa, 2014; Wigley, 1989). Besides, many studies passed the ambiguity of word, and found technological clarifications in the fields of architecture and landscape architecture and soundscape design (Cerwén, Wingren, & Qviström, 2017; Steele, Luka, & Guastavino, 2012).

The purpose of this chapter is revealing the extent of the term "ambiguity" in general and in the fields of urban design principally. This chapter also highlights why the word "atmospheres" is an unambiguous concept in urban design. Therefore, additional studies are necessary to explain further each term in the field of architectural practice. The author used two methods to achieve these objectives: 1) A literature review related to the different theoretical and empirical thoughts; 2) direct communication with experts in several fields of practice.

Accordingly, in the following, this chapter includes six sections: 1) *Background*, which provides an overview of the ambiguity of the term atmospheres in modern thought; 2) *Methodology*, which illustrates the methodology the author adopted in this study; 3) *Historical Extrapolation: Literature Review*, which highlights the progression of the definitions and concepts of the word "atmospheres," based on analytical data; 4) *The Deductive Survey*, which illustrates the aim and features of the questionnaire the author used in this study; 5) *Discussing the Findings*, which provides a two-fold discussion of atmospheres, based on the literature review, and on the findings; 6) *Conclusion*, which sums up the work and its outcomes.

BACKGROUND

Prior research suggests that the ambiguity of the term "atmospheres" in the modern global thought starts from Schmitz's (1978) work. According to Böhme (1993), Schmitz understands atmospheres in the full sense through an emotional impression resulting from the impact of the atmospheres on the body. In this regard, Schmitz is one of the theorists who deal with atmospheres or the general feeling as previous or daily experience in the theoretical sense (Böhme, 1993). Depending on the concept of phenomenology, based on Böhme, Schmitz does not provide a definition of atmospheres, but refers to everyday experiences, such as the experience of a tense atmosphere in a room, a repressive atmosphere or a tranquil atmosphere in a garden. Schmitz's school is characterized by intuition; it is interested in studying the sensitivity of what is a phenomenon in people's conscious experience (Böhme, 1993).

Contrarily, Böhme (1993) sees that atmospheres are indeterminate about their "ontological status, or [...] cosmic reality" (114). Besides, he said that it is uncertain whether he should attribute them to the objects, organisms or environments from which they originate "or to the subjects who experience them" (p. 114). Moreover, Böhme is unsure of where atmospheres are going. Further, Böhme describes atmospheres in several ways, including "fill the space with a certain tone of feeling like a haze, [...] an expression for something vague" (p. 114), which does not mean that an atmosphere "is itself vague" (p. 118). Besides, the knowledge of the production of atmospheres is "very seldom explicit, it is destroyed by the subject-object dichotomy, an aesthetic concept" (p. 123). Atmospheres also seem "a colloquial term, [...]due to the ambiguity of its usage and experienced as an emotional effect" (Böhme, 2014, pp. 43, 46).

In another vein, the term "atmospheres" in the field of geography seems to express something vague, and already used vaguely and interchangeably with mood, feeling, ambience, tone and other ways of naming collective affects. Each of these words has "different etymology" (Anderson, 2009, p. 78). However, Anderson (2009) adds that "the referent for the term atmosphere is multiple: epochs, societies, rooms, landscape, couples, artworks, and much more are all said to possess atmosphere (or be possessed by them)" (p. 78). He describes atmospheres as being "impersonal or transpersonal intensity, [...] the transmission of the other's feeling, ... qualified aura, ... tone in literature, [...] mimetic waves of sentiment, [...] more broadly a sense of place" (p. 78). Anderson also summarises atmospheres are as follows:

1. The atmosphere holds a series of opposites–presence and absence, materiality and ideality, definite and indefinite, singularity and generality–in a relation of tension.
2. According to Karl Marx, "Atmospheres are singular affective qualities that emanate from but exceed the assembling of bodies" (p. 80)
3. Atmospheres have a characteristic spatial form–diffusion within a sphere.
4. According to Deleuze and Guattari, Atmospheres are some form of "envelopment"; they are generated by bodies–of multiple types–affecting one another
5. Based on the concept of everyday situations, atmospheres in the arts and sciences "become the 'object' of explication and intervention" (p. 80).

These situations emanate from the effective qualities of human bodies, discursive bodies, nonhuman bodies, and all the other bodies.

Seminal contributions have been made by contemporary global scholars, such as Hermann Schmitz (1978; Schmitz, Owen, & Slaby, 2011), Peter Zumthor (1998, 2006), Mark Wigley (1998), Ben Anderson (2009), Luke Oliver Jaaniste (2010), Juhani Pallasmaa (2014), Gernot Böhme (1993, 1998, 2000, 2013, 2014), and Beata Labuhn (2016). Those researchers noticed the ambiguity of the term "atmospheres," whether on the level of meanings, definitions or usage. Even though they agreed with what many dictionaries stated (Abrams & Harpham, 2011; Algeo, 2010; "Ambience," n.d.; Simpson & Proffitt, 1997), they recognised that the term "atmospheres" describes humans' emotions toward a place or a person or a landscape or artwork. They see that the issue of the atmospheres is mainly related to the nature of creating feelings. This term describes how an individual finds himself/herself in a place, feels joy and pleasure or monotony and depression, and this sense frequently occurs swiftly and suddenly.

METHODOLOGY

This chapter is classified among the descriptive, correlational, and exploratory studies. It searches in literature and historical artworks, side by side with the reality, to discuss the ambiguity of the word "atmospheres" in the field of practice, and to provide precise guidelines for the use of this term in urban design. This investigation revolves around two precise questions: What is the certainty of the ambiguity of the word atmospheres in general and in the fields of specialization? What are the justifications (plausibility) for its appropriating in urban design? The methods for collecting and analysing data to address these questions include two steps.

The first step includes the use of gaps analysis of the relevant research literature (Cooper, 2014). This method focuses on developing the necessary information. It provides a way to identify suboptimal concepts, processes, and practices, and then recommends initiatives that will help the fields of urban design meet their goals. An in-depth literature review can contribute to obtaining beneficial synthesis about the term "atmospheres." The author divided the results into four phases, which allowed him to discuss the progression of the term "atmospheres" in the period 1600-2017, as follows:

1. **First Phase:** In the period 1600-1700, so over 100 years.
2. **Second Phase:** In the period 1800-1900, which includes four philosophers and two architects.
3. **Third Phase:** In the 19th century, which includes five philosophers and three architects.
4. **Fourth Phase:** In the first two of decades of the 21st century, which includes five philosophers and five architects.

The second step is the deductive survey based on the hypothesis that the word "atmosphere" is different from the term "sense of place." The author created the survey in SurveyMonkey and collected the data for three months (from October 5, 2018, to January 5, 2019). The respondents were 58 participants of the International Ambiances Network ("The ambiances network," n.d.); they would answer six questions and could also add a comment. The questions revolved around the ambiguity of the word "atmospheres" in general, and in the field of urban design, in particular. This questionnaire addressed the effects of the places of the city on people's satisfaction, the similarity between the idea of atmospheres and the concepts of sense of place, affective atmosphere, urban emotion, and urban atmosphere. It also converged on how one can see the linkage between atmospheres and dimensions of urban design.

HISTORICAL EXTRAPOLATION: LITERATURE REVIEW

The word "atmospheres" focus on the ambiguity of this word on its levels of explanation or usage (Abusaada, 2019; Algeo, 2010; Anderson, 2009; Böhme, 2014; Gandy, 2017; Jaaniste, 2010). The phenomenon of atmospheres has been widely observed in many fields of practice connected with the fields of architecture (Figure 1).

Figure 1. Areas of interest of the term "atmospheres"

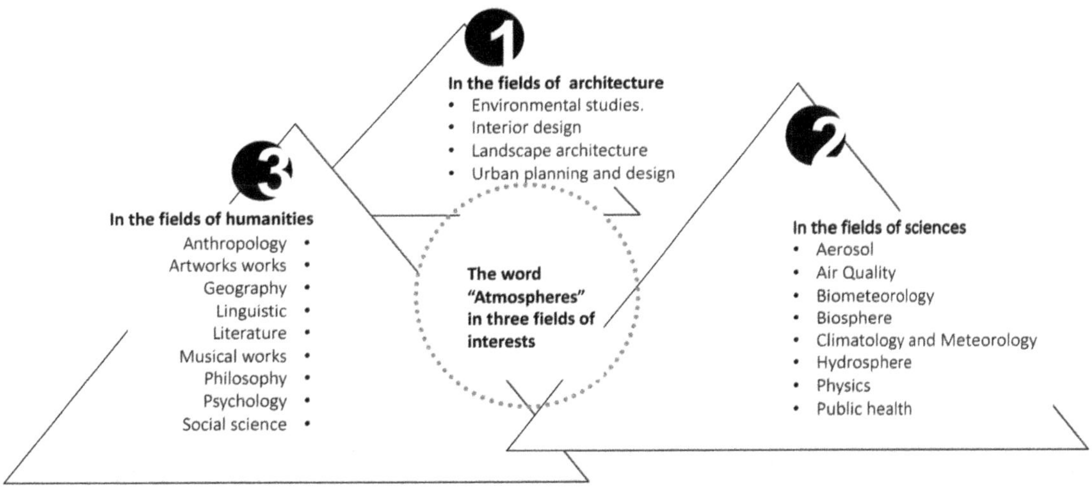

Notably, the term "atmospheres" has a widely different use in each field. This section commences with a short review of the literature regarding the use of this term. It has been discussed through 27 elected thinkers throughout 417 years, in many fields of practices. Figure 2 shows these thinkers related their date of birth and the time when they mentioned the word "atmospheres" in their works.

More than 400 years between the old and new phenomenon, and the development of the term "atmospheres" passed through three phases of progression. The next subsections illustrate the three phases of progression in the use of the term "atmospheres" (Figure 3).

The First Phase of Progression

In the 17th and 18th centuries, William Shakespeare, as an English poet, playwright, and actor, used the term "atmospheres" to show the tone of the dialogue between the characters of his novels. John Wilkins, as an English Anglican clergyman, and Newton, as a physicist, astronomer, and theologian, discussed atmospheres a scientific tract. The first phase shows the early usage of the word atmospheres, which

Figure 2. The 27 thinkers in the period 1600-2020

The fields of scholars specialisation	The scholars	Life period	Date of used the word "atmospheres"	
An English poet, playwright, and actor.	William Shakespeare	1564 –1616	1609	
An English Anglican clergyman, natural philosopher and author.	John Wilkins	1614 – 1672	In the 17th century	
An English mathematician, physicist, astronomer, theologian, and author.	Isaac Newton	1642 – 1726	In the 18th century	
An English poet, literary critic, philosopher and theologian.	Samuel Coleridge	1772 – 1834	1816	In the 19th
A German architect, art critic, and professor of architecture.	Gottfried Semper	1803 – 1879	1860-62	centuries
A British writer on art and fashion.	Mary Merrifield	1804 –1889	1869	
An American writer, editor, and literary critic.	Edgar Allan Poe	1809 – 1849	1842	
A German philosopher, economist, historian, sociologist, political theorist, journalist and socialist revolutionary.	Karl Marx	1818 – 1883		
A German architect .	Richard Lucae	1829 – 1877	1869	
An English novelist and poet.	Thomas Hardy	1840 – 1928	1878	
				In the 20th
An Austrian-born American music composer for theatre and films.	Max Steiner	1888 – 1971	1933	century
An Irish novelist and short story write.	Elizabeth Bowen	1899 – 1973	1948	
A French philosopher and aesthetician.	Mikel Dufrenne	1910 – 1995	1966	
A German philosopher.	Hermann Schmitz	1928	1978	
A Finnish architect and professor of architecture.	Juhani Pallasmaa	1936	2014	•
A German philosopher and musicology and doctor of urban studies.	Jean-François Augoyard	1937	2007	•
A German philosopher and author: aesthetics, ethics and anthropology.	Gernot Böhme	1937	1993 1998; 2001;	•
A British linguist, academic and author.	David Crystal	1941	2013; 2014	•
		1943	1990	
A Swiss architect.	Peter Zumthor	1949	2006	
A German geographer, educator and urban researchers.	Jürgen Hasse	1956	2014	
A New Zealand architect and author.	Mark Wigley	1958	1998	•
An Italian professor of Aesthetics.	Tonino Griffero	1960	2014	•
A French sociologist, urban planner, developer, and researcher.	Jean-Paul Thibaud	1956	2002	•
A French human geographer interested in public spaces, everyday aesthetics and sensory geographies.	Rainer Kazig		2016	•
An English geographer and urbanist.	Matthew Gandy	1965	2017	•
An Austrian artist: architect and urban planner.	Dietmar Offenhuber	1973	2008	
An English cultural-political geographer.	Ben Anderson	1973	2009	

• The indication shows the ten thinkers who used the term atmospheres in the two first decades of the 21st century.

The indication shows the ten specialists in urban studies, including the fields of architecture.

emerged in the 17th century ("Ambience," 2018; Stevenson, 2010; Wilkins, 2018). According to Gandy (2017), this word derived from the Latin word *atmosphaera,* which included a set of meanings that have nothing to do with each other, such as vapor, sphere, breath, life or soul. In the same decade, Isaac Newton used this word in the field of physics, between 1642 and 1727. The basic definition emerged originally as a scientific term; it expressed on the vast void of the universe that comprises the planets, stars, and galaxies through which orbits pass (Jaaniste, 2010).

The Second Phase of Progression

This phase extended for 100 years during the 19th century, and includes two stages. The first stage of progression occurred in the fields of social sciences and humanities around 1800-1850. Four authors drove the further development of understanding this term mainly in the development of culture, poetry, drama, art, fashion, aesthetic discourse, and architecture (Ford, 2012; Jenner, 2013; Semper, 2004). Research has provided evidence of the use of the word "atmospheres" in the domain of sensation, in Samuel Coleridge's poetry in 1816 (Weigelt, 2010), and in the domain of situations, in Edgar Allan Poe's short stories in 1842 (Poe, 2017). Coleridge induces a combination of religious and superstitious terror by his description of the first scene in the narrative poem *Christabel* (1816),

Figure 3. Three phases of progression over 400 years—in the period 1600-2000

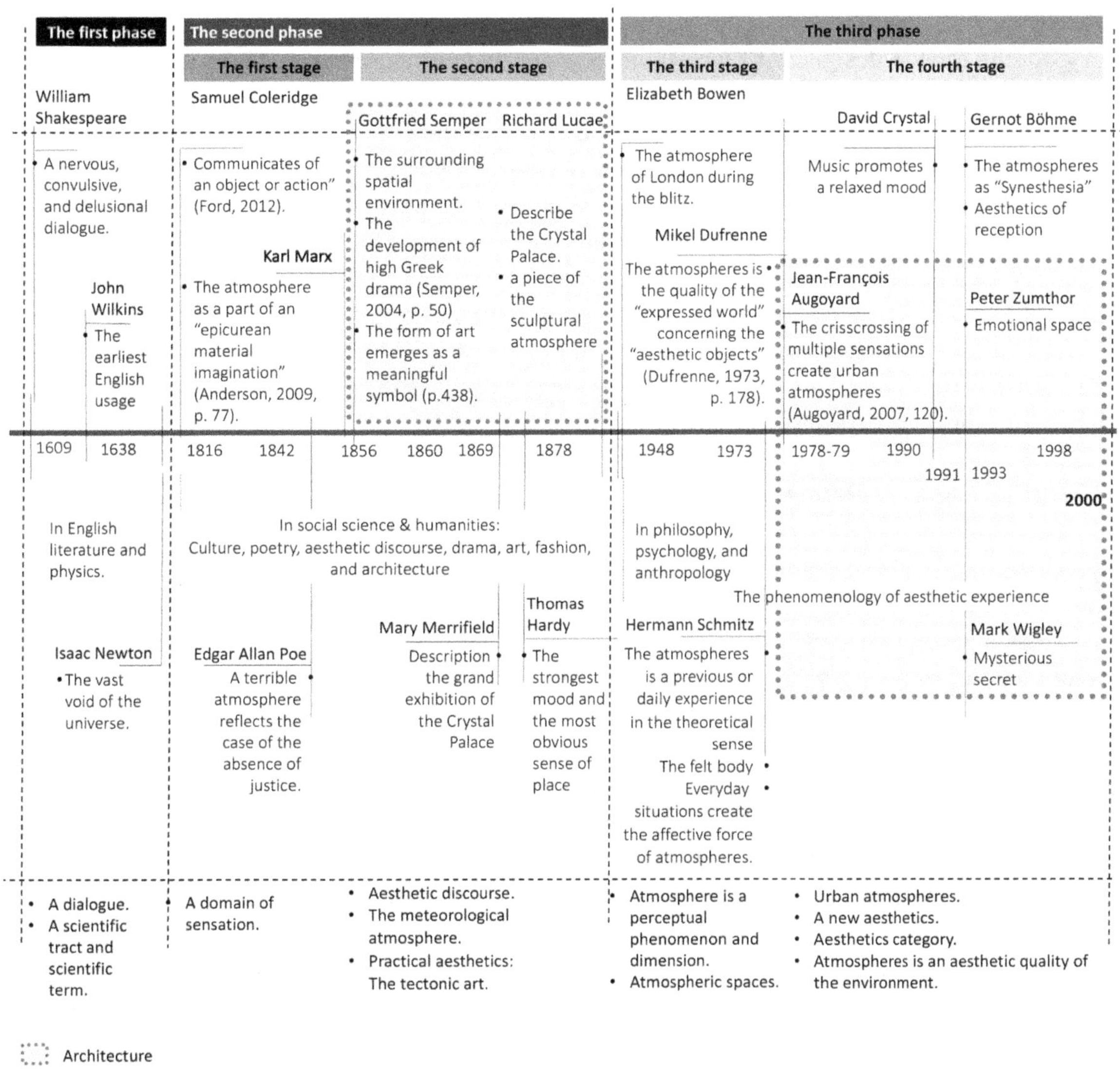

In1856, the term revolutionary atmosphere has used for the first time by Karl Marx concerning the idea of meteorological atmosphere. In this regard, Anderson (2009) classified this usage in everyday speech and aesthetic discourse. He classified this usage as turbulent materialism, which described that "life is imagined through a combination of different elements and different states" (p. 77). Moreover, Anderson believes that Marx's classification put the atmosphere in the category of phenomena, but "they are not necessarily sensible phenomena" (p. 78).

In the field of artworks, drama, and architecture, Semper (2004) used the word "atmospheres" in the realm of culture as a tone of an occasion or setting and an environing mood (e.g., the development of Greek drama regarding the birth of a classical Greek temple). Besides, atmospheres show the power of a *petit bourgeois* in two examples. The first is that porphyry (igneous rock) as a material creates atmospheres of sovereignty. The second is that granite expresses atmospheres of the homeland. In another vein, Semper (2004) sees that "the human culture has always been enchanted by the veil or mask, the haze of carnival candles is the real atmospheres of art" (Mallgrave, 1996, p. 202). At the mid of the 19th century, Joseph Paxton designed the Crystal Palace in London in 1869 as an "icon of Empire exhibits par excellence" (Sturm & Turner, 2012, p. 26).

Mary Merrifield used the word "atmosphere" to describe the grand exhibition of the Crystal Palace (Jenner, 2013). Merrifield described the atmospheres saying that the parts of the building were covered with blue-coloured fog, as if open to the air, contrasting with the warm colour of the cloth and the ceiling with light blue beams. Also, the blue sky above the church wing produced a very satisfactory and, at the same time, very natural appearance. It is difficult to distinguish where art begins and nature finishes. In Jenner's (2013) view, Richard Lucae, the architect of the Frankfurt Opera House, described the Crystal Palace as "a piece of the sculptural atmosphere" in London and compared the floating elements of the Palace with Turner's painting–"in both, the effect was achieved through a humid atmosphere which dematerialises the landscape and dissolves it into infinity" (p. 14).

The Third Phase of Progression

Many existing studies have examined atmospheres in the third phase of progression as the sociological imagination in the literature (Crystal, 2014; Dufrenne, 1973; Wyndham-Davies, 1997). Atmospheres were used as a perceptual dimension in new ephemeral phenomenology, which characterizes the sensory realm in philosophy, psychology, and anthropology (Schmitz, 1978; Schmitz et al., 2011). In *The Return of the Native* (1878), Thomas Hardy use of atmospheres makes Egdon Heath an immense and brooding presence to show the futility of the human's struggle for happiness. Afterwards, Elizabeth Bowen (1948), Mikel Dufrenne (1973), and David Crystal (2014) used atmospheres in various forms. In the film *The Heat of the Day* (Wyndham-Davies, 1997) the viewer can capture "the atmosphere of London during the blitz" (McCrum, 2015). In 1966, Mikel Dufrenne used ballet dancers and music to create atmospheres to make the spectator aware. He used the atmospheres around the dancers intending to revive what he called the ballet spirit (Dufrenne, 1973, p. 76); it is an aesthetic thing that dancers bring to life. These atmospheres exist and are tangible even in pure dance, where the expression is not driven by or fortified with a subject. Dufrenne believed that, while seeing what in this world is not easy, people are at once sensitive to its existence as an aesthetic object that has much to offer them. Besides, in the art of acting, it seems the use of changes in the atmosphere contributes to how one interprets the world. Atmospheres is a unique affective quality of an expressed world, which concerns the aesthetic objects. In the art of acting, David Crystal (2014) gives an example to the music that "promotes a relaxed mood," like "chill-out music" (p. 233).

In philosophy, psychology, and anthropology, three terms were used concerning atmospheres as a dimension of perception, namely the affective force of atmospheres (Schmitz, 1978), atmospheric spaces (Jenner, 2013; Kazig, 2016; Schmitz, 2019), and urban atmosphere (Augoyard, 2007). Concerning these three terms, the concepts of atmospheres ranged between three different aspects. The first is that atmospheres are intimate. The second is atmospheres refer to compositional process of gestures and worlds concerning dwelling in space (Stewart, 2011). These two terms are related to the human body as the focal point of analytical studies of the effect of atmospheres. The third term is that atmospheres do not occupy a physical space, but a surface-less space around the conscious subject (Schmitz, 2019; Schmitz et al., 2011).

Later, the felt body experience, bodily sensations, and physical reactions have been considered to measure the impacts of atmospheres. Each person has an independent cognitive field or body-based experience, as Schmitz (1978) mentioned in his ten-part German-language literature from 1964-1980 *System of Philosophy [System der Philosophie]*. In this vein, Schmitz sees the effects of emotional atmospheres based on the feeling of the body, not only due to the prior experiential knowledge. In this regard, atmospheres are classified as objective facts and quasi-objective quality (Kazig, 2016; Schmitz et al., 2011).

Other studies address atmospheres as a phenomenology of aesthetic experience, in the fields of philosophy of science, theory of time, geography, and architecture (Augoyard, 2007; Böhme, 1993; Zumthor, 1998; Wigley, 1998). With a short review of the literature, this subsection follows atmospheres as an aesthetic category in the last decade of the 21st century. In 1979, Augoyard (2007) saw the crisscrossing of multiple sensations create urban atmospheres. In this respect, atmospheres mean emotions; they focus on the context of specific cultures of sensory perception for atmospheric spaces. Thus, the affective force of atmospheres is due to the catalytic role of everyday situations, individual or collective. Emotions are not a purely psychological category; they are atmospheres. According to Labuhn (2016), a wide choice of visions is available in the literature of atmospheres concerning atmospheres as an intellectual approach, operational approach, and unconscious approach (Böhme, 1993; Wigley, 1998; Zumthor, 1998). These approaches discuss atmospheres as a concept of an aesthetic of reception based on the intellectual approach (Böhme, 1993), an aesthetic category and emotional space (Zumthor, 1998), and a mysterious secret based on the unconscious approach (Wigley, 1998).

The Fourth Phase of Progression

Over the first couple of decades of the 21st century, some scholars drove the progression of the term "atmospheres" in many fields of practice (Anderson, 2009; Böhme, 2000, 2013, 2014; Griffero, 2014, 2016; Stewart, 2011). Five architects participated in the intellectual discourse about atmospheres (Hasse, 2007, 2014; Offenhuber, 2008; Pallasmaa, 2014; Thibaud, 2002; Zumthor, 2006) (Figure 4).

According to the scope of an aesthetic architecture, the term "atmospheres" emerged as an ecological aesthetics concept (Böhme, 2000). Based on the praxeological approach to the perception of human behavior, the atmosphere is classified as the third term of perception and immediate surroundings of something (Thibaud, 2002). Additionally, atmospheres are an aesthetic category, creating based on the spiritual quality of designing the buildings and the conceptual and transcendental dimension of meaning in architecture. It is a "singular density and mood, this feeling of presence, well-being, harmony and beauty" (Zumthor, 2006, p. 11). It follows the concept of sensuous materiality (Offenhuber, 2008).

Many early studies, as well as current work, focus on atmospheres, which follow the everyday life of urban experience (Böhme, 2013; Gandy, 2017; Hasse, 2014; Labuhn, 2014; Pallasmaa, 2014, Stefansdottir, 2017). This experience is based on the "mindful physical presence in space" (Böhme, 2013, pp. 21, 27), "the multi-sensory in its very essence" (Pallasmaa, 2014, p. 231), and the places and situations in the cities (Hasse, 2014). As a result, atmospheres are influential and significant, but are not expressed in a visible, countable or measurable shape. Besides, Gandy (2017) sees that atmospheres originate in "a spontaneous-affective realm of grounded sensation that lies (outside of the individual human subject)" (p. 365). Hass (2014) studied atmospheres based on the new phenomenology to the understanding of urban space, expressions of medial power, and manifestations of urban life. In this debate, based on Bohm's (1991) words, atmospheres as *Synesthesia*, (Labuhn's (2016) recent study concluded that atmospheres as multisensorial experiences of the subject and the various characteristics of the object are intrinsically inter-

Figure 4. The progression of the term "atmospheres" in the last two decades—around 2000-2020

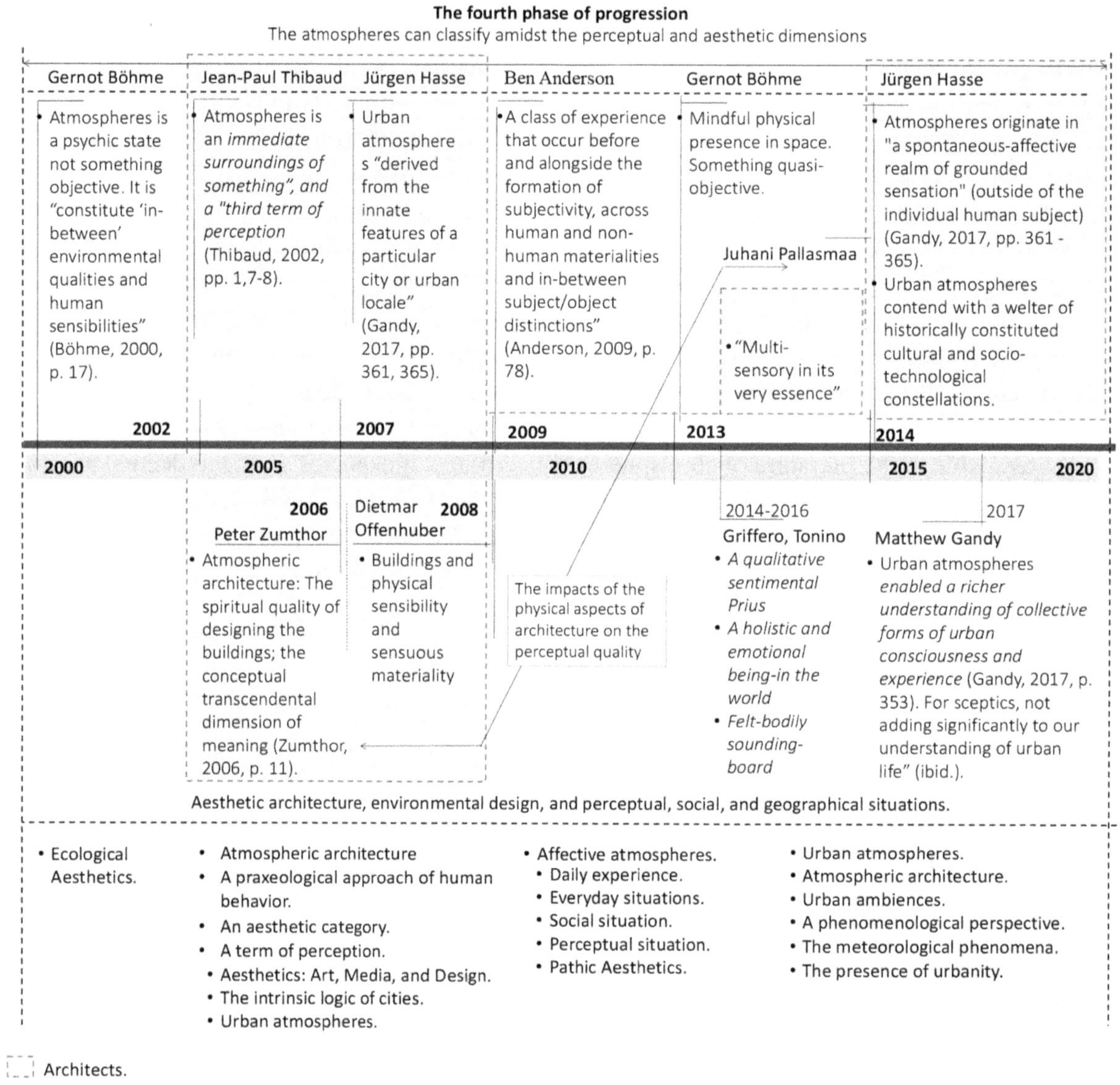

related. However, Labuhn believes that the city's inhabitants are the producers of atmospheres. Moreover, many researchers used the principles of pathic aesthetics in the intellectual discourse of atmospheres. The atmosphere is "a qualitative sentimental Prius (Griffero, 2014, p. 5), it is "a holistic and emotional being-in the world" (p.15). Furthermore, Griffero (2016) sees that "an atmosphere finds in the perceiver a more or less clearly defined felt-bodily sounding-board" (p. 27). Atmospheres support different modes of interpretation related to geometries of perception (Gandy, 2017): (a) The sensory characteristics of corporeal experience, and (b) the "manipulated atmospheres" of music festivals, shopping malls or street lighting. In this phase, many researchers considered a consistent terminology (Anderson, 2009, Buser, 2014; Böhme, 2014; Gandy, 2017; Hasse, 2014; Kazig, 2016; Stefansdottir, 2017).

Atmospheric Spaces

Atmospheric (airy) spaces relate to a phenomenology of the felt body (Schmitz, 2016), such as the room-filling phenomenon (Schmitz et al., 2011). Coherently, atmospheres are classified as objective facts and quasi-objective quality (Schmitz et al., 2011). In another vein, atmospheric spaces emerge in two concepts: Void space and apparent emptiness (or empty space) (Jenner, 2013). The first concept depends on the aerial quality, which indicates

the penetration of the light and air between gaps, grilles, frames, screens in combination with the lifting of elements in compositional play by the lightest of supports. (Jenner, 2013, p. 19)

Moreover, according to (Jenner, 2013), Carla Zanini's works focus on the aerial composition (an expression of abstract art), an impressionist sensibility, and a desire for fusion with nature. The second concept is apparent emptiness, relates to the abstraction, in her work *Spazio Vuoto* to make float the "plastic solitude" of a new world of objects.

Affective Atmospheres

Many scholars in human geography and beyond have used the term affective atmospheres (Abusaada, 2019; Ash, 2013; Bissell, 2010; Buser, 2014; McCormack, 2008; Michels, 2015; Michels & Steyaert, 2017), which are "perceived and sensed through the body" (Bissell, 2010, p. 272). This term is defined as "a quality of environmental immersion;" it is something which diffuses in the air (ethereal) and registers by using the sensing bodies (McCormack, 2008, p. 413). McCormack mentioned three ways to design affective atmospheres through affect, feeling, and emotion. These ways are a prepersonal field of intensity, a sensing body, and a socioculturally recognizable form. Afterwards, this term was used as "a class of experience that occur before and alongside the formation of subjectivity, across human and non-human materialities, and in-between subject/object distinctions" (Anderson, 2009, p. 78).

About restricting particular practices in everyday experience, affective atmospheres became central conduct on the move they lead to precipitate particular structures of feeling (Bissell, 2010). Thus, it is defined as relevant to space as "the relational potential for things to act or change" (Bissell, 2010, p. 273). Further, this term used to focus on how it "shapes humans' emotions and capacities, [...] produced and shaped due to the technical and non-human objects and structures" (Ash, 2013, p. 22). Furthermore, affective atmospheres explore the changes in the characteristics of a space or the bodies within it, based on the emotional experience. Afterwards, Buser (2014) developed the notion of affective atmospheres to indicate the variety "of collective affects," which "produced through dynamic, relational place encounters–to consider the imperceptible and the affective sensations situated in place experiences and relations" (p. 228).

Place experience plays a vital role in the concept of affective atmospheres. Affective atmospheres are dynamic spatial patterns radiating outwards; they emerge and shift through bodily interactions. They are also transpersonal, quasi-autonomous, and characterised by collective/shared qualities (Buser, 2014). This term links the material and spatial processes of organising with the sensual and the affective which are interrelated to them (Michels, 2015). Affective atmospheres consider the "key concept for the critical study of affect in organization" (Michels & Steyaert, 2017, p. 82). It focuses on the unshifting of organizational spaces, by reorganizing a site's constellation of material and immaterial elements based on the affectual composition through aesthetic experimentation (Michels & Steyaert, 2017).

Urban Ambience

Ambiance "involves a socio-aesthetic approach that enables us to grasp everyday urban atmospheres" (Thibaud, 2002, p. 203). Urban ambiances address the issue of everyday perception; "it implies a reconsideration of everyday life" (Thibaud, 2002, p. 203). Urban ambience deals the "interactivity, surveillance and signalling" for people of the city, which represents the city as "a dialectic" in itself, and people feel then connected and estranged go together (Jaaniste, 2010). Additionally, today, this concept is not just related to how people learn to avoid or love the places or the harsh industrial noises. "Urban ambiance" is the term that has French tradition and is comparable to the urban atmosphere as it is experienced (Thibaud, 2015). This term discusses the role of the effect of atmospheres in everyday urban situations and daily social activities (Stefansdottir, 2017).

Urban Atmospheres

Urban atmosphere manifests as a pervasive quality, which ties the components of any situation into a consistent whole and gives each situation a specific character (Dewey, 2005; Thibaud, 2002). This quality emerged due to "the crisscrossing of multiple sensations" (Augoyard, 2007, p. 120), as an individual and indivisible situation focuses on the objects and events in immediacy experienced, before being conceptualised or analysed. It then refers to the characteristics of any situation, which

involves the prereflective dimension of the experience, an intuitive way of grasping reality. It belongs to the realm of immediate feeling and bodily sensation and gives primacy to the aesthetic side of ordinary experience, [...] is precisely what gives internal consistency to the situation, conferring meaning and a clear direction, [...] involves a course of action, [...] applies to the situation itself, [...] a domain defined by the interaction between an organism and its environment. (Thibaud, 2002, p.6)

Urban atmospheres are related to the "urban consciousness and experience" (Gandy, 2017, p. 353), as well as "the physical characteristics of the urban fabric and the activities within the city and its sociality" (Stefansdottir, 2017, p. 319). Urban atmospheres focus on urban consciousness, urban life, and two categories of experience: The historical human subject (e.g., in streets and quarters) and the generic, impersonal modes of social interaction (e.g., in airports and highways) (Gandy, 2017). Additionally, urban atmospheres "contend with a welter of historically constituted cultural and socio-technological constellations, [...] cut across diverse ontological realms, [...] encompasses mundane interventions such as the dimming of lights as well as threats of violence or impending meteorological events" (Gandy, 2010, pp. 365-370).

The characteristics of urban atmospheres play an essential role in the uality of life for citizens; this quality might be influencing on the perceived well-being at different locations in a city. These atmospheres are a generator to a valuable opportunity to interpret the meaning of experiencing characteristics of places as totalities, including social interaction, memories about places, cultural influence, locational perspectives, and other situational aspects (Stefansdottir, 2017, p. 320-235).

Limitations, Questions, and Research to Be Explored

This subsection points out some of the problems the author encountered in the extant research. Gandy (2017) sees that there are many sceptics, considering that the concept of atmosphere has remained somewhat "fuzzy" and ill-defined. He sees the idea of "atmosphere" is still conflated with that of "affect," or it does not add significantly to people's understanding of urban life. A closer look at the literature on the most recurrent definitions and concepts, however, reveals a few gaps and shortcomings. Historically, there has been a great deal of confusion in the literature regarding the word "atmosphere." It should be noted that one of the solutions is adding the words urban or affective (Anderson, 2009; Böhme, 2014; Gandy, 2017). The previous studies reveal that atmospheres as a perceptual phenomenon and an aesthetic category are usually the most problematic to most of the scholars (Augoyard 1990; Böhme, 1993, 2017; Griffero, 2014; Thibaud, 2002, Wigley, 1998, Zumthor, 1998). One of the tough challenges for all researchers in this domain is the presence of many stable terms that explore emotions in the field of urban design. These terms are a sense of place, *genius loci*, and spirit of place (Buser, 2014). For instance, a sense of place as a perceptual term has strong linkage with the emotional impacts in the field of urban design (Guthey et al., 2014; Jackson, 1994; Jorgensen & Stedman, 2001). Additional studies to understand more completely the critical tenets of these concepts with terms such as affective atmospheres and urban atmospheres are required. The next section highlights that the limitation of studying these concepts have strong relations with the field of urban design, which is so far lacking in the literature of atmospheres.

THE DEDUCTIVE SURVEY

This section discusses the inherent ambiguity when interpreting the word "atmospheres" in the general meaning and the fields of urban design through the online questionnaire. The author developed this questionnaire to discuss two main issues. The first is the ambiguity of atmospheres in general and in urban design mainly. The second is the differences between atmospheres and sense of place. The main results appear as follows:

1. The respondents who selected "approved" as an indicator of the ambiguity of the word "atmospheres" in the general meaning represented 60.34% % of the total, and 56.90% of respondents selected "agree" about the ambiguity of the concept in the field of urban design. Regarding the ambiguity in general, 24.14% of respondents selected "neither approve or disapprove", while 15.52% of respondents indicated "disapproved". In the same vein, 22.41% of respondents selected "neither agree or disagree," while 20.69% of the respondents indicated "disagree".
2. In urban design, the study revealed that 34% of the respondents selected that affective atmosphere and sense of place are the appropriate expressions to explain the effects of the city's places on people's satisfaction. The other dimensions ranged as follows: Urban ambience 26%, urban atmosphere 24%, urban emotion 12%, and others 10%.
3. The respondents who selected the word "likely" to express the strong link between "atmosphere" and "sense of place" represented 42%. In another vein, 22% of the respondents selected "likely or unlikely," while 32% of the respondents rejected the existance of this dependency at all; they saw the correlation as "unlikely;" finally, 4% of the respondents selecte: "I do not know."

4. Concerning urban design dimensions, the study revealed that 66% of the respondents selected that the perceptual is the dimension which is most related to change the atmosphere of the city's places. The other aspects ranged as follows: 2) Environmental 40%, 3) functional, 4) temporal 30%, 5) ecological 28%, and 6) visual 24%.
5. The standard deviation of the use of the word "atmosphere" to describe the change of feelings was due to the cinematic works (2.24%), the musical works (2.20%), and the architecture of the city (1.82%).
6. The study revealed that 55.32% of the respondents chose atmosphere is the first element which the urban designer uses to change the feeling. The other elements ranged as follow: The behavioural settings are the second element (53.19%), the sense of place and kinetic perception (42.55%) are the third, and the last element is architectural style (40.43%).

DISCUSSION OF FINDINGS

Throughout 400 years, since the year 1600 till 2020, amongst 27 theorists that have used the term "atmospheres" from Austria, France, Finland, Germany, New Zealand, United Kingdom, and United States, 10 scholars worked in the fields of architecture and urban planners. The results of the term "atmospheres" in the first phase showed that the term had been used at first as a scientific tract in the meteorological field, which refers to vapor, sphere, breath, life or soul. It also turned to be a scientific term referring to the great void of the universe, in the fields of physics. Afterwards, this term emerged in poetry to describe the objects or actions, occasion or setting, tone, aura, or mode of the surrounding environment. Furthermore, it has been used in political discourse to describe the epicurean material imagination, in drama, and to describe the surrounding spatial environment, which refers to the chromatic status of the environment.

The results of the relationship between architecture and atmospheres show or at least hint that this relationship stems from creating meaningful symbols. For examples, related to Semper, materials create atmospheres in: An igneous rock creates an atmosphere of sovereignty, the granite expresses the atmosphere of the homeland, and the haze of carnival candles represents the authentic atmosphere of art. Mary Merrifield and Richard Lucae used atmospheres to describe the relationship between art and nature in a place, for instance in the design of the grand exhibition of the Crystal Palace.

The author showed that the term "atmospheres" was used to express on the perceptual dimension based on the new ephemeral phenomenology, which characterizes the sensory realm in literature, philosophy, psychology, and anthropology. Another promising finding was that atmospheres are an aesthetic experience, mainly in the fields of aesthetics, geography, and architecture. The implications of these findings discussed atmospheres in a novel of *The Return of the Native* (1878), by Tomas Hardy, and *The Heat of the Day* (1948), by Elizabeth Bowen, as well as in reflecting on ballet dancers in Mikel Dufrenne's (1966) work. In the novel and musical works, atmospheres used to promote a relaxed mood.

In the last two decades of the 20th century, the results of an analytical review of the two philosophers Hermann Schmitz and Gernot Böhme, besides to the three architects Jean-François Augoyard, Mark Wigley, and Peter Zumthor, demonstrate an essential finding in the understanding of atmospheres through two things: First, as a perceptual phenomenon and conceptual dimension based on the feeling of the body and the everyday situations; second, as an aesthetic quality of the environment based on the quality of the expressed world, which is related to the mysterious secret, emotional space, and aesthetics of reception.

From the short review above, for most of the scholars in the first two decades in the 21st century, key findings show that most of them admit that this term is ambiguous. During the first two decades in the 21st century, this study confirmed the results of five architects, namely Thibaud (2002), Zumthor (2006), Offenhuber (2008), Pallasmaa (2014), and Hasse (2014). This finding sees atmospheres can classify amidst the perceptual and aesthetic dimensions. This progression can classify within three main interests, namely aesthetic architecture, environmental design, and perceptual, social, and geographical situations. Since 2013, the author's results have cast a new light on the situations as a determinant to the effects of atmospheres. It emerged through such themes as daily experience, everyday situations, social situation, and perceptual situation. The comparison between the author's results and those of the newest studies points out that the words "atmospheres" are used in many researches in urban planning and design (Abusaada & Elshater, 2019a; Abusaada & Elshater, 2019b; Ash, 2013; Bissell, 2010; Buser, 2014; Dovey, 2010; Hasse, 2014).

From the results, many concepts lead the way to deal with the term "atmospheres" in all fields of practice. These terms are affective atmospheres, urban atmospheres, urban ambiences, atmospheric architecture or architectural atmospheres, and atmospheric spaces. A difference between affective atmospheres and urban atmospheres can only be attributable to the scope of interest. The first term expresses the perceived structures of feeling, while the second discusses the situations which encompass feeling. Nonetheless, the author believes that it is well justified to synthesize affective and urban atmospheres together in one concept. From these results, it is clear that affective atmospheres is the most relevant expression to explain the effects of the city's places on people's satisfaction.

Another promising finding was that the term "affective atmospheres" has been connected to several different environmental, human, and spatial aspects. These aspects ranged as follows (Anderson, 2009; Ash, 2013; Bissell, 2010; McCormack, 2008; Michels, 2015):

1. The ethereal space that is affecting the body.
2. The daily human experience that has a socially and culturally recognizable form.
3. The human behaviour that shapes human feelings and abilities.
4. The characteristics of space or objects within it via physical and immaterial structures.
5. The aesthetic abilities.
6. The physical and immaterial dimensions.
7. The processes of physical and spatial organization with sensory and emotional attachment.

This result highlights that urban atmospheres are little known, which emerges as an emotionally perceived quality of a situation and manifests as pervasive quality. The situation is a consistent whole and has a character respecting the following six issues (Augoyard, 2007; Dewey, 2005; Gandy, 2017; Stefansdottir, 2017):

1. The collective forms of urban consciousness.
2. Two categories of experience, namely a historical human subject and impersonal modes of social interaction.
3. The constituted cultural and sociotechnological constellations.
4. The physical characteristics of the urban fabric; the activities within the city and its sociality.
5. Different locations in a city; The nonphysical characteristics of citizens' quality of life and well-being.

6. The meaning of experiencing characteristics of places, including social interaction, memories about places, cultural influence, locational perspectives, and other situational aspects.

The results of the online questionnaire showed that the word "atmospheres" is generally understood, but it seems that the method of using in urban design looks still ambiguous. More than 50% of the respondents could see that the atmosphere is like the sense of place. Besides, the study revealed that most of the respondents preferred to use the perceptual dimension related to change of the atmosphere of the city's places and to add the atmosphere is the first element that the urban designer uses to improve the feeling. At this stage of understanding, several questions remain unanswered: Can atmospheres be used instead of a sense of place? Can atmospheres be used as aesthetic and perceptual dimensions in the urban design?

Previous studies indicated that a sense of place is a cognitive perception held by people toward a natural or built environment, and depends on rational thinking to interpret the current physical experiences, individual activities, and socially formed appreciations. It is a criterion to measure and reconstruct the quality of the environment through fosters memories, unique ambience, identity, and character. A sense of place is "an incorporeal, complex and irreducible entity, at the surface of things" (Deleuze, 1990, pp. 19-20). It is an awkward and ambiguous translation of the Latin term *genius loci* (Jackson, 1994, p. 158). It refers to the two things: First, what might more accurately be called the spirit of a place (Hough, 1990; Relph, 1976; Steele et al., 2012), and, second, the personal meaning built subjective qualities (Hay, 1998). The interaction between people and place involves mediation, representation, and human activity, and it fosters memories, unique environmental ambience, character, and the qualities of natural and cultural assets, as well as bioregional context, and physical properties, in other words, the landscape and urban history (Cresswell, 2009; Holt-Jensen, 2009; Relph, 1976). A sense of place is "an assemblage" (Dovey, 2010, p. 25), which means rapprochements between "physical experiences" that are including using, encountering, looking, hearing, and breathing. It is also indicating to the "imaginative constructions that give meanings and values [...] through individual activity and socially formed appreciations" (Healey, 2010, p. 34).

Though it seems that the term "atmospheres" is still ambiguous in urban design, it is not. It implies the immediate context of the built environment and should be addressed more in architecture. It is the sensory experience of a place that profoundly affects the way people relate to and behave around others and the built environment. Therefore, the whole sensorial apparatus should be considered when designing for the term "atmospheres." This consideration is respecting not only the visual and aesthetic sensations, but also sounds/acoustics, smells, touch sensations, and even the hidden sensations, such as electromagnetic or acoustic waves beyond the conscious threshold.

Besides, the spiritual aspects are relating to the intuitive and unconsciousness self-experience, which focuses on the kinetic perception for one in a permanent place (stroll), or a sequential movement (following events). Atmospheres focus on previous daily experience in the theoretical sense, an aesthetic concept, singular affective qualities, and the "object" of explication and intervention in everyday situations. Therefore, it remains unclear whether the term "affective urban atmospheres" is preferred to the term "sense of place"—this may explain why the field of urban design needs more studies to show the extent of similarity or difference between these two terms (i.e., affective atmospheres and a sense of place). This similarity or difference may raise concerns about what they can add to the term "affective urban atmospheres," which can be addressed by more analytical and empirical studies.

Affective atmospheres are an emotionally perceived quality toward a specific situation, and focus on the emotional and sensual experiences, which can be considered to be equivalent in a particular and indivisible situation (has a specific character). The quality perceived and sensed through the body occurs due to the crisscrossing of the multisensorial experiences of the subject, various characteristics of the object (the physical and non-physical structures) in surrounding space. This perceptual phase occurs via restricting practices in everyday experience (collective forms of urban consciousness in each situation), which leads to shaping humans' emotions and changing structures of people's feeling and capacities.

CONCLUSION

More generally, the primary findings of this study are consistent with research showing that the term "atmospheres" not confined to a specific field of practice. In the areas that related to the physical sciences, the term "atmospheres" emerged in the fields of architecture and environmental design, mainly associated with the issues such as climate change, soundscape experiences, and lighting design. This chapter argued that the areas that are most relevant to urban design concerning the fields of humanities are urban geography, environmental psychology, and literary artworks, namely the novel, cinema, and music, which discussed atmospheres precisely. This study represents the systematic beginnings of review of the definitions and concepts of atmospheres in the fields of humanities and physical sciences. It also focused on how to create atmospheres in some fields of architectural practice, most notably urban design.

The data of the deductive survey indicate (via online questionnaire) that there is still a long way to identify the factors and tools to explore atmospheres, whether in the built environment or that still under construction and design. It is hard to understand at any previous results the reasons why urban design experts are reluctant to use the term "atmospheres" in their research works. Indeed, they tend to use terms such as a sense of place, a spirit of place, and *genius loci.*

These findings confirm that the experts in the fields of architecture deal with atmospheres as a perceptual phenomenon and aesthetic category. Meanwhile, the experts of urban planning and design use the term "affective" or "urban atmospheres" to measure the city microclimates and the people's emotional perceptual quality of life in a situation. Importantly, the author's results provide evidence that the study of affective atmospheres represents an added value in the field of urban planning and design.

In this respect, affective atmospheres differ from the other perceptual dimensions in the potentiality to consider the indicators, which refers to the change in peoples' feelings towards physical and intangible properties. These indicators emerged from the elements of the natural and built environment as quantitative and qualitative measurements.

These indicators emerged from the elements of the natural and built environment as quantitative and qualitative measurements.

The apparent implication of this study is it confirmed there is a difference between affective atmospheres and sense of place. This difference can be attributed related to how the perceptual quality could be perceived or sensed of the urban environments. Moreover, this difference also is related to the nature of the place which should examine

The first relies on detecting the atmosphere during the idea of the felt body, which takes place through people's reactions related to the prereflective dimension of the experience in the situation(s). The second depends on building people's feeling towards the environment based on rational thinking, which focuses on the post-analysis evaluation of the actual experiences in the natural and built environments. It aims at interpreting the reflection of these experiences on people-place and vice versa.

Future research should consider the likely effects of the impacts of atmospheres more carefully; for instance, recognising the physical and nonphysical characteristics for the situation(s). Future research on the meaning of the situation(s) might provide extensive explanations about how atmospheres can enhance people's feelings. In future work, examining the indicators related to people's interactions, according to the idea of a felt body, might prove valuable. A vast investigation is needed to delimitate, according to the perceptual qualities of affective urban atmospheres, the aesthetic and perceptual dimensions of urban design.

Accordingly, prospective inquiry requires further practical prospects on urban design guidelines concerning the concept of affective atmospheres —the newest digital method to measure the indicators of people's reactions warrants also further investigation. Furthermore, searching about the participation between experts outside the fields of urban design represents an added value, which offers the outset for debate and new research.

Developing the technical data needed for gap analysis in the fields of architecture and urban design concerning the term "atmospheres" is an enormous task. The gap analysis in the future should focus on three issues associated with using the term "atmospheres" in the fields of urban planning and design, namely theoretical literature, applied research, and professional practice.

REFERENCES

Abrams, M., & Harpham, G. (2011). *A Glossary of Literary Terms*. Cengage Learning.

Abusaada, H. (2019). Revisiting the Word "Atmospheres" in the Urban Design Academic Field Based on the Artworks. *International Journal of Sustainable Entrepreneurship and Corporate Social Responsibility*, *4*(2), 22–43. doi:10.4018/IJPPPHCE.2019070102

Abusaada, H., & Elshater, A. (2019a). Internet of Things Interpreting a Myth: Internally Displaced Persons Transform Cities of Hardship into Chaos Cities. In H. Abusaada, C. Voulguth, & A. Elshater (Eds.), *Handbook of Research on Digital Research Methods and Architectural Tools in Urban Planning and Design* (pp. 47–81). IGI Global. doi:10.4018/978-1-5225-9238-9.ch004

Abusaada, H., & Elshater, A. (2019b). A Crowdsourcing Approach in Urban Design: A bibliographic review of cities of singularity. In H. Abusaada, C. Voulguth, & A. Elshater (Eds.), *Handbook of Research on Digital Research Methods and Architectural Tools in Urban Planning and Design* (pp. 1–24). IGI Global. doi:10.4018/978-1-5225-9238-9.ch001

Abusaada, H., & Elshater, A. (2020). Effects of urban atmospheres on changing attitudes of crowded public places: An action plan. *International Journal of Community Well-Being*, *3*(2), 109–159. doi:10.100742413-019-00042-w

Adams, D., Smith, M., Larkham, P., & Abidin, J. (2019). Encounters with a future past: Navigating the shifting urban atmospheres of place. Journal of Urban Design. *Journal of Urban Design*, *25*(3), 1–20. doi:10.1080/13574809.2019.1592668

Algeo, J. (2010). The Origins and Development of the English Language (6th ed.). Wadsworth, Cengage Learning.

Ambiances. (n.d.). *The Ambiances Network.* Retrieved from https://www.ambiances.net/network/about-the-international-ambiances-network.html

Ambience. (2018). In *Meriam Webester's dictionary.* Retrieved from https://www.merriam-webster.com/dictionary/ambience

Anderson, B. (2009). Affective atmospheres. *Emotion, Space and Society*, *2*(2), 77–81. doi:10.1016/j.emospa.2009.08.005

Ash, J. (2013). Rethinking affective atmospheres: Technology, perturbation and space times of the non-human. *Geoforum*, *49*, 20–28. doi:10.1016/j.geoforum.2013.05.006

Atmosphere. (n.d.). Retrieved from http://www.literarydevices.com/atmosphere/

Augoyard, J.-F. (2007). *Step by Step: Everyday Walks in a French Urban Housing Project.* University of Minnesota Press.

Backer, M. D. (2019). "Being different together" in public space: Young people, everyday cosmopolitanism and parochial atmospheres. *Social & Cultural Geography*, 1–20. doi:10.1080/14649365.2019.1594352

Beggs, P. J., Šikoparija, B., & Smith, M. (2017). Aerobiology in the International Journal of Biometeorology, 1957–2017. *International Journal of Biometeorology*, *61*(S1), 51–58. doi:10.100700484-017-1374-5 PMID:28607999

Bissell, D. (2010). Passenger mobilities: Affective atmospheres and the sociality of public transport. *Environment and Planning. D, Society & Space*, *28*(2), 270–289. doi:10.1068/d3909

Böhme, G. (1993). Atmosphere as the Fundamental Concept of a New Aesthetics. *Thesis Eleven, 36*, 113-126.

Böhme, G. (2013). Atmosphere as Mindful Physical Presence in Space. *OASE Journal for Architecture*, *91*, 21–32.

Böhme, G. (2014). Urban atmospheres: Charting new directions for architecture and urban planning. In C. Borch (Ed.), *Architecture, architectural atmospheres – On the experience and politics of architecture* (pp. 42–59). doi:10.1515/9783038211785.42

Buser, M. (2014). Thinking through non-representational and affective atmospheres in planning theory and practice. *Planning Theory*, *13*(3), 227–243. doi:10.1177/1473095213491744

Cerwén, G., Wingren, C., & Qviström, M. (2017). Soundscape actions: A tool for noise treatment based on three workshops in landscape architecture. *Frontiers of Architectural Research*, *6*(4), 505–518. doi:10.1016/j.foar.2017.10.002

Cooper, C. (2014). *Wiley Encyclopedia of Management.* John Wiley & Sons, Ltd.

Cresswell, T. (2009). Place. In N. Thrift & R. Kitchen (Eds.), *International Encyclopedia of Human Geography* (Vol. 8, pp. 169–177). Elsevier. doi:10.1016/B978-008044910-4.00310-2

Crystal, D. (2014). *Words in Time and Place: Exploring Language Through the Historical Thesaurus of the Oxford English Dictionary*. OUP Oxford.

Deleuze, G. (1990). *The Logic of Sense*. Columbia University Press.

Dewey, J. (2005). *Art as experience* (1934 ed.). Penguin.

Dovey, K. (2010). *Becoming Places, Urbanism/Architecture/Identity/Power*. Routledge.

Dufrenne, M. (1973). *The Phenomenology of Aesthetic Experience*. Northwestern University Press.

Egan, K. (2014). Film Production Design: Case Study of The Great Gatsby. *Elon Journal of Undergraduate Research in Communications*, *5*(1), 5–14.

Ford, T. (2012, December 6). *Nineteenth-Century Climate Change: Atmosphere, Culture and Romanticism*. Retrieved from http://rsha.cass.anu.edu.au: http://rsha.cass.anu.edu.au/nineteenth-century-climate-change-atmosphere-culture-and-romanticism

Gandy, M. (2017). Urban Atmospheres. *Cultural Geographies*, *24*(3), 353–374. doi:10.1177/1474474017712995 PMID:29278257

Griffero, T. (2014). *Atmospheres: Aesthetics of Emotional Spaces* (S. D. Sanctis, Trans.). Ashgate Publishing Limited.

Griffero, T. (2016). Atmospheres and felt-bodily resonances. *Studi di Estetica*, *XLIX*(IV), 1–49.

Guthey, G. T., Whiteman, G., & Elmes, M. (2014). Place and Sense of Place: Implications for Organizational Studies of Sustainability. *Journal of Management Inquiry*, *23*(3), 254–265. doi:10.1177/1056492613517511

Hasse, J. (2014). Atmosphere as Expressions of Medial Power, Understanding Atmospheres in Urban Governance and Under Self-Guidance. *Lebenswelt*, *1*(4), 214–229.

Hay, R. B. (1998). Sense of place in a developmental context. *Journal of Environmental Psychology*, *18*(1), 5–29. doi:10.1006/jevp.1997.0060

Healey, P. (2010). *Making Better Places*. Palgrave Macmillan. doi:10.1007/978-1-137-01379-8

Holt-Jensen, A. (2009). *Geography: History and Concepts* (4th ed.). SAGE.

Hough, M. (1990). *Out of Place: Restoring Identity to the Regional Landscape*. Yale University Press.

Hsu, H. L. (2017). Literary Atmospherics. *Literary Geographies*, *3*(1), 1–5.

Jaaniste, L. O. (2010). The Ambience of Ambience. *A Journal of Media and Culture, 13*(2). Retrieved from http://www.journal.media-culture.org.au/index.php/mcjournal/article/view/238

Jackson, J. B. (1994). *A Sense of Place, a Sense of Time*. Walker Art Center.

Jenner, R. (2013). Ambient atmospheres: Exhibiting the immaterial in works by Italian Rationalists Edoardo Persico and Franco Albini. *Interstices: Journal of Architecture and Related Arts*, 13-24.

Jorgensen, B., & Stedman, R. (2001). Sense of Place as an Attitude: Lakeshore Owners' Attitudes Toward Their Properties. *Journal of Environmental Psychology*, *21*(3), 233–248. doi:10.1006/jevp.2001.0226

Kazig, R. (2016, April 27). *Presentation of Hermann Schmitz' paper, "Atmospheric Spaces"*. Retrieved April 30, 2017, from Ambiances [En ligne]: http://ambiances.revues.org/709

Kemmis, D. (1995). *The Good City and the Good Life*. Houghton Mifflin Co.

Labuhn, B. (2016). Conceptualisations of Atmosphere in Text and in Image in Architectural Journalism (1991 - 2013). *Ambiances, tomorrow*. In *Proceedings of 3rd International Congress on Ambiances* (pp. 289-294). Volos, Greece: International Network Ambiances, University of Thessaly.

Mallgrave, H. (1996). *Otto Wagner: Reflections on the Raiment of Modernity*. Getty Publications.

McCormack, D. P. (2008). Engineering affective atmospheres on the moving geographies of the 1897 Andrée expedition. *Cultural Geographies*, *15*(4), 413–430. doi:10.1177/1474474008094314

McCrum, R. (2015). *The 100 best novels: No 69 – The Heat of the Day by Elizabeth Bowen (1948)*. Retrieved from https://www.theguardian.com/books/2015/jan/12/100-best-novels-elizabeth-bowen-heat-of-the-day-blitz

Michels, C. (2015). Researching affective atmospheres. *Geographica Helvetica*, *70*(70), 255–263. doi:10.5194/gh-70-255-2015

Michels, C., & Steyaert, C. (2017). By accident and by design: Composing affective atmospheres in an urban art intervention. *Organization*, *24*(1), 79–104. doi:10.1177/1350508416668190

Pallasmaa, J. (2014). Space, Place, and Atmosphere: Peripheral Perception in Existential Experience. In C. Borch (Ed.), *Architectural Atmospheres: On the Experience and Politics of Architecture* (pp. 18–41). Birkhauser. doi:10.1515/9783038211785.18

Poe, E. A. (2017). *Great Horror Stories*. Dover Publication.

Reichler, T. (2009). Changes in the Atmospheric Circulation as Indicator of Climate Change. In T. M. Letcher (Ed.), *Climate Change: Observed* (pp. 145–165). Elsevier., doi:10.1016/B978-0-444-53301-2.00007-5

Relph, E. (1976). *Place and Placelessness*. Pion.

Sandercock, L., & Attili, G. (2009). *Where Strangers Become Neighbours*. Integrating Immigrants in Vancouver. doi:10.1007/978-1-4020-9035-6

Scannell, L., & Gifford, R. (2010). Defining place attachment: A Tripartite Organizing Framework. *Journal of Environmental Psychology*, *30*(1), 1–10. doi:10.1016/j.jenvp.2009.09.006

Schmitz, H. (1978). *System der Philosophie*. Bouvier.

Schmitz, H. (2016, April 27). *Atmospheric Spaces, Ambiances*. doi:10.4000/ambiances.711

Schmitz, H. (2019). New Phenomenology: A Brief Introduction (Atmospheric Spaces) (T. Griffero, Ed.). Academic Press.

Schmitz, H., Owen, R., & Slaby, M. (2011). Emotions outside the box—The new phenomenology of feeling and corporeality. *Phenomenology and the Cognitive Sciences*, *10*(2), 241–259. doi:10.100711097-011-9195-1

Semper, G. (2004). *Style in the Technical and Tectonic Arts, Or, Practical Aesthetics*. Getty Research Institute.

Shaw, R. (2014). Beyond night-time economy: Affective atmospheres of the urban night. *Geoforum*, *51*(1), 87–95. doi:10.1016/j.geoforum.2013.10.005

Simpson, J., & Proffitt, M. (1997). *Oxford English Dictionary Additions Series* (Vol. 3). Oxford University Press.

Steele, D., Luka, N., & Guastavino, C. (2012). Constructing ideal soundscapes: A practical study on closing the gaps between soundscape studies and urban design. In *Proceedings of the Acoustics 2012 Nantes Conference* (pp. 2163-2168). Nantes, France: Société Française D'acoustique. 10.1121/1.4709108

Stefansdottir, H. (2017). The role of urban atmosphere for non-work activity locations. *Journal of Urban Design*, *23*(3), 319–335. doi:10.1080/13574809.2017.1383150

Stevenson, A. (2010). *Oxford Dictionary of English*. OUP Oxford.

Stewart, K. (2011). Atmospheric Attunements. *Environment and Planning. D, Society & Space*, *29*(3), 445–453. doi:10.1068/d9109

Sturm, S., & Turner, S. (2012). "Built pedagogy": The University of Auckland Business School as crystal palace. *Interstices*, *12*, 23–34.

Tawa, M. (2017). Consilient Discrepancy: Porosity and Atmosphere in Cinema and Architecture. *Architecture_media_politics_society*, *11*(13), 1-17. doi:10.14324/111.444.amps.2017v11i3.001

Thibaud, J.-P. (2002). From Situated Perception to Urban Ambiences. In *Proceedings of the First International Workshop on Architectural and Urban Ambient Environment* (pp. 1-11). Nantes, France: Centre de recherche méthodologique Learn to pronounce d'architecture, Ecole d'architecture de Nantes.

Thibaud, J.-P. (2015). The backstage of urban ambiances: When atmospheres pervade. *Emotion, Space and Society*, *1*(8), 39–46. doi:10.1016/j.emospa.2014.07.001

Weigelt, R. (2010, Dec 16). *Samuel Taylor Coleridge, "Christabel"*. Retrieved from wiki.uiowa.edu: https://wiki.uiowa.edu/pages/viewpage.action?pageId=41878932

Wigley, M. (1998). Architecture of Atmosphere. *Daidalos*, *68*, 18–27.

Wilkins, J. (2018). *The Discovery of a World in the Moone*. CreateSpace Independent Publishing Platform.

Wyndham-Davies, J. (Producer), & Morahan, C. (Director). (1997). *The Heat of the Day* [Motion Picture]. Retrieved from https://www.amazon.com/Heat-Day-Masterpiece-Theatre-VHS/dp/156442667X

Zumthor, P. (1998). *Thinking Architecture*. Lars Müller.

Zumthor, P. (2006). *Atmospheres: Architectural Environments - Surrounding Objects*. Birkhauser.

Chapter 2
The Concept of Atmospheres:
From Goethe to Bratton – How Atmosphere Is Key to Creating Smart Cities

Diana Soeiro
Centre for Socioeconomic and Territorial Studies (DINÂMIA'CET), ISCTE, University of Lisbon, Portugal

ABSTRACT

Architects and urban planners currently face the challenge of the Fourth Industrial Revolution (4IR) that quickly permeates urban environments. In an effort to understand this event as one that goes beyond technology use and technological innovation, the authors investigate how the concept of atmosphere is key to envisioning the future of smart cities. In order to clarify the understanding of the concept, this chapter presents an original alignment of three clusters of authors (Goethe and Wittgenstein, Böhme and Schmitz, and Bratton). The goal is to demonstrate that atmospheres and urban ambiances are fundamental urban design elements. They have the ability to positively shape technology use in cities as decisive elements to promote sustainable smart cities.

INTRODUCTION

The concept of 'atmosphere' is a mysterious one with it being hard to define its borders. Though this is fully in tune with the nature of the phenomenon itself, the main motivation in this chapter is to investigate the concept of atmosphere in-depth in order to better clarify why and how it is relevant to our experience of the built environment. It is important to do this in order to demand a more humanising use of technology in urban environments. We, therefore, identify three clusters of reference authors in order to trace how the concept has been developing. To present a genealogy of the concept of atmosphere is, therefore, our main goal and less attention will be given to smart cities literature – a rapidly increasing research area with an extensive body of literature. The goal is to identify the concept of 'atmosphere' as a foundational concept of smart cities. Becoming aware of this will allow us to determine how to better orient technology design and use towards smart cities where the human element prevails over technology.

DOI: 10.4018/978-1-7998-3856-2.ch002

We start by analysing the work of three pioneers. In *Theory of Colours* (1810) by Johann W. von Goethe (1749-1832), we can find a key contribution to the concept of 'atmosphere'. Which elements can we find in Goethe's proposal that is relevant to a better understanding of what is at stake in the concept of 'atmosphere' and, more importantly, how is his method of investigation on colour relevant to the inquiry into the concept's qualities? In order to answer these questions, we claim that Goethe anticipated the phenomenological method as a primary form of examination and that this method is vital to grasp the concept of 'atmosphere'. To show this, we link his method to the definition of phenomenology given by Edmund Husserl (1859-1938), who originally coined and defined the approach (1913; Escoubas, 2003). Later on, around 1950, Ludwig Wittgenstein wrote about Goethe's work on colours (1979) giving his own contribution to the theme. He worked on it during the last eighteen months of his life. These authors make us aware that in order to understand colour, it is necessary to study 'atmosphere' and due to the nature of the phenomenon, the methodology that is used to grasp the concept is crucial.

The second cluster of authors makes the connection between 'atmosphere' and architecture. We will first consider the work of philosopher Gernot Böhme (*b.*1937) who, considering Goethe's work, is one of the most relevant thinkers on the concept in the world today. We will specifically address four of his works (1995, 2006, 2017, 2017a) tackling the concept of 'atmosphere', the relation between architecture and 'atmosphere' and one of a broader scope denoting the built environment. Secondly, we will evaluate the work of philosopher Hermann Schmitz (*b.*1928), also deeply influenced by Goethe, who is better known as the founder of the movement New Phenomenology (*Neuen Phänomenologie*). Schmitz has written about the relation between space, body and feeling, which is a combination that simultaneously broadens and deepens the scope of the concept of 'atmosphere'. Through Schmitz's perspective, we become aware that the concept goes beyond sight, beyond our five senses, and this is what the philosopher aims to convey in his work (2004, 2009, 2014).

Benjamin H. Bratton (*b.*1968), the American sociologist, architectural and design theorist, in his seminal work *The Stack* (2015), develops the concept of 'atmosphere' under the concept of globalisation and the Fourth Industrial Revolution (4IR). Bratton's main reference is French philosopher Gilles Deleuze (1925-1995); however, the book's title allows him to forge his own identity, presenting a powerful perspective on today's world. The concept of 'atmosphere' is a valuable one to understand Bratton's viewpoint, which is described and scrutinised in different areas (six 'layers') including the 'city layer'. However, all layers intertwine and are embedded in technology, creating an indivisible space between reality and the virtual. What architecture and urban ambience can be created when the visible and invisible worlds are becoming one in the built environment surrounding us? The architecture of the virtual is everywhere, and it is exceeding the visible built environment, creating an added layer that has different rules. How can both be harmonised in our surrounding environment, and what impact does it have in the way we experience and feel space? The advent of smart cities is at the centre of this conflict. By reviewing recent literature on the impact of technology use in the built environment we conclude that cities must be sustainable in order to be smart. (Yigitcanlar et al., 2019). This implies a broader understanding of what is at stake when we say 'sustainability', humanising the way we conceive and use technology. We end by proposing our own concept of a smart city where we claim that a deeper understanding of the concept of 'atmosphere' can contribute greatly to both its sustainability and our wellbeing.

BACKGROUND

Ranging from the 18th century to nowadays, this chapter aims to present a historical background, establishing the genealogy of the concept of atmosphere. Over time, it is possible to observe that the development of the concept of atmosphere slowly incorporates a broad-ranging spatial component that becomes vital to better grasp what is at stake in the Fourth Industrial Revolution (4IR), reframing our understanding of the growing role of technology in urban environments.

GENEALOGY OF THE CONCEPT OF ATMOSPHERE

The Pioneers: Goethe and Wittgenstein

Johann W. von Goethe (1749-1832) is widely recognised as a German writer and statesman. However, only by a few is he acknowledged or even referenced when it comes to his scientific writings. The main reason for this is that although he wrote about scientific matters, his methodological approach did not follow the prevailing scientific system then or now.

In the eighteenth century, a *Naturforscher* (a student of nature; a naturalist) was someone who took an interest in *Naturgeschichte* (Natural History). This area encompassed various research subjects, predominantly *Naturwissenschaften* (Natural Sciences), which included an interest in organisms (*Lebewesen*) studied in Biology (Botanics and Zoology), Palaeontology, Ecology and Biochemistry, and Archaeology, Geology, Astronomy, Physics and Meteorology as well. In his scientific writings, Goethe investigated and reflected on all these subjects at some point, which confirms him as a *Naturforscher*. According to von Engelhardt (2009), *Naturforschung* cannot be identified with *Naturphilosophie* around 1800 because at the time it itself was not homogeneous. Though there are many divergences between the individual *Naturforscher* (and although Goethe never considered himself a Romantic (Uberoi, 1984; Knight, 2009, p. 17)), the one thing they had in common was that "in the interest of a meaningful and beneficial investigation into nature" (von Engelhardt, 2009, p. 56) one must not deny the interrelation between physics and metaphysics. In the eighteenth century, the big shift is that nature is perceived as subject to change. (von Engelhardt, 2009, p. 56)

Goethe's perspective is that nature's immutable character and its laws appear to us through careful observation of its infinite variations. Consequently, this implies that one should observe first and only then dare to formulate a theory (instead of formulating a theory and then conveniently accommodating what is observed to fit it); and secondly, that the possibility of claiming to know something scientifically is perhaps never possible since one is never fully aware of all of nature's variations. Science is a constant dialogue between the ability to carefully observe, experiment and formulate an idea that once established is questioned again by careful observation, leading to experimentation, and so on. (Goethe, 1989, p. 220-232)

The incompatible nature of his scientific interests with his literary pursuits was addressed by Goethe himself, particularly in relation to the reception of his work *The Metamorphosis of Plants* (Die Metamorphose der Pflanzen) published in 1790 (e.g. Goethe, 1989, pp. 167-176). In his defence, he states unapologetically: "[…] a man of lively intellect feels that he exists not only for the public's sake, but for his own" (Goethe, 1989, pp. 169).

Out of Goethe's three seminal scientific achievements, the first is his discovery of the "intermaxillary bone", now known as the "premaxilla", a pair of cranial bones that are located at the front of the upper jaw and bear the incisors in animals that have these teeth. It was published in 1784 and later given the title *An Intermaxillary Bone of the Maxilla Is to Be Ascribed to the Human as well as to Animals* (Dem Menschen wie den Tieren ist ein Zwischenknochen der obern Kinnlade zuzuschreiben). Up until then, anatomists had only recognised this structure in animals and Goethe demonstrated that it also existed in humans. Whether or not this was truly a Goethian discovery, what Feigenbaum (2015) interestingly discusses is that this event (and the full awareness that Goethe has of its importance) is a fundamental step against the anthropocentric ontology that grounded the natural sciences of the time. Moreover, it lays the ground for a non-anthropocentric vision of nature that will prevail in the future. In 1786, Goethe states that he was chiefly occupied with comparative anatomy, "taking enormous pains [...] to interest others in [his] conviction that the intermediary bone was undeniably present in man" (Goethe, 1989, p. 245). That same year, Goethe decides to take a two-year trip to Italy. During that time, "by means of active observations and creative imagination" (Goethe, 1989, p. 245), he pursues the versatility of Nature in the plant world. The result is the publication of *The Metamorphosis of Plants* (Die Metamorphose der Pflanzen) in 1790. His third innovative work is *Theory of Colours* (Farbenlehre), published in 1810.

The effort of organizing his observations on plants to be published as a book led him to formulate a particular method, alternative to that of traditional science. This method is now briefly presented as it is relevant to understanding Goethe's *Theory of Colours* and, consequently, his understanding of 'atmosphere'.

Goethe's method is the genetic method, and the realm of his scientific approach is specific enough for him to boldly propose a new discipline that conducts an investigation of a given object according to itself – morphology. This means that morphology frames the object, which is then observed focusing on the question 'how' and not so much on the 'why'. According to Goethe, the observation of science should not be teleological because to him 'natural system' is "a contradiction in terms. Nature has no system; she has, she *is* life and its progress from an unknown centre toward an unknowable goal" (Goethe, 1989, p. 116). Progress is a centrifugal force towards infinity, responsible for nature's constant state of metamorphosis. The concept of metamorphosis is central in Goethe's concept of science as it is responsible for identifying scientific research as endless. However, Goethe warns, it can be a highly dangerous concept leading to formlessness, destroying knowledge and disintegrating it (Goethe, 1989, p. 116). So that progress does not lose itself in the infinite, a centripetal force of specification must be provided in order to act as a counterbalance. This centripetal force that cannot be disturbed in its deepest nature by anything external (Goethe, 1989, p. 116) is polarity. Progress and polarity are identified as the two main 'activating forces' in Nature. "Polarity is a property of matter insofar as we conceive it as material; progression is a property of spirit, insofar as we conceive it as spiritual" (Goethe, 1989, p. 245). According to Goethe, "the life force" [...] creates just as well *from* the outside as well as *toward* the outside" (Goethe, 1989, p. 83) and "Nature is molded *from* all directions and *toward* all directions" (Goethe, 1989, p. 84).

The metamorphosis of plants draws attention to a twofold law: "(1) the law of inner nature, whereby the plant has been constituted; (2) the law of environment, whereby the plant has been modified" (Goethe, 1989, p. 85). In this formulation, Spinoza's influence on Goethe can be observed where the latter's proposal of nature as a creative power and as created product translates into Spinoza's terminology as *Natura naturans* ("nature naturing") and as *Natura naturata* ("nature natured"). (Goethe, 1790, p. 106)

According to this view, morphology is the science that frames an object of study. Goethe considered that natural history was restricted to the outward appearance of forms, regarding them as a whole, while anatomy was aimed at the knowledge of their inner structure, of the parts of an organic body, privileging the human body (Goethe, 1989, p. 87, 89). Morphology should be regarded as an independent science, auxiliary to physiological science ("Consideration of form both in its parts and as a whole, the conformities and deviations, apart from all other considerations."). His proposed definition of its object was "Consideration of the organic whole by visualizing and linking all these considerations through mental processes" (Goethe, 1989, p. 88). The considerations at stake were those taken by other sciences: Natural History, Natural Philosophy, Anatomy, Chemistry, Zoonomy and Physiology.

Morphology should include a theory of form, formation and transformation of organic natures (Goethe, 1989, p. 88). Theory of form is the consideration through mental processes that visualise the organic whole and link all considerations from other disciplines. The formation and transformation of organic natures corresponds to a careful observation of both the centripetal and centrifugal aspects (metamorphosis).

The theory of form is achieved both by observation and intuition, and it is described in his genetic method as practised by those who find a balance between fact-finding and contemplation. The process of finding that balance is described thus:

When I see before me something which has already taken shape, inquire about its origin and trace back the process as far as I can follow, I become aware of a series of stages. Naturally, these cannot be observed side by side with the physical eye but must be pictured mentally as a certain ideal whole.

Inclined at first to postulate certain stages, I am finally compelled, since Nature never proceeds by skips and jumps, to regard the sequence of uninterrupted activity as a whole, annulling individual details so as not to destroy the total impression (Goethe, 1989, p. 93).

Attempting to find intermediate points in the process and reflecting upon the results of these attempts, "one sees eventually fact-finding must be terminated, contemplation of a growing thing must begin, and ultimately an idea must be expressed" (Goethe, 1989, p. 93). At this point he specifically quotes "a city as a work of a man" as an example of this process. Goethe's privileged example of the theory of form, along with metamorphosis, is a city. The exercise of envisioning a city, just like the exercise of (scientifically) knowing an object, implies contemplation, polarity and progression. Polarity requires a continual attraction and repulsion; progression is constant upward striving. The idea resulting from the back and forth between contemplation and fact-finding is what Goethe will designate as 'original phenomenon' (Urphänomen). (The idea stems from an initial formulation from his plant observation, 'originary plant' (Urpflanze)). This is a "virtual inner identity" (Goethe, 1989, p. 85) not fully perceived by the senses and only possible to be recognised also through an inner sense. (p. 90) Only this explains why we are able to know a given object since we cannot regard an organic nature as a unit for long. The empirical unity is recognised with our senses but the ability to acknowledge all stages of development (past, present and future) as having originated from an idea archetype (*from* and *towards* a specific form) concerns the contemplative skill. This is again why no natural entity can be imagined without an environment. (Goethe, 1989, p. 90, 94, 95)

In *Theory of Colours* (1810) Goethe puts his scientific approach (Morphology) into practice taking colour as his privileged research subject. The choice is a bold one since, according to traditional science and particularly Newton (on the polemic between both authors, see Sepper, 1988), colour is perceived by nature as a formless phenomenon. It is investigated as an isolated phenomenon occurring independently of other phenomena and objects – as if having its own nature apart from Nature. Goethe's point of departure is the opposite: in order to understand any phenomenon where colour is present in Nature, it is necessary to understand the phenomenon of colour itself. This is why those who take an interest in the subject of colour are usually severely disappointed when reading Goethe's *Theory of Colours*. The work is an assemblage of the most relevant careful observations about colour (its formations and transformations, i.e. metamorphosis) occurring in Nature. More significantly, the book's main research area is epistemology and not aesthetics. It is the most solid example of Goethe's proposed science of Morphology with all its advantages and disadvantages. Among the former are: a careful observation of the phenomenon not in a laboratory setting but as we witness it daily in Nature; observations are described in a simple and clear manner; and facts are not twisted in order to formulate a theory that the author previously had in mind. Among its disadvantages, owing to its object of study (colour), the work presents itself as the ultimate test for the possibility of Morphology as science and when evaluated as a scientific method, it fails. Why? The book is in fact an infinite book that could assemble many other observations but Goethe strongly resists ending it, remaining the never-ending fact-finder and never accomplishing the contemplative idea.[1] What is colour's 'originary phenomenon' (Urphänomen)? Perhaps light.[2] In an epistemological sense, perhaps truth itself.

Adopting a very similar method later on, the philosopher Edmund Husserl proposes phenomenology where he urges those who think and aim at producing scientific thought to go back to things themselves. Colour is not perceived as a secondary property added to things but as inherent to the thing itself being an intrinsic part of its form and identity[3] (1913; Escoubas 2003). Also, what we know is partially known by our senses but because we cannot fully trust them, Husserl aims at describing the underlying mental processes at stake. The rupture with traditional science (and positivism) that Husserl proposed was anticipated by Goethe, who in essence proposed a similar method.

Around 1950, the philosopher Ludwig Wittgenstein wrote about Goethe's work on colours (1979) adding his own contribution to the subject and, having worked on it during the last eighteen months of his life, the book was published posthumously. Wittgenstein takes Goethe's book as his main reference and quotes it several times. The main concept that plays a central role in Wittgenstein, clearly reflecting Goethe's influence, is that of 'context'. Any colour can only be known according to its context. And because context frequently varies, it is hard to establish scientific knowledge about the phenomenon colour. It is the ultimate phenomenon that constantly escapes being crystallised, resisting having an established identity. Its identity is always virtual and temporary. The main thing both authors have in common is that their priority is not to observe facts to be able to establish laws as quickly as possible. The gaps – to delve more deeply into the deviations and what escapes being established as a law – are what enable them to get to know the phenomenon better. This strategy is similar to the phenomenological method where every act of knowing a given object starts by identifying what it is not and in which the gaps between things become extremely relevant to establish an object's identity. The main difference between Goethe's and Wittgenstein's approach is that to Goethe, colour is an expression of an infinite nature, of light, and simultaneously the phenomenon that best illustrates the way we know things. To Wittgenstein, colour is approached through the concepts of 'context' and 'language games' – what makes us recognise colour as a specific colour, how do we learn what a colour is, how are we able to name a given colour, and, ultimately, what allows us to have a shared understanding about colour when we denote it. Both authors tackle the subject in order to address epistemological concerns. Still, Wittgenstein, instead of framing his investigation in Nature, frames it in establishing an analogy with the way we use language. (Soeiro, 2011)

In a given space, along with a set of objects, colour and light play a key role in establishing an atmosphere. Although objects can be identified, a particular colour chosen and a certain light selected, it is not easy to determine what exactly generates a specific atmosphere. All elements work as a whole, each contributing with its individuality but fully depending on all the others to appear to us as a whole. If you add living organisms, e.g. people, to this, whether in a private or public place, it becomes even harder to identify what determines an atmosphere.

In urban environments, the presence of technology is rapidly growing. Therefore, when envisioning the future of the urban ambiences we aim to design and implement in cities, it is an extra layer to consider. Objects, colour, light and people are all highly volatile elements and context-dependent phenomena. The technological aspect has the advantage of being a measurable and easy to reproduce element, but given technology's tendency towards artificial intelligence where machines can be flexible enough to take their own decisions based on context instead of mere reproducible mechanisms, they themselves have their status upgraded to that of a living organism. As Goethe writes, taking Spinoza as his reference: "A living thing cannot be measured by something external to itself; if it must be measured, it must provide its own gauge. This gauge, however, is highly spiritual, and cannot be found through the senses." (Goethe, 1988, p. 8)

With Goethe and Wittgenstein, through the phenomenological method anticipated by Goethe's Morphology, we establish that something highly spiritual that cannot be found exclusively through the senses corresponds to the concept of atmosphere.

Böhme and Schmitz: New Phenomenology

Contemporary philosophers Gernot Böhme and Hermann Schmitz have both studied Goethe, including his scientific writings and his book on colours. Böhme's work relies on Schmitz's proposed New Phenomenology (*Neuen Phänomenologie*) which takes body, space and feeling as its main elements.

Böhme's main historical references are Plato, Kant and Goethe. Having first addressed the concept of atmosphere in 1989 (*Für eine ökologische Naturästhetik*), in 1995 he dedicates an entire book to the subject, *Atmosphäre: Essays zur neuen Ästhetik* [Atmosphere: Essays on a new aesthetics]. Then, in 2006, he publishes *Architektur und Atmosphäre* (translated into English in 2017 as *Atmospheric Architectures: The Aesthetics of Felt Spaces*). Also, in 2017 two other volumes are published, *Critique of Aesthetic Capitalism* (a translation of a work previously published in 2016) and an edited volume that is a great introduction to Böhme's work on the concept of atmosphere for English speakers titled *The Aesthetics of Atmospheres* (2017c).

Since his first book on the subject, Böhme has developed the concept of atmosphere from an idea that fitted his ecological aesthetics of nature into a key element of his overall aesthetic, i.e. he conceived it as a theory of sensory perception (Thibaud, 2017, p. 17). To Böhme, "what mediates objective factors of the environment with aesthetic feelings of a human being is what we call atmosphere. The atmosphere of a certain environment is responsible for the way we feel about ourselves in that environment" (Böhme, 2017c, p. 26). Following Schmitz, atmospheres are nothing without a subject feeling them. Böhme's development of the concept is mainly in the field of aesthetics. Over time he used it to formulate a critical theory of everyday life and the aestheticization of our lifeworld, aiming at contributing to the ongoing democratisation of culture (Böhme, 2017c, p. 41, 42). Böhme's general theory of perception states that what we first and immediately perceive are "neither sensations nor shapes or objects or their constellations, as gestalt psychology thought, but atmospheres, against whose background the analytic regard distinguishes such things as objects, forms, colors etc." (Böhme, 2017c, p. 82). In particular, atmosphere "is something between the subject and the object; therefore, aesthetics of atmosphere must also mediate between the aesthetics of reception and the aesthetics of the product or of production" (Böhme, 2017c, p. 88). It can therefore be experienced as a passive phenomenon or as an active phenomenon. Architecture and urban environments allow us to produce atmospheres in the sense that in essence these presuppose that someone creates them. It is not simply a matter of designing an object but also of creating the conditions for its appearance (Böhme, 2017c, p. 231, 232). When it comes to the atmosphere of a city, it is important to distinguish between atmosphere and image. "The image of a city is the consciously projected self-portrait, and the sum of its advantages, that an outsider might enjoy. (…) The atmosphere of a city is precisely the way life unfolds within the city" (Böhme, 2017c, p. 383, 384). The concept of atmosphere to think about a city is useful for three reasons. First, it frees us from the restriction of being bound to the visual and the symbolic. Secondly, instead of the city being judged in terms of its aesthetics or cultural aspects, the priority becomes to understand how we feel in the city. The third and final advantage is the ability to focus both on the subject, who exposes himself to an atmosphere, and also on the object, the means by which an atmosphere is created (Böhme, 2017c, p. 384-388; pp. 404-422). The ability to create atmospheres using light and sound is also a topic dealt with by Böhme (2017c, p. 500-640).

Hermann Schmitz's dissertation, which he finished in 1955 and later published in 1959, was on Goethe and his work strongly influenced the philosopher's lifework. According to Böhme, Schmitz's concept of atmosphere uncouples the phenomenon away from things, developing in its place the spatial character of atmospheres. "Atmospheres are always spatially 'without borders, disseminated and yet without place that is, not localizable.' They are affective powers of feeling, spatial bearers of moods." (Böhme, 2017c, p. 63)

On the spatial character of atmospheres, Schmitz states: "Determining location […] presupposes that the reference objects are stationary." This implies stillness (or vice-versa) as experienced in

area-less spaces, e.g. as still water, calm silence, evening repose, resting when tired. One must start out from such morphic experiences of stillness in order to find reference objects through which one can set up a locational space with positions and distances; in this, these reference objects can then be determined themselves by locations, and then secondarily the stillness of the reference objects can count as remaining at these locations (Schmitz, 2016, p. 3).

These examples are excellent to contextualise, in the realm of architecture and urbanism, elements that can actively be used to create an atmosphere. Relating body, space and feeling, Schimtz proceeds to identify as the two most important types of area-less spaces the space of the *felt body* and the space of emotions as atmospheres.

The *felt body* of a person is everything that he himself can perceive as

belonging to himself, in the vicinity – not always within the boundaries – of his material body [Körper], without making use of the five senses of sight, hearing, touch, smell, taste and of the perceptive material body schema (the habitual image of a person's own material body) that has been obtained from his or her experiences, particularly those of sight and touch (Schmitz, 2016, p. 3).

One may assume that if the five senses are not considered to investigate the concept of atmosphere, one is left with nothing. On the contrary, following Goethe's approach, focusing on the inner sense and going beyond the five senses, one unveils the true potential of the concept not only to aesthetics but also as an ontological operational concept. This is the highly spiritual aspect of living things that has its own measure, a highly spiritual one that goes beyond the senses – which Goethe mentioned having Spinoza in mind. According to Schmitz, what is included in atmosphere that goes beyond the five senses is the purely *corporeal* stirrings (such as fright, anxiety, pain, hunger, thirst, itching, sharp pains, disgust, vigour or fatigue); the *corporeal* stirrings which are affective marks of emotions (joy, sadness, anger, shame, fear, courage, compassion, contentedness and despair); the perceived voluntary and involuntary movements (such as walking, grasping, jumping, dancing, shivering, twitching or swallowing) and irreversible *corporeal* directions, sometimes occurring without movement such as when one gazes and sometimes tied to movements as in breathing out and swallowing. All these cannot be perceived by one's own *felt body*. "With the corporeal stirrings, we encounter atmospheres for the first time" (Schmitz, 2016, pp. 3, 4).

Schmitz presents his own definition: "an *atmosphere* is a total or partial, but in any case comprehensive, occupation of an area-less space in the sphere of that which is experienced as being present. I speak of 'occupation' rather than filling, so that an atmosphere of emptiness can be taken into account also" (Schmitz, 2016, p. 5).

To the philosopher, the area-less emotional space has three layers. The base layer is formed by two pure moods: contentedness and despair. The second layer consists of pure stimuli, e.g. abstract joy and melancholy, apprehension and longing, discontent and mistrust, which are just looking for triggers and are characterised by presentiment, a sense of anticipation. The third layer is formed by the thematically centred emotions in which the directed stimuli are pulled together around a theme (Schmitz, 2016, p. 5).

What is achieved by dwelling is the introduction of atmospheres into locational space which can go just as far as a person preparing his own emotional space from the emotions that have been made available to him at particular locations. This is why a dwelling must be delimited so that a given atmosphere does not escape from the person's control. "Dwelling is more than satisfying physical needs through walls to protect against wind and weather, namely the art of capturing atmospheres which are emotions and designing them such that the person can attune to them in harmony with his corporeal mood" (Schmitz, 2016, p. 9).

This approach is entirely in tune with the goal of New Phenomenology, founded by Schmitz, that is to be able to conceptualise the involuntary lived experience. (Schmitz, 2018, p. 52) The concept of atmospheres takes a central role in the sense that in it, contrary to any already established system, feelings are perceived as moods of the soul belonging to the private sphere and having no spatial dimension. Schmitz argues "feelings are atmospheres that spread in space and forces that grasp carnally" (Schmitz, 2018, p. 51).

This understanding of a phenomenon in Schmitz's New Phenomenology is different from that of Husserl's ("back to things themselves") and even from that of Heidegger's (for whom a phenomenon is "what is revealed"). "A phenomenon for someone at a time is a state of affairs of which the person cannot in earnest deny that it is a fact." (Schmitz, 2019, 1144) Two main features of this understanding of phenomenon are at stake: (1) A phenomenon respects 'someone', 'at a particular point in time' having a more dynamic and moment to moment basis instead of being presented as an apodictic certainty; (2) The concept state of affairs should replace that of a thing when addressing a phenomenon because it contains "the aspect of the thing by determining something as a case of something." (Schmitz, 2019, 1158) New Phenomenology's understanding of knowing a given object/state of affairs as it is presented to someone at a particular point in time brings the phenomenological method even closer to Goethe's Morphology. The infinite possibility of different situations that resist being grasped by a single concept or idea is only possible to access through contemplation as it belongs to the realm of spirituality and not accessible through the five senses.

Going back to its original sense, an atmosphere is not only related to the five senses, objects, sound, light and colour. To New Phenomenology, the concept of atmosphere is again central since the phenomenon is perceived as a highly volatile element similar to that of moods or weather conditions ("being under the weather").[4] All elements from the realm of the visible and invisible world, being temporary state of affairs, influence each other constantly creating and renovating previously established atmospheres. To Schmitz, the relation between body, space and feeling is what allows him to design a philosophy of the body based on the lived experience and grounded in the concept of atmosphere that is strategically located in between the visible and invisible world, allowing for an ever-dynamic/virtual identity foundation of ontology.

As for Goethe and Wittgenstein, the concept of atmosphere is deeply intertwined with epistemological and ontological questions in Böhme's and Schmitz's work as well. However, in this second cluster of authors, particularly in Schmitz, we establish that the concept of atmosphere is undoubtedly connected with our lived experience of space – the space that we dwell in and the space that affects, and is affected by, our moods through our body and mostly through feelings. The future of architecture and urbanism and what is at stake when it comes to creating new atmospheres in increasingly technological urban environments becomes clearer.

Schmitz affirms: "Mankind is helplessly carried away with technological progress and social connectedness by an unavoidable amount of offers without having salient conceptual criteria for self-contemplation and thus asserting itself. New Phenomenology aims to aid humans by equipping them with poignant but elegant concepts that enable involved self-contemplation." (Schmitz, 2019, 1212-1219) Along with Goethe, Schmitz aims at contributing towards a better balance between fact-finding and contemplation endorsing a more radical understanding of contemplation, i.e. self-contemplation. New atmospheres and urban ambiences should consider the body, its moods and feelings, as the key element for designing new solutions.

Bratton

American sociologist Benjamin Bratton has produced a significant amount of literature on digital and visual culture, media and technology. Here we choose to focus on his 2015 book, titled *The Stack: On Software and Sovereignty*, which is about architecture, aesthetics and the impact of the cultural implications of computing and globalisation. Strongly influenced by the work of French philosopher Gilles Deleuze, Bratton creates an original work where he establishes an analogy between the architecture of technology and the architecture of space, arguing that both are converging into one and the same object. This is relevant to understanding how the Fourth Industrial Revolution (4IR) can impact urban environments and how we can better humanise the future of smart cities. This implies becoming aware of the path being traced by technology and artificial intelligence in order to, hopefully, find elements and solutions that allow for quality lived experiences in urban environments, in particular in smart public places. Atmospheres go far beyond the five senses. As we have established, they are crucial to our understanding of ourselves as humans, deeply impacting the way we know, what we are and the way we feel.

Bratton's pivotal work presents the concept of "The Stack" and this is the book's gravitational centre. The Stack is a kind of planetary-scale computing system that expresses the new architecture in which we divide up the world into different sovereign spaces. He aims at presenting new vocabulary and a new interdisciplinary model that crosses political philosophy, architectural theory and software studies. This is a book of design theory where the presented model "does not put technology as a phenomenon taking place 'inside' a 'society', but sees a technological totality as the armature of the social itself". (Bratton, 2015, p. xviii) His perspective is not pro- or anti- technology, pro- or anti- "The Stack". The claim is that "The Stack" is a multi-layered structure of deeply intertwined software protocols that establish a global vertical order creating a different world to the one before where computation becomes governance. Bratton's aim is to present the state-of-the-art of this new order and present some solutions so that we can redesign and replace the Stack-we-have for the Stack-we-want. This calls for a geopolitics of design that delves in computation but also explores new possibilities when it comes to vertical systems of designation and decision. "The Stack" makes it clear that the way software systems are used nowadays, owing to their global scale and widespread use in every area, strongly impacts cultural, institutional and technical systems. At this point, technology cannot be investigated as if existing in a niche. Correspondingly, cultural and institutional changes occurring in the last decade or so cannot be explained without a deep understanding of technology's growing role, which greatly shapes and impacts them. *The Stack* (2015) is a brave attempt to map out these all-encompassing changes and megastructures.

Bratton identifies six layers that correspond to different forms at different scales of planetary-scale computation: (1) energy and mineral sourcing and grids (Earth layer); (2) subterranean cloud infrastructure (Cloud layer); (3) urban software and public service privatisation (City layer); (4) massive universal addressing systems (Address layer); (5) interfaces drawn by the augmentation of the hand, the eye or dissolved objects (Interface layer); (6) users both over-outlined by self-quantification and also exploded by the arrival of legions of sensors, algorithms and robots (User layer). (Bratton, 2015, pp. 4, 5)

Being aware that "The Stack" describes a vertical megastructure because each layer is somewhat independent, let us now consider the "City layer". The goal is to briefly present Bratton's description so that we can proceed and explore the current role of technology in urban environments, having in mind how atmospheres can play a significant role in shaping their future.

Bratton's definition of a city is an interesting one: "While it captures humans and consoles them within walls, it also provides boulevards of escape toward other urban interiors". (Bratton, 2015, p. 148) Being part of "The Stack" also allows for other layers to become tangible infrastructures, for example, opening and closing urban spaces to different *Users* as mobilisation (acceleration) is a key feature in cities, contrasting with that of settlement (enclosure). Physical objects engage side by side with data (accessed through Apps using the Cloud-devices that allow us to access an array of modern itineraries). This allows us to identify two ideal models of a city: "one is a city of partitions, permanent centres, and enveloped populations, and the other a city of movement expressed through nomadic landscapes, shifting perspectives, and impermanent networks" (Bratton, 2015, p. 150). As happens in any other platform, the grid as experienced in an urban environment is what allows us to describe a city (through its rigidity) and also what keeps it permeable through its open channels of circulation.

"The Stack" is designed "to disseminate a repertoire of accidents that frustrate the more unwanted forms of centripetal incarceration" (Bratton, 2015, p. 153). Therefore, the privileged way of creating a sense of bounded universality is to promote the generic *User* subject position. In order for this to be possible, the *Earth* layer provides the energy; the *Cloud* layer provides access to data centres that have the ability to create urban envelopes that are able to be accessed through *Interfaces* and finally the *Ad-*

dress layer provides a network presence for any "thing" within the urban landscape and for the potential of communication among them.

What is it like to build landscapes at Stack level? Megastructures aim at organising human habitation and urban flow. The result seems to be not so much the neutralisation of placefulness but rather "a total architecture withdrawn from the public city and bound by its own structural borders, gates, walls, and skins, gardens, introverted from its immediate environment so as better to connect to external economies on its own terms. Enclaves inside of enclaves digest one another all the way down" (Bratton, 2015, p. 155). Ideally, no discipline has more expertise in interface design than architecture. (Bratton, 2015, p. 166) This demands that university curricula catch up with a rapidly changing landscape in order to be up to the task.

Taking architect Aleandro Zaera-Polo's essay "The Politics of the Envelope" as a reference, he uses the term 'envelope' to describe buildings as envelopes that, more than their material expression, embody a particular link between spatial typologies and political modalities. Software envelopes also conform into typological envelopes and in cities physical and virtual envelopes overlap.

Concerning physical and virtual envelopes, Bratton conducts a relevant thought experiment where all architects and urbanists are divided into two groups. The first group should stop designing new buildings and new developments immediately and focus on designing and programming software that provides for the better use of structures and systems we already have. The second group should continue as before. What would very likely take place is that the first group would prioritise the design of mobile handsets, i.e. mobile control apparatus, essential to fulfilling a function that is crucial to the city – that of proximity to people, markets, goods, transport and information (Bratton, 2015, p. 168).

Bratton's proposal is that the design questions, and solutions, should not be addressed from a perspective where artificial and physical envelopes are separated. Software and urban programming should reorient to mix and merge "in more clever ways and, through this, to express emergent sovereignties more explicitly. (…) [O]ur design concern should be with how a mixed interfacial regime could support a loquacious multiplication of outcomes impossible for either hard or soft systems alone" (Bratton, 2015, pp. 169, 170).

Examples given by Bratton where it is clear that hard systems shaped physical architecture are: the *Googleplex* (by Clive Wilkinson and others) and its upgrade to *Google 2.0* by architects Bjarke Ingels and Thomas Heatherwick (in North Bayshore); Frank Gehry's early proposal for a new Facebook headquarters in Menlo Park, a company town designed to promote contact between employees in motion; *Campus 2* in Cupertino, as proposed by Apple and Sir Norman Foster during Steve Jobs' last years where a giant spaceship seems to have landed among apricot groves in an apparent pre-launch position; and Amazon's biosphere dome (South Lake Union, Seattle), a city within a city. (Bratton, 2015, pp. 176-189) These examples show how the utopian can rapidly convert into the dystopian. These *Cloud* platform megastructures, as Bratton calls them, draw "economies of the flesh, information, energy, and symbolisation into a web of settlement and displacement as vast as it is uneven and asymmetrical" (Bratton, 2015, p. 189).

There are also megastructures all around the world that are architecture of and for things, conceived to "accommodate and express the algorithmic resorting and distribution of inanimate objects, commodities, and cargo from sites of extraction, to assembly, to consumption, and back into landfills or the production cycle. (…) this object-oriented economy of molecular logistics is expressed in 'planetary surfaces'" (Bratton, 2015, p. 188). He proceeds to mention several examples of what is at stake such as warehouses that are so large their floors have been laser-levelled against the curvature of the Earth;

bar codes, RFID chips and scanners that replace windows and walls; and robotic platforms, shelves and stockers that can easily lift over a ton of goods at once.

Along with Bratton, the authors of this chapter claim that the way forward should encompass a solution where design is the key feature, allowing architects and urbanists to develop solutions that mix and merge hard and soft systems.

The concept of atmospheres, bridging physical and virtual elements, is vital to lead the way on how to do this. Also, owing to its balance between fact-finding (experiment) and contemplation, it ensures that whatever environments will be designed, they will be human and suited for humans.

TECHNOLOGY AND URBAN ENVIRONMENTS

Having identified a genealogy of the concept of atmosphere, let us briefly sum up how each of the approaches presented can contribute to promoting the role of the senses in smart cities. The main element common to all the approaches is that from Goethe to Bratton they all successfully intertwine physical and virtual elements going beyond the five senses.

Goethe and Wittgenstein demonstrate that colour is a phenomenon which in order to be truly explained cannot be perceived as an isolated phenomenon. Its behaviour and dynamics can only be grasped when putting into context, when placed side by side with all other surrounding elements. In that sense, to understand one colour, it is necessary to understand all other colours, expressed by a given amount of elements at a given time in a given place. This ground-breaking perspective demonstrates that colour goes beyond the sense of sight.

Having studied both Goethe's and Wittgenstein's work on colour, Böhme and Schmitz go a step further and explicitly address the concept of atmosphere contextualising it inbuilt space. The pioneering aspect is identifying the body as a fundamental element to define the concept of atmosphere. Now we are far beyond the sense of sight and, particularly with Schmitz, and we are being the five senses. The body itself is perceived as an element that is able to generate its own atmosphere. Goethe and Wittgenstein showed that the eyes are not merely passive receptacles of light and that there was more at stake. Böhme and Schmitz argue that the body is not merely a passive receptacle of its surrounding environment and atmospheres as it has the ability to actively generate its own atmosphere.

Bratton, in *The Stack*, claims that not only our bodies but also our minds are deeply embedded *in* and embodied *with* technology use, which is particularly experienced in urban environments. The architecture of technology is increasingly hard to separate from the architecture of the mind and the architecture of the body, influencing us beyond the five traditional senses. Examples of other senses at stake, up until now neglected, have recently been identified by Eccleston (2016): balance, movement, pressure, breathing, fatigue, pain, itch, temperature, appetite and expulsion. Decisions should be taken to conceive, design and implement technology in the realm of urban environments that do not hinder but strongly contribute to promoting the highest level of performance of the senses among the orientations for smart cities.

Smart cities encompass the challenge of integrating the use of Artificial Intelligence (AI) in their design. This includes incorporating in the urban fabric decentralised consensus, nanotechnology, robotics and autonomous vehicles. These elements are framed within the event of the Fourth Industrial Revolution. Professor Klaus Schwab (Founder and Executive Chairman of the World Economic Forum, the International Organization for Public-Private Cooperation) indicates that its beginnings took place after the Internet (in the 1990s). The 4IR is still very connected with digital technology, but a mobile internet, a smaller scale and a lower cost provide the opportunity both for worldwide massification/accessibility and for the integration of systems. This is what enables it to propel artificial intelligence and machine learning. (Schwab, 2017) Luciano Floridi (2014) also claims that we are living in the 4IR era but uses cultural events to determine its advent. According to him, the 4IR was initiated by Alan Turing's (1912-1954) work which revolutionised computer science, a task that once belonged to humans but since then has belonged to the machine.

City planning increasingly involves physical and virtual elements, and the Internet of Things (IoT) becomes a key element affecting the way we design buildings, connectivity, data, energy, governance and transport. (Artificial Intelligence and Life in 2030, 2016; Tegmark, 2017; Acuto *et al.*, 2019; Allam *et al.*, 2019; Park, 2019; Pelton, 2019) However, as Bratton argues, it is not the case that the virtual element is permeating the physical realm. At this point, we are beyond that. There is already a hierarchy put in place where the virtual element, the architecture of the virtual, is commanding how the architecture of physical things should be shaped from the large scale (the *Earth* scale) to the scale of the *User*.

What then are the shapes of buildings and cities to come? How can we find our place within them?

SOLUTIONS, RECOMMENDATIONS AND FUTURE RESEARCH DIRECTIONS

The concept of atmosphere, in the sense the authors have expanded on in this chapter, is the solution for the reconstruction of urban ambiences in smart public spaces. It is the element that allows both architects and urban planners to find the balance between hard and soft systems, between the architecture of software (virtual) and the architecture of physical things.

In practical terms, this implies conceiving urban design solutions that are formulated and implemented by interdisciplinary teams who are able to go beyond data (Lisdorf, 2020). For this to become a reality there are four significant obstacles that need to be overcome in the meantime. First, university curricula need to keep up to speed with rapid on-going changes. It is the case that in many universities all over the world, architecture and urbanism are taught within a framework that has long been overtaken by the reality of 4IR. For solutions to be found, professionals need to be educated and given the tools to be able to do this. Second, decision-makers need to be better informed and aware of the crucial impact that 4IR has on urban environments (for better or worse). Omission, neglect or taking a passive role when it comes to this subject can severely impact a city's economy and well-being. Third, private companies and other social actors should also become aware of the subject and actively contribute to designing and implementing solutions that promote human and higher quality living urban environments. (Soeiro, in press) Fourth, innovative governance design systems need to be conceived and implemented.

To sum up, urban design should envision solutions that aim to go beyond the five senses using a broad-scope understanding of atmosphere by considering how its design in urban settings can positively impact living organisms, the human body, mood and feelings. On this, more research is called for, and the work of Schmitz, Griffero (2014, 2017, 2019), Fuchs (2012, 2019) and Summa (*et al* 2012[5]) is at the forefront.

CONCLUSION

With no accurate diagnosis, it is not possible to formulate the best possible solution. Therefore, when it comes to investigating the role of urban ambiences in smart public places, it is important to recognise the role of technology and 4IR in today's world and, in particular, how the architecture of software is converging with the architecture of things.

This chapter presents a clear diagnosis of the subject, claiming that the concept of atmospheres has a leading role in humanising cities. In order to prove this, we go back in time and trace an original succession of authors who have paved the way to this solution. First Goethe and Wittgenstein, through a reflection on the subject of colour, show us that any given object, in order to be known, demands that it be considered in context. Analogously, atmospheres also depend on context as they are acknowledged as highly volatile, temporary events. The research method put in place to investigate both colours and atmospheres is simultaneously the subject of research and the best possible example to describe the way we know a phenomenon and how atmospheres in particular inform philosophical anthropology. Böhme and Schmitz are contemporary philosophers who, along the lines of New Phenomenology and with a strong connection to Goethe's body of work, build on the concept of atmospheres. Particularly in Schmitz, the concept is framed in an all-encompassing ontology where body, space and feeling play the leading role in his phenomenology of the body. The relation between atmospheres and cities is specifically addressed. Bratton declares that there is a growing convergence between the architecture of software and the architecture of physical things. He aims at providing a new vocabulary that enables us to describe the current state of affairs and effectively elaborates a diagnosis, pointing at possible future scenarios.

As the city is a layer of convergence of different software platforms, in order to promote a sustainable smart city concept, hard and soft systems should be balanced in order to positively impact cultural, institutional and technical systems. In this process, atmospheres play a key role in shaping solutions that include and positively stimulate the traditional five senses and other neglected senses. Furthermore, the body and the mind should both be perceived as elements that are simultaneously passive and active generators of atmospheres themselves. We claim that this expanded view of what is understood by 'senses' is a crucial argument to adopt the concept of atmospheres as foundational in order to orient the future of smart cities, incorporating humankind's spiritual side and creating urban environments where technology thrives and the human element prevails.

ACKNOWLEDGMENT

This research received no specific grant from any funding agency in the public, commercial or not-for-profit sectors.

REFERENCES

Acuto, M., Steenmans, K., Iwaszuk, E., & Ortega-Garza, L. (2019). Informing urban governance? Boundary-spanning organisations and the ecosystem of urban data. *Area*, *51*(1), 94–103. doi:10.1111/area.12430

Adler, J. (2009). Goethe's use of chemical theory in his *Elective Affinities*. In A. Cunningham & N. Jardine (Eds.), *Romanticism and the sciences* (pp. 263–279). Cambridge University Press.

Allam, Z., & Dhunny, Z. A. (2019). On big data, artificial intelligence and smart cities. *Cities (London, England)*, *89*, 80–91. doi:10.1016/j.cities.2019.01.032

Artificial Intelligence and Life in 2030. (2016), *One Hundred Year Study on Artificial Intelligence: Report of the 2015-2016 Study Panel*. Stanford, CA: Stanford University. Retrieved from https://ai100.stanford.edu/2016-report

Böhme, G. (1995). *Atmosphäre: Essays zur neuen Ästhetik* [Atmosphere: Essays on a new aesthetics]. Suhrkamp.

Böhme, G. (2006). *Architektur und Atmosphäre* [Architecture and Atmosphere]. Wilhelm Fink Verlag.

Böhme, G. (2017a). Atmospheric Architectures: The Aesthetics of Felt Spaces. London: Bloomsbury.

Böhme, G. (2017b). *Critique of Aesthetic Capitalism*. Mimesis.

Böhme, G. (2017c). *The Aesthetics of Atmospheres* (J.-P. Thibaud, Ed.). Routledge.

Bratton, B. H. (2015). *The Stack: On Software and Sovereignty*. MIT Press.

Eberhard-Kaechele, M. (2012). Memory, metaphor, and mirroring in movement therapy with trauma patients. In S. C. Koch, T. Fuchs, M. Summa, & C. Müller (Eds.), *Body Memory, Metaphor and Movement* (pp. 267–287). John Benjamins Publishing Company.

Eccleston, E. (2016). *Embodied: The psychology of physical sensation*. Oxford University Press.

Escoubas, E. (2003). La Bildung des Couleurs (M. Elie, Trans.). In *Goethe, Matériaux pour l'histoire de la théorie des couleurs* (pp. 9–23). Presses Universitaires du Mirail.

Feigenbaum, R. (2015). Toward a Nonanthropocentric Vision of Nature: Goethe's Discovery of the Intermaxillary Bone. *Goethe Yearbook*, *22*, 73–93. doi:10.1353/gyr.2015.0012

Floridi, L. (2014). *The 4th Revolution: How the infosphere is reshaping human reality*. Oxford University Press.

Fuchs, T. (2012). The phenomenology of body memory. In S. C. Koch, T. Fuchs, M. Summa, & C. Müller (Eds.), *Body Memory, Metaphor and Movement* (pp. 9–22). John Benjamins Publishing Company.

Fuchs, T. (2019). The Uncanny as Atmosphere. In G. Francesetti & T. Griffero (Eds.), *Psychopathology and Atmospheres: Neither Inside nor Outside* (pp. 101–118). Cambridge Scholars Publishing.

Griffero, T. (2014). *Atmospheres: Aesthetics of Emotional Spaces*. Ashgate.

Griffero, T. (2017). *Quasi-things: the paradigm of atmospheres* (S. De Sanctis, Trans.). State University of New York.

Griffero, T. (2019). The invasion of felt-bodily atmospheres: Between pathic aesthetics and psychopathology. In G. Francesetti & T. Griffero (Eds.), *Psychopathology and Atmospheres: Neither Inside nor Outside* (pp. 6–34). Cambridge Scholars Publishing.

Husserl, E. (1913). Ideas Pertaining to a Pure Phenomenology and to a Phenomenological Philosophy – First Book: General Introduction to a Pure Phenomenology (F. Kersten, Trans.). The Hague: Nijhoff.

Knight, D. (2009). Romanticism and the sciences. In A. Cunningham & N. Jardine (Eds.), *Romanticism and the sciences* (pp. 13–24). Cambridge University Press.

Koch, S. C. (2012). Testing Fuchs' taxonomy of body memory: A content analysis of interview data. In S. C. Koch, T. Fuchs, M. Summa, & C. Müller (Eds.), *Body Memory, Metaphor and Movement* (pp. 171–186). John Benjamins Publishing Company.

Lisdorf, A. (2020). *Demystifying Smart Cities: Practical Perspectives on How Cities Can Leverage the Potential of New Technologies*. Apress/Springer Nature.

Park, J. H. (2019). Advances in future internet and the industrial internet of things. *Symmetry*, *11*(2), 244. doi:10.3390ym11020244

Pelton, J. N. (2019). *Preparing for the Next Cyber Revolution: How our world will be radically transformed – again!* Springer.

Schmitz, H. (2007). *Der Leib, der Raum und die Gefühle* [Body, Space and Feeling]. Aisthesis.

Schmitz, H. (2014). *Atmosphären* [Atmospheres]. Freiburg: Alber.

Schmitz, H. (2016). Atmospheric Spaces/Espaces atmosphériques. *Ambiances*. doi:10.4000/ambiances.711

Schmitz, H. (2018). Les sentiments comme atmosphères [Feelings as atmospheres]. *Communications*, *1*(102), 51–66.

Schmitz, H. (2019). *New Phenomenology: A brief introduction* (R. O. Müllan, Trans.). Milan: *Mimesis*.

Schwab, K. (2017). *The Fourth Industrial Revolution*. Penguin.

Sepper, D. L. (1988). Goethe contra Newton: Polemics and the Project for a New Science of Colour. Cambridge University Press.

Soeiro, D. (2011). *A cor como abrigo: a arquitectura como cuidado* [Colour as shelter: architecure as care] (PhD Thesis). Department of Philosophy. Repository of Universidade Nova de Lisboa. http://hdl.handle.net/10362/14113

Soeiro, D. (in press). Smart Cities, Well-being and Good Business: The 2030 Agenda and the role of knowledge in the era of Industry 4.0. In F. Matos, V. Vilarinhos, I. Salavisa, L. Edvinsson, & M. Massaro (Eds.), *Knowledge, People, and Digital Transformation: Approaches for a Sustainable Future* (pp. 41–53). Springer Nature.

Summa, M., Koch, S. C., Fuchs, T., & Müller, C. (2012). Body memory: An integration. In S. C. Koch, T. Fuchs, M. Summa, & C. Müller (Eds.), *Body Memory, Metaphor and Movement* (pp. 417–444). John Benjamins.

Tegmark, M. (2017). *Life 3.0: Being human in the age of artificial intelligence*. Alfred A. Knopf.

Thibaud, J.-P. (2017). Forward. In The Aesthetics of Atmospheres (pp. 17-23). New York: Routledge.

Uberoi, J. P. S. (1984). *The Other Mind of Europe – Goethe as a Scientist.* Oxford University Press.

von Engelhardt, D. (2009). Historical Consciousness in the German Romantic *Naturforschung*. In A. Cunningham & N. Jardine (Eds.), *Romanticism and the sciences* (pp. 13–24). Cambridge University Press.

von Goethe, J. W. (1790). The Metamorphosis of Plants. Cambridge, MA: MIT Press.

von Goethe, J. W. (1970). Theory of Colours (C.H. Eastlake, Trans.). MIT Press. (Original publication 1810)

von Goethe, J. W. (1988). *Scientific Studies* (D. Miller, Ed. & Trans.). Suhrkamp.

von Goethe, J. W. (1989). Goethe's Botanical Writings (B. Mueller, Trans.). Woodbridge, Canada: Ox Bow Press.

Wittgenstein, L. (1979). Bemerkungen über die Farben/Remarks on Colour (G.E.M. Anscombe, Ed.; L.L. McAlister & M. Schättle, Trans.). Hoboken, NJ: Blackwell.

Yigitcanlar, T., Kamruzzaman, M., Foth, M., Sabatini-Marques, J., da Costa, E., & Ioppolo, G. (2019). Can cities become smart without being sustainable? A systematic review of the literature. *Sustainable Cities and Society*, *45*, 348–365. doi:10.1016/j.scs.2018.11.033

ENDNOTES

1 In an essay based on Spinoza (Goethe, 1988, p. 9), he states: "The things which enter our consciousness are vast in number, and their relations—to the extent the mind can grasp them—are extraordinarily complex. Minds with the inner power to grow will begin to establish an order so that knowledge becomes easier; they will begin to satisfy themselves by finding coherence and connection. Thus all of existence and totality must be made finite in our minds so that it conforms to our nature and our way of thinking and feeling. Only then will we say that we understand something, or enjoy it."

2 "The colours are acts of light; its active and passive modifications: thus considered we may expect from them some explanation respecting light itself" (Goethe, 1810, Preface, p. xxxvii). Light is therefore a foundational element of colour and not a consequence (Escoubas, 2003, p. 16).

3 The relation between Goethe and Husserl can be found in the works of Gernot Böhme, Hermann Schmitz and Elaine Escoubas.

4 See Adler (2009).

5 See also Eberhard-Kaechele, M. (2012) and Koch (2012).

Chapter 3
Affective Atmospheres, Essence of Architecture, and Spirit of Place

Hisham Abusaada
https://orcid.org/0000-0001-6530-7714
Housing and Building National Research Center (HBRC), Egypt

Abeer Elshater
https://orcid.org/0000-0002-5061-6861
Ain Shams University, Egypt

ABSTRACT

This chapter reviews the implications of using the words "essences" and "spirit" in urban studies and their link with the concept of affective atmospheres in the realms of architecture. Two assumptions are valid when this matter is addressed. The first is that, despite affective atmospheres being considered as the fifth dimension in architecture, they are not their essence or spirit. The second is that these atmospheres impersonate a crucial role in reconstructing different urban environments, which are based on the perceptual dimension. The purpose of this chapter is to provide an informed view bibliographically and conceptually about distinguishing between essence, spirit, and affective atmospheres. The chapter also provides an analysis of the concept of affective atmospheres to verify the hypothesis. The conclusion is latent in the possibility that the expression "affective atmospheres," instead of "essence" or "spirit," can be used for referring to people's emotional impressions in urban environments as a fifth dimension.

INTRODUCTION

All cities in the world are built on the same structural elements of buildings, open spaces, gardens, roads, and streets, together for people. Nevertheless, each city has an entirely different impression in its outdoor places, even though some cities in the world lack these impressions. In everything that God created and in some of what man has done throughout history, everywhere in an urban environment, people who

DOI: 10.4018/978-1-7998-3856-2.ch003

visit some places for the first time always feel hidden or explicit things that suddenly impact on them (Ash, 2013; Thrift, 2004). For instance, if a person crosses a place for the first time, whether intentionally or unintentionally, he/she will feel a sudden sensory reaction. Sometimes, the atmosphere affects them with complacency and acceptance; on the contrary, it causes them to have a mood [1] disorder and change their attitudes (Abusaada & Elshater, 2020).

In this regard, Dewey (2008) expressed his total overwhelming impression, which would not only affect his body, but also arouse passion and change his mood. This immediate response describes an emotional moment that occurs slowly or quickly, violently or quietly, but it must happen, and no matter how different the human samples. Indeed, the authors refer to people's age, gender, educational, social, and cultural status, as well as economic level, in addition to the differences in human nature. An emotional shock is always instantaneous and occurs when the senses receive the influences of the natural or human-made elements that exist in a place. These effects vary among taste, smell, touch, sight, and sound. Besides, the results of all these immediate impacts that excite emotions create effects on a person anywhere.

According to some researchers, this change in people's impressions has presumably occurred as a result of multiple characteristics in the spirit of a place; some theorists have called it a spirit of a place (Jackson, 1994; Norberg-Schulz, 1980; Relph, 1976). In another context, Gandhi (2017) called these effects–from an urban geography perspective–an urban atmosphere. From a metaphorical perspective, Anderson (2009) coined the term affective atmospheres. These researchers, along with other researchers, see the emotional influences of a person or place that make that hidden spirit active (Anderson, 2009; Ferreira, 2018; Norberg-Schulz, 1980; Terranova & Tromble, 2016). Moreover, many commentators linked the words of essence and spirit to understand precisely the atmospheres of the urban environments (Jackson, 1994; Norberg-Schulz, 1971, 1980).

Therefore, this problem is synonymous with the classical problem of using terms such as essence, spirit of place or affective atmospheres as an indication of people's changing impressions toward a place. This issue can be defined as the problem of estimating the validity of referring to affective atmospheres as the essence or spirit of architecture. However, doubts remain on reliability of such study, due to the difficulties of naming affective atmospheres as the fifth dimension (Abusaada & Elshater, 2018). Is this the same situation as when somebody sees time or intuition as a fourth or sixth dimension, respectively?

This research seeks to investigate the discrimination between essence (substance), spirit, and affective atmospheres. It aims to explore and explain the practical reasons for establishing affective atmospheres that activates the hidden spirit that people feel towards a person or place. Besides, it attempts to show urban planners and designers the importance of adopting the term affective atmospheres as a fifth dimension of the field of architecture, with time (the fourth dimension) and intuition (the sixth dimension). This fifth dimension addresses and directs hidden and mystical things and topics related to the spiritual dimension and human experience, which are also necessary to learn how to assess them in urban environments. There are two reasons for this. The first is related to the terminology that focuses on dealing with hidden meanings such as spirit, essence, and atmospheres. The second tries to find links between these three terms and the professional terms that have been in use in the field of architecture in general and urban planning and design. These terminologies are a sense of place and *genius loci*. This research endeavours to indicate these links between these concepts in urban studies. Besides, it focuses on the qualities of the dimensions of perceptual dimension.

METHODOLOGY

The authors conducted the literature review through a bibliographic approach (Boon, 2017), with a general inductive approach based on the process of qualitative raw data in the fields of phenomenology and epistemology, urban geography, architecture, urban planning, and urban design. This method enabled to condense extensive and varied raw text data into a brief, and to derive the results. This approach answered the key question: Are affective atmospheres the essence of architecture or spirit of place?

Consequently, this study explores the concept of essence of architecture and spirit of a place in general and mainly in the fields of architecture. It evolves to search for the concept of affective atmospheres as an absent dimension. Besides, it provides a review of the spiritual dimension as a dominant dimension from the perceptual dimensions. This review also allowed to answer the main research question and prove that an affective urban atmosphere is not the essence of architecture, but may constitute its fifth dimension. Besides, this dimension ensures that architecture is for people as well as people are for architecture. Finally, this fifth dimension is related to the spiritual aspects of the perceptual dimension.

BACKGROUND

According to Sartre (1993), “essences and facts are incommensurable, and one who begins his/her inquiry with facts will never arrive at essences;” he/she must even recognize that “essences alone permit us to classify and inspect the facts” (p. 9). In general, the literature that has discussed essence is restricted to books and periodical articles published between 1975 and 2014. Some examples are theorists Charles Taylor (1975), Joan Kung (1977), Norman Dahl (1997), Karin Dahlberg (2006), Lorenza Saitta and Jean-Daniel Zucker (2013), Lucas Angioni (2014), and Milica Muminovic (2015).

In the past decades, many commentators discussed the space as the essence of architecture; these were August Schmarsow, Alois Riegl, Theodor Lipps, Herman Sörgel, and Adolf Hildebrand (Stanek, 2012). In the mid-twentieth century, architects began to see space as the essence of architecture. Among these architects were Le Corbusier (1948), Geoffrey Scott (1954), Louis Kahn (1957), Jane Rendell (2006), Andrea Sauchelli (2012), and Johanna Gullberg (2016). Notably, in the 1980s, Charles Jenkins (2013), Chris Abel (1982), and Michael Herzfeld and Margot Lenhart (1982) rejected the idea that architecture has an essence. Instead, the theorists who discussed the essence and the spirit of place in the fields that have strong relationships with architecture are Edward Relph (1976), Edward Casey (1997), and Tim Cresswell (2009).

The gap in the literature emerges with regard to the perceptual qualities related to affective atmospheres in the areas of urban studies. Principally, this lack is in the definition of their relationship with the concept of essence of architecture and spirit of a place. A search on Google engine indicated that a limited number of articles combined the three terms essence of architecture, spirit of place, and affective atmospheres. However, many researchers linked essence and spirit to understand precisely the atmospheres of the urban environments. In the mid-twentieth century, Christian Norberg-Schulz (1971, 1980) and John Brinckerhoff Jackson (1994) did so. Besides, some of them linked the spirit of place, the sense of place, and affective atmospheres, such as John Brinckerhoff Jackson (1994).

This spirit of a place, which "retained through such change, is 'subtle', 'nebulous', not easily analysed in formal and conceptual terms, but nonetheless extremely obvious (Carmona, Heath, Oc, and Tiesdell (2003, p. 96, as cited in Relph, 1976, p. 99) stated. They added these characteristics might vary to become global brands for each city. According to Relph (1976),

the spirit of a place involves topography and appearance, economic functions and social activities, and particular significance deriving from past events and present situations–but it differs from the simple summation of these. Spirit of place can persist in spite of profound changes in the basic components of identity. (p. 106)

The spirit in a place inspires love and joy, while there is another spirit hidden in another place that gives rise to depression and sadness (Hough, 1990; Relph, 1976; Steele, Luka, & Guastavino, 2012).

Böhme (2002) and Zumthor (2008) discussed the spiritual view of the impacts of atmospheres. Now, the question is: Can essence, spirit, and atmospheres be used as synonyms? In this respect, are the perceptual qualities of affective atmospheres understood as the same qualities of the other two terms?

This chapter includes two phases. The first phase shows the similarity between essence, spirit of place, and affective atmospheres. This is an attempt to prove that affective atmospheres are still an absent dimension in some areas of urban studies, mainly in architecture, urban planning, and urban design. Moreover, can affective atmospheres be considered the fifth dimension in architecture, urban planning, and urban design? Then, in the second phase, the authors discuss a controversial issue about people for architecture or architecture for people. In this, the effect of the sixth dimension, which relates to intuition, the sixth sense, or the innate cognitive experience of people, appears. The discussion of the findings highlights how influencing people's emotions and feelings in urban environments is related to changing affective atmospheres as a fifth dimension and not their essence.

BETWEEN THE ESSENCE, SPIRIT, AND ATMOSPHERES

The definition of essence in general considers a confusing and vague concept, which continuously needs reasonable interpretations (Dahlberg, 2006; Kung, 1977; Muminovic, 2015). Essence is an independent entity, an object on what it is or something that exists evermore. Its descriptions differ: Being as substance, an independent entity, something that exists evermore, an ideal model, a perfect form of something, and the ultimate nature of a thing. Instead, essence belongs to the realm of the transcendental, and lies between abstract and tangible ideas giving things their identity. It is the origin of something, not its appearance. Essence is the domain in which one sees things not just by themselves, "immediately," but as founded on an underlying basis (Taylor, 1975). It is

the realm of mediacy [...] inescapably mediates, that is, we can only get to it via another: We come to essence by reflecting on Being, seeing that it does not suffice to itself, and hence referring back beyond it to what underlies it." (p. 258)

In this regard, Hadot (1995) pointed out that essence is "amongst the most confused and confusing notions" (p. 76); it seems that a definitive explanation of the concept of essence is still needed. This chapter focuses on interpreting the meaning of essence in general and in the fields of architecture. Therefore, the next decree carefully follows the concepts of essence and spirit, which are based on affective atmospheres.

A recent review found out that essence is derived from the Greek adjective *ousia*, referring to "substance," which is "the essence of a thing," according to Aristotle (Dahl, 1997, p. 241). The essence of things is what underlies the difference between abstract and concrete ideas. According to Plato, abstract ideas are the ideal entities that capture the essence of things (Saitta & Zucker, 2013). This essence is also what decides how things will behave or what a thing will be, while things without essence will lose their identity. In philosophy, based on Aristotelian thought, in relation to the existence of changing qualities of things, essence is the holder, the ideal model, and the necessary condition for changing attributes without changing it (Angioni, 2014).

The concept of essence is still an internal, implicit, and invisible object reflected in one's behaviour and everyday attitudes. Moreover, it is "an almost occult quality, a hiddenness of things" (Dahlberg, 2006, p. 11). Besides, tracking the experience of someone's life may reveal his/her essence, and his/her personality can be judged based on this knowledge. This offers the possibility of highlighting his/her nature in seriousness and humour, in work and rest, and in the relationships of love and friendship. In addition, some literature indicates that the essence is "a phenomenon's style, its way of being, and thus the essence cannot be separated from the phenomenon that it is the essence of things" (Dahlberg, 2006, p. 18). It is not possible to talk about essence as a characteristic in the same way as one talks about the properties of characteristics[2] (Dahlberg, 2006). Moreover, essence uses to make a distinction between things, and its nature in this matter is "universal, abstract, invariant, eternal, ideal, and belonging to the realm of transcendental. [...] determines what the thing is. [...] will define the point at which that thing stops to exist or transforms into something else" (Muminovic, 2015, p. 302).

In geographical perspectives, there have been previous attempts to distinguish between terms relevant to the concepts of space, place, and essence. In geography, space seems to have more abstract meanings, while place is more objective and subjective. Lukermann (1964) believed that the word "place" "has more substance than the word location suggests: It is a unique entity, a 'special ensemble'" (p. 70). On the other hand, Relph (1976) sustained that the places are "profound centres of human existence" (p. 17). In addition, the essence of the place is latent in this "unselfconscious intentionality" (p.169).

In Germany, since the 19th century, it seems the idea that "space is the essence of architecture" has dominated in the field of architecture. It was been based on August Schmarsow, Alois Riegl, Theodor Lipps, Herman Sörgel, and Adolf Hildebrand (Abusaada, 2017, p. xii; Stanek, 2012). Schmarsow (1893), Van der Rohe (1908), and Berlage (1908) sustained that architectural history should be seen across the history of the sense of space and the relation between the body and what surrounds it. Besides, ample space is a room with no function, and universal space is an architecture-independent part of a specific function. Further, the aim of man's creations is the art of space, which somebody sees as the essence of architecture. (Gullberg, 2016; Krohn, 2014). Afterwards, many researches discussed the intellectual trend to perceive space as the essence of architecture (Gullberg, 2016; Kahn, 1957; LeCorbusier, 1948; Rendell, 2006; Sauchelli, 2012). In this trend, space plays an influential role, in one way or another, on the construction of buildings in general. Besides, history proves man shaping space, a spatial practice, an experience "of space" (generally) and "of enclosed space" (particularly) (Rendell, 2006; Sauchelli, 2012).

Specifically, between the 1980s and 2010s, some critics linked the essence of place with its spirit to understand the atmospheres of urban environments. These critics were Christian Norberg-Schulz (1971, 1980) and John Brinckerhoff Jackson (1994). In addition, the spirit of the place "gives life to people and places" (Norberg-Schulz, 1971, p. 160). As Norberg-Schulz (1980) believes that this spirit not only "accompanies them from birth to death," but also determines "their character and essence" (p. 18). In this regard, the term *genius loci*, which Norberg-Schulz developed, is based on the concept of genius and is used to describe the place according to its essence, wherein "the meanings which are gathered by a place constitute its *genius loci*" (p. 170). In addition, he used the concept of "eternal essence" to describe the place and ways of building, and to maintain the qualities of a location that has become a place[3]. In another vein, Jackson (1994) believed that a sense of place depends "on awareness of the familiar environment, a ritual repetition, a sense of fellowship," which is based on "a shared experience" (p. 161). Moreover, he used the term *genius loci* to describe the atmospheres of a place in the perspective of the quality of its environment. In these critics' view, the continuity of the spirit of the place not only refers to the built environment, but also reflects its atmospheres, which can be found in the natural characteristics of the location prior to the act of building.

The Atmospheres and Sense of Place

The word "atmospheres" is still a vague concept, and its usage is ambiguous (Anderson, 2009; Böhme, 1993). For instance, in the past several decades, the term atmospheres referred to "impersonal or transpersonal intensity, [...] environment, or the transmission of the other's feeling, [...] qualified aura, [...] tone in literature, [...] mimetic waves of sentiment, [...] more broadly a sense of place" (Anderson, 2009, p. 78). Indeed, the two terms "sense of place" and "atmospheres" seem similar in terms of maintaining the quality of the urban environments. In addition, both reflect a sense of time, people's attitude towards their environment, and studies of sustainability (Guthey, Whiteman, & Elmes, 2014; Jackson, 1994; Jorgensen & Stedman, 2001).

However, crucial differences exist between those two terms. Several research groups have been working on defining the sense of place by referring to what might more accurately be called the spirit of a place (Hough, 1990; Relph, 1976; Steele, 1981). Nevertheless, affective atmospheres refer to mindful physical presence, sense in bodily presence by human beings, a property of objects and subjects, coalescence of sensuous elements, felt-bodily resonances, and multi-sensory experiences (Abusaada, 2020; Anderson, 2009; Böhme, 2013b; Dufrenne, 1973; Griffero, 2016; Schmitz, Owen, & Slaby, 2011). Sense of places refers to "the personal meaning built for subjective qualities" (Hay, 1998, p. 7), while affective atmospheres are

something thinglike, belonging to the thing in that things articulate their presence through qualities–conceived as ecstasies. [...] and subjectlike, belonging to subjects in that they are sense in bodily presence by human beings, and this sensing is at the same time a bodily state of being of subjects in space. (Böhme, 1993, p. 122)

Today, the term "affective atmospheres" is not linked to how people learn to avoid or love places or harsh industrial noises. Besides, affective atmospheres are not spoken in a legible, countable or measurable shape, but the expression is still powerful and significant (Hasse, 2014). However, these atmospheres can be included within the visual, aesthetic, and spiritual aspects of perceptual dimension.

At the beginning of the predesign stage, the specialists start thinking about how architectural products must touch the spirits of people before how the spirit of a place affects people's feeling (Barrie, Bermudez, & Bermudez, 2015). Nevertheless, the spirit of a place seems to receive much research in the realm of architecture, mainly in the issues of emotions and the atmospheres (Kazig, 2016). On the contrary, in some cases of practice and application, the spirit of people has not been considered as the spirit of a place, even though many researchers (Griffero, 2016; Hasse, 2014; Schmitz et al., 2011) have recently debated these concepts. Besides, many scholars assumed that the place has a sense of place (Cresswell, 2009; Deleuze, 1990; Dovey, 2010; Hay, 1998; Jorgensen & Stedman, 2001), a spirit of a place (Anderson, 2009; Ash, 2013; Dewey, 2008; Thrift, 2004), and *genius loci* (Norberg-Schulz, 1980; Pallasmaa, 2014).

Are Affective Atmospheres as an Absent Dimension?

Arguably, several studies have examined architecture as a living organism that carries the spirit (i.e., the unknown psychological essence) and the body (i.e., the physical essence), which are the characteristics of the four physical dimensions (i.e., length, width, height, and time) (Ferreira, 2018; Norberg-Schulz, 1980; Pallasmaa, 2013; Terranova & Tromble, 2016). The dimensions of this research are related to urban environments (Figure 1). The first and second are the horizontal dimensions, which are the length and width of a specific area of the land, while the height represents the third dimension. These three dimensions are measurable physical dimensions. Time is the fourth dimension in architecture; it is the time necessary for movement within a specific space (Lynch, 1972; Shakibamanesh & Ghorbani, 2017). It is a nonphysical dimension and is not measurable directly, but its effects on people can be measured. In this research, the authors assume that two dimensions resemble time in their indirect effects on people. The first dimension is intuition (considering that it is the sixth sense after the five senses), while the second is affective atmospheres, which relates to what fills the spaces of urban environments with things that are not tangible, but perceived by the feeling of the body (Schmitz et al., 2011).

Now, the question is: Are affective atmospheres an absent dimension in urban studies? In this regard, can the atmospheres become the fifth dimension in the fields of architecture, particularly in urban planning and design?

ARCHITECTURE FOR PEOPLE AND PEOPLE FOR ARCHITECTURE

The primary purpose of this section is to highlights the impacts of the spiritual aspects of urban environments. In this chapter, the authors have linked these aspects to the concept of self-awareness, which is primarily related to the human body as the first portal to generate the human feeling (Schmitz et al., 2011). In terms of affective atmospheres, this work specifically concentrates on their spiritual qualities (Böhme, 2013a, 2013b; Zumthor, 2006). Consequently, this orientation implies two prime objectives: The first focuses on the assumption of supporting people's self-awareness through their bodies, via focusing on the idea of affectivity; the second promotes the idea that perceptual qualities of urban planning and design could create affective atmospheres for a place.

Figure 1. The six dimensions of architecture

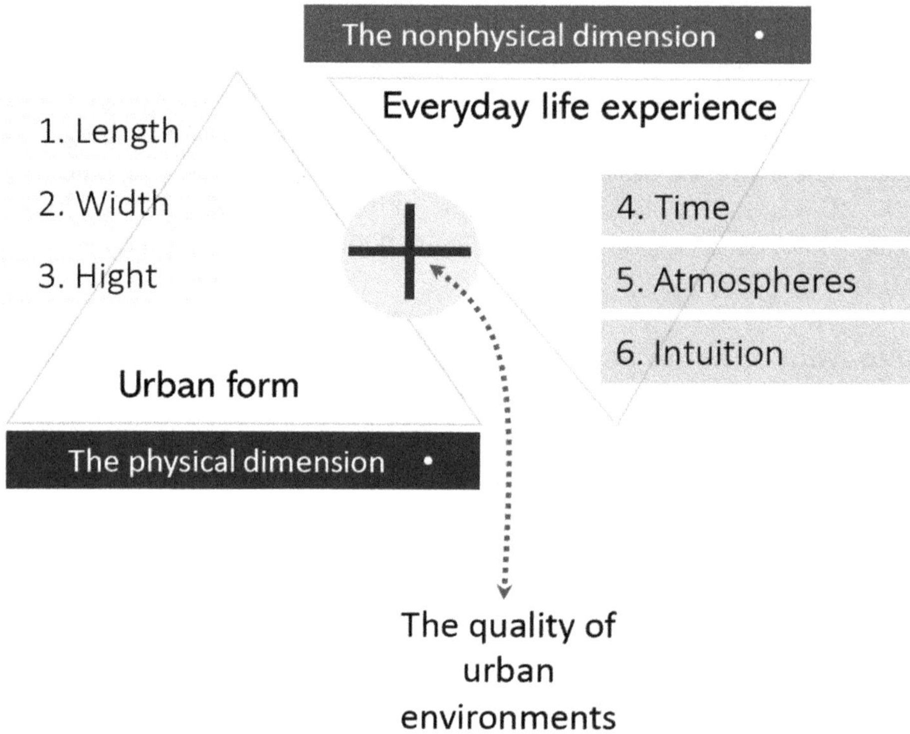

Supporting People's Self-Awareness

Vlad Ionescu (2016) indicated that two famous vanguard historians, Wölfflin (1864-1945) and Schmarsow (1835-1936), described the architecture movements related to the intuitive concerning affectivity and empathy.

Heinrich Wölfflin (1888) thought that the spatial form of architecture focused on the understanding of the change of situations of human awareness. It did not only rely on the sense through the vision, but reflected its natural sensation according to the self-experiences of the human body that obtained over time. Consequently, Ionescu believes that Wölfflin dealt with "the corporeal sensory impact as an expression of a corporeal state" and that he "reinterpreted architecture as a visual regime that appealed to the mood" (Ionescu, 2016, p. 1). In this regard, Wölfflin's questions were:

1. How can architectural forms be an expression of a mood?
2. How is it possible that an inanimate object (e.g., a building) conveys an impression that is felt like an expression?

In this respect, Wölfflin referred *to* "architectural structures" regarding the concept of "lived experience." The analysis of tectonic structures is subordinated to their impression on the viewer's mood. Architecture becomes a correlate for a lived experience as opposed to a set of quantifiable mathematical proportions. As Ionescu (2016) explained, Wölfflin presented the expression "vital feeling" because the beauty of architectural space is not a response to the forms seen on the retina, but is reflected by the human body's feeling of an open relationship between "the perceived forms and the vital feeling of our body" (p.8). The vital feeling refers to "a constancy of energy that is felt during the unconstrained movement of the body" (p.8). Each human body has its own constitution that helps to feel, perceive, and create "the experience of architectonic structures;" it relies on what Wölfflin called the "symbolic mediation," which means

the relation between bodily proportions and architectural structures mediate a feeling. [...] to move geometric relationships from one domain to another. Consequently, if architectural structures are felt as beautiful, it is because they affirm the feeling that the body has of its own 'well-being', a notion too rapidly and too often translated as "pleasure." (Ionescu, 2016, pp. 3-4)

Besides, according to Ionescu (2016), Wölfflin's theory was founded on the intuition that architectural forms mediate between people's felt bodily structure and space, namely architecture as structures subordinated to the human bodily constitution. The scale of the lived environment should follow the human scale. The symmetry, proportion, and vertical position of the body are felt as a harmonious affective disposition called "mood." Pleasurable responses to space are the sign of the congruence between visual forms and vital feeling. Wölfflin dealt with the body as a criterion for the experience of space; it is experienced as mood.

August Schmarsow was the first architectural historian to specialize in the idea of space, in1893. He formulated a comprehensive theory of "architecture to discuss the spatial creation at the frontiers of the paradigm of perceptual empiricism" (Schwarzer & Schmarsow, 1991, pp. 50-52). Schmarsow believed that the concept of architectural space should be understood based on the perspective of a cognitive process. In his view, a distinct link exists between self-awareness and spatial awareness; the second originates based on the first. As a result, the designing of architectural forms should follow and depends on the bodily structure, sensation, and movement. In other words, the designer should understand the structure, sensation, and movement of the human body to take all the results to be a guide to design the architectural form and formation. Therefore, the identification of the human gestures and motions help the architect to design the spatial shape of the building. Therefore, in order to build architecture, the designer must understand how man uses space and that the architect does not impose on people the design of space. Architecture is created organically and not geometrically. Schmarsow (1994) believes that "what is truly essential can only start in the mind of the artist and end in the mind of the observer" (p. 283). He described architecture as "creatress of space," "the ideal forms of the human intuition of space, and 'spatial construct' built around the phenomenological human axis." (pp. 287-289),

The Perceptual Dimension and Spiritual Qualities

One enters a specific place and he/she suddenly feels steam emotions towards that place; these emotions appear without his/her conscious mind controlling, but rather the result of pent-up inner awareness or emotion. The idea here revolves around the immediate and sudden effect on people as a result of relying on the feeling of the body, and the result is steam feelings toward the place, whether positive or negative. This influence is often linked to each individual's self-awareness, and its implications are beyond what people experience over time in their daily life. This experience is made up of individual and collective senses, is rooted in the conscience, and becomes the spiritual roots that control the emotions without the individual's interference. The next section plays a preliminary role in exploring spirituality in architecture by re-adjusting the physical and intangible conditions in urban environments.

Over the years, this challenge has reached a mature stage in phenomenological approaches (Dufrenne, 1973; Merleau-Ponty, 2002; Schmitz et al., 2011) and urban geography (Anderson, 2009; Gandy, 2017). In practice, the areas of study have become strongly linked with the fields of architecture (Böhme, 2005, 2017; Pallasmaa, 2011; Schonhammer, 2018) and urban planning and design (Abusaada, 2019; Abusaada & Elshater, 2020; Stefansdottir, 2018). According to the relationship between environmental psychology, architecture, and spatial design, in this research perceptual dimension is closely linked with some recent theoretical terms, such as affective atmospheres (Anderson, 2009; Shaw, 2014) and urban atmospheres (Böhme, 2014; Gandy, 2017). From this perspective, several groups of research addressed the impact of affective atmospheres on people's feelings. The main objective here is to investigate the toolkit that enhances the use of the affective (or emotional) dimension through the ideas of body material and daily experience in the predesign stage (Böhme, 2013a; Dewey, 1984a; Schmitz, 2005, 2017; Schmitz et al., 2011), which is based on the visual, aesthetic, and spiritual dimensions.

The spiritual dimension focuses on the tactile perception of the spatio-temporal atmospheres. This perception happens through the corporeal presence of people, which relies on the emotional sensibility as a form of perception that works incredibly quickly (Zumthor, 1998). The corporeal presence of space is based on local relationships, skins, and topological space (Böhme, 2013b). These topological spaces, which have geometric characteristics and spatial relations, are not influenced by the constant change in shape or size of shapes, but include the directions, above and below, cantering, contraction, and expansion (Schmitz et al., 2011).

Böhme's spiritual view of the changing feelings towards the place emerged as a response to an "aesthetic of reception" (Böhme, 1993) and due to the physical presence (Böhme, 2013a). Instead, Zumthor (2006, 1998) focused on the semantics and his interest in buildings, physical sensibility, and attention to the place, which led him to concentrate on the public sphere, building facades, and physical sensibility. Böhme (2013b) sees atmospheres as an aesthetic category, while Zumthor (2006) defined architectural atmospheres as "singular density and mood," (p. 11). Besides, Zumthor seeks to create emotional space that implies the property of emotional feeling and spiritual property to realize a structure free of symbolism or everything related to experience[4].

According to the spiritual dimension, Zumthor (2006) discusses the conceptual dimension in designing the buildings in the architectural discipline through the transcendental aspect of meaning. This dimension focuses on the materials used in the architectural construction to be the entrance to the establishment of the affective atmospheres[5]. Thus, the synesthetic qualities of buildings constitute the sensitive state that the body engenders, which enables the users to perceive the materials through the affective impact of their corporal feelings. This aspect is based on changing the nature and properties of the materials used and the way they are used to change the impact on the human body and then of the individual's feelings, where the corporeal feeling drives to synesthetic experience. According to Zumthor (1998), the sensation can be reached by adapting the materials which he uses in design to go beyond all rules of composition, as he processes them to take advantage of their tangibility, smell, and acoustic qualities.

FINDINGS AND DESCUSSION

The findings of this study highlight the importance of accepting the challenge to acknowledging the relationships between affective atmospheres, essence of architecture, and spirit of place. Although affective atmospheres may be the fifth dimension, inevitably they cannot be the essence of architecture or spirit of place, for three reasons:

1. Essence may have the ability to change, but it is not changeable. On the other hand, both architecture and atmospheres can make a change, but they can also be changed.
2. Essence cannot draw from explicit connotations (characteristics), which can be monitored by vision, hearing, taste, touch, and smell, as well as through intuition. Instead, atmospheres reflect how the senses affect the felt body and based on the prior experiences of each person. This matter is not simple, mostly because the essence of architecture and affective atmospheres are always linked, in one way or another, with the consciousness and emotions of the human (Ferreira, 2018; Pallasmaa, 2014).
3. Atmospheres, not the essence, may be merely considered for their impact on the spirit of place, which creates sensory impressions by controlling the characteristics of the place. Accordingly, affective atmospheres can add to the spiritual dimension that follows the perceptual and cognitive dimensions, as well as their qualities could classify through the six senses.

The authors showed that the essence of the place is latent in the unselfconscious intentionality of human existence, but the place itself has no steadfast essence (Casey, 1997; Relph, 1976). Besides, the natural characteristics of any geographical location create the spirit of the place (i.e., the essence), which, accordingly, reflects its atmospheres. In this regard, as the authors mentioned above, Norberg-Schulz (1971, 1980) developed, and Jackson (1994) used, the term *genius loci*, to describe the essence of the place, and "eternal essence," to describe how to maintain the qualities of a location associated with a sense of place.

As such, importantly, environmental awareness, a ritual repetition, and a sense of place that is based on a shared experience will create the essence and reconstruct the atmospheres. Some findings suggest that essence cannot be separated from any phenomenon. Besides, phenomena represent the essence of things, which lies beyond abstract and tangible ideas together. Essence also decides how things will behave, controls their identity and characteristics of changes, as well as distinguishes between things. Instead, the place is a complex entity with a uniform and harmonious performance. It is inherent in the unconscious, and its meaning is inspired by people's daily practices, experiences, and feelings. This is what differentiates man-made places from spaces, which are cosmic realities without specific meaning.

There was a time when specialists knew that space was the essence of architecture. The impact of the atmospheres on the built environment has been understood thanks to many critics. These critics have linked the essence of the place with its spirit. They believe that the spirit of the place reflects the atmosphere, which can be identified through the natural characteristics of the location that existed before the construction of the building. It is this spirit that gives life to the place and people. This spirit derives its vitality from the meanings that every place reflects and through which its genius is formed. Environmental quality is associated with a sense of place and related to the atmospheres of the place.

Although some thinkers point out that phenomena are the essence of things, in the authors' view, the affective urban atmosphere as a human phenomenon is not the essence of architecture. The spirit and essence of the place are connected in the sense that the place reflects. This meaning is related to the physical and nonphysical characteristics of the site, on one side, and to people's feelings and experiences, on the other. Moreover, based on Dahlberg's (2006) view, knowing the characteristics of a phenomenon (or something related to this phenomenon) does not mean knowing the essence of the phenomenon[6]. The essence is continuously changing; it changes with the change of shape, background, and any additions that occur every time and change the style of the phenomenon (or thing). As to architecture and affective atmospheres, they cannot be considered as separate entities.

Architecture consists of many things; each of these things has its characteristics and then its independent essence. Further, the characteristics of every element alone or together with other elements, whether physical or nonphysical, change constantly, and then its essence changes. Accordingly, the essence of architecture is nothing more than an abstract idea that cannot reach its essence. However, it is necessary for specialists to know the causes and potentialities of the effects on the users by tracking the characteristics. Besides, they should know the variables that affect these characteristics. Therefore, this research sees that affective atmosphere is one of those influences that change the characteristics of the architecture and construction of the place, which change according to the change of people's feelings and emotions; then, their impressions change. In this research, affective atmospheres are invisible external effects that generate varying sensory impressions based on the felt body, multisensory experiences, and the hidden self-awareness (Abusaada, 2020). They cannot be measured, but their generated effects can measure as indicators that appear through facial expressions and body movement.

Accordingly, specialists do not require to define the essence of architecture as much as they should know the characteristics and the reasons for changing these characteristics. Concerning affective atmospheres, one of the reasons for the change of the characteristics affecting feelings and emotions follow the change of perceptual qualities, namely visual, aesthetic, and spiritual. Therefore, it became clear that the study of these dimensions and the knowledge of their influencing and changing characteristics allowed to identify the causes of people's changing impressions. Consequently, the authors should determine the impressions that are desired to achieve reflections in an environment. Afterwards, they should use the tools necessary to do so and apply the perceptual attributes necessary to make the properties that reflect those impressions.

The concept of affective atmospheres exceeds the concept of the spirit of place—the atmospheres depend on conscious and unconscious experiences; nevertheless. Atmospheres focus on perceptual qualities, which can measure their causes and review their effects. The spiritual dimension is related to the experience of each individual's self-awareness. Despite the stability of the physical elements that influence the experience, based on the five senses, hidden experiences can add to these senses. These hidden experiences arise from one's own experiences as per his/her beliefs and genesis, as well as per other elements built on his/her self-awareness. This practice has been the subject of extensive research in the fields of humanities, including urban studies. As to the new in this research, it relates to how to link the outputs of the humanities that discuss creating an individual subjective experience, on one side, and to track the impact of this subjective experience on the felt body, facial reactions, and body movements, on the other side. The addition here is to indicate the needs of making use of new digital research methods to study spiritual aspects through simulation.

The fifth dimension serves the aspect related to the spirit of architecture (Abusaada & Elshater, 2018). Alternatively, perhaps this dimension may imply the sense of the term atmospheres, which depends on prior knowledge and intuition. One will discover this sensation through the human senses, before consciousness occurs through human thinking. The authors' second assumption deals with essence as the main reason behind this rapprochement between man and everything else in life, whether natural or human-made. Whenever essence predominates, the reason for rejection or acceptance becomes more transparent for ordinary people. The recommendation is to follow the five-phases conceptual framework below for reconstructing affective atmospheres of urban environments in the future, based on the perceptual dimension:

1. People's reactions and responses to the physical characteristics of urban environments (urban form) and nonphysical characteristics of urban environments (everyday life experience emerge according to the individual and combined senses).
2. Affective atmospheres are reconstructed based on the multisensory experiences, which involves addressing the combined effects of the senses (i.e., vision, hearing, smell, touch, and taste), without neglecting the importance of intuition (i.e., the sixth sense: the spiritualities).
3. The spiritual dimension derives its effectiveness from the meanings that urban environments reflect and the visitors these environments receive through an individual's senses, then through the influence of the combined senses, and through everyone's own experience.
4. Some qualities of the perceptual dimensions (spiritual aspects) depend on designing the building blocks, the public realm, context, and the material. The design should consider people's impressions through the indicators of facial reactions and body movement.
5. The best digital tools can be used to develop the conceptual framework.

CONCLUSION

The authors conclude by arguing that the concept of affective atmospheres is the fifth dimension in architecture, as time and intuition are the fourth and six dimensions, respectively. Although all three represent nonphysical and invisible dimensions, they play an influential role in the design of urban environments. Besides, by the literature review, the authors tested that affective atmospheres are not the essence of architecture. The proof is that positive and negative effects of architecture have come not only throughout changing the situations of the built environment, but also based on two couple of streams of emotional effect and affect. The first focuses on the characteristics of the built environment. The second is related to the changes in the nonphysical indicators that form people's emotional responses, which emerge due to the qualities of the perceptual dimension in the design of urban environments.

These findings confirm that affective atmospheres should follow the perceptual dimension in urban planning and design, mainly under the aspects of spiritual investigations. By using the theoretical, analytical review, the authors tested the hypothesis that the fifth dimension of architecture represents a crucial role in reconstructing the different affective atmospheres. This issue may allow the conclusion of the plausibility of dealing with affective atmospheres as a fundamental dimension in relevant studies.

Accordingly, while reshaping a traditional city or designing a new one, the urban planner and designer should refer to the guidelines that not only respect the presence of atmospheres, but also help the designer to create different atmospheres. Moreover, the possibility of using affective atmospheres as a fifth dimension in the field of architecture, both in urban planning and design, warrants further investigation. Perhaps this investigation requires a more empirical effort focused on reviewing the concept of affective atmospheres through a thorough study of perceptual qualities, particularly of the spiritual dimension, and a study of their mutual effects within the urban environments.

The achievements in the field can be summarized as studying the impacts of affective atmospheres on architectural products which touch people's spirits, before thinking about how the spirit of a place affects people. Besides, the contributions of this chapter have applicability to reconstruct affective atmospheres in public places, based on the bodily sensation and daily or lived experiences. Moreover, much remains to be done to strengthen the authors' suggestions. Intuition, atmospheres, spiritual qualities, and spirit of place require further empirical attention.

Ultimately, this research presented a theoretical framework for exploration based on the concept of affective atmospheres. This framework depended on establishing the impact of the characteristics of the urban environments and the indicators that emerge from people's emotional response to these characteristics. Further research is needed to confirm this finding related to the morphological aspects of urban design. The stresses in these future studies, first and foremost, should focus on the urban form and everyday life experience in the urban environments. These studies should focus on testing the quality of urban environments based on the perceptual qualities of the visual-aesthetic and spiritual dimension.

REFERENCES

Abel, C. (1982). Architecture as Identity, I: the Essence of Architecture. In M. Herzfeld & M. D. Lenhart (Eds.), *Semiotiqs 1980* (pp. 1–12). Plenum Press. doi:10.1007/978-1-4684-9137-1_1

Abusaada, H. (2017). Architecture Evermore: The World of Architecture via the Western Thought an Extended Term (A. Elshater, Ed.). Partridge Africa, A Penguin Random House Company.

Abusaada, H. (2019). Revisiting the Word "Atmospheres" in the Urban Design Academic Field Based on the Artworks. *International Journal of Sustainable Entrepreneurship and Corporate Social Responsibility*, *3*(2), 22–42. doi:10.4018/IJPPPHCE.2019070102

Abusaada, H., & Elshater, A. (2018). *The Fifth Dimension: Urban design - The atmospheres of the City*. Academic Bookshop. (in Arabic)

Abusaada, H. (2020). Strengthening the affectivity of atmospheres in urban environments: The toolkit of multi-sensory experience. *Archnet-IJAR: International Journal of Architectural Research*. doi:10.1108/ARCH-03-2020-0039

Abusaada, H., & Elshater, A. (2020). Effects of urban atmospheres on changing attitudes of crowded public places: An action plan. *International Journal of Community Well-Being*, *3*(2), 109–159. doi:10.100742413-019-00042-w

Anderson, B. (2009). Affective atmospheres. *Emotion, Space and Society*, *2*(2), 77–81. doi:10.1016/j.emospa.2009.08.005

Angioni, L. (2014). Definition and Essence in Aristotle's Metaphysics vii 4. *Ancient Philosophy (Garland Publishing)*, *34*(1), 75–100. doi:10.5840/ancientphil20143416

Ash, J. (2013). Rethinking affective atmospheres: Technology, perturbation and space times of the non-human. *Geoforum, 49*, 20–28. doi:10.1016/j.geoforum.2013.05.006

Barrie, T., Bermudez, J., & Bermudez, J. (Eds.). (2015). *Architecture, Culture & Spirituality*. Ashgate Publisher.

Böhme, G. (1993). Atmosphere as the Fundamental Concept of a New Aesthetics. *Thesis Eleven, 36*, 113-126. doi:10.1177/072551369303600107

Böhme, G. (2005). Atmosphere as the subject matter of architecture. In H. Meuron & P. Ursprung (Eds.), Natural History (pp. 398–407). London: Springer Science & Business Media.

Böhme, G. (2013a). Atmosphere as Mindful Physical Presence in Space. *OASE Journal for Architecture*, 21-32.

Böhme, G. (2013b). Encountering atmospheres: A reflection on the concept of atmosphere in the work of Juhani Pallasmaa and Peter Zumthor. *Building atmosphere. OASE, 91*, 93–100.

Böhme, G. (2017). *Atmospheric architectures: The aesthetics of felt spaces* (T. Engels-Schwarzpaul Ed. & Trans.). Bloomsbury Academic.

Boon, B. (2017). Bibliographic research. In M. Allen (Ed.), *The SAGE Encyclopedia of communication research methods*. SAGE Publications, Inc., doi:10.4135/9781483381411.n37

Carmona, M., Heath, T., Oc, T., & Tiesdell, S. (2003). *Public places-urban spaces: The dimensions of urban design*. Architectural Press.

Casey, E. S. (1997). *The fate of place: A philosophical history*. University of California Press.

Cresswell, T. (2009). Place. In N. Thrift & R. Kitchen (Eds.), *International Encyclopedia of Human Geography* (Vol. 8, pp. 169–177). Elsevier. doi:10.1016/B978-008044910-4.00310-2

Dahl, N. O. (1997). Two kinds of essence in Aristotle: A pale man is not the same as his essence. *The Philosophical Review, 106*(2), 233-265. doi:phr1997106215

Dahlberg, K. (2006). The essence of essences – the search for meaning structures in phenomenological analysis of lifeworld phenomena. *International Journal of Qualitative Studies on Health and Well-being, 1*(1), 11–19. doi:10.1080/17482620500478405

Deleuze, G. (1990). *The logic of sense*. Columbia University Press.

Dewey, J. (2008). *The later works of John Dewey, 1925-1953: 1934, Art as experience* (A. Boydston, Ed., Vol. 10). Southern Illinois University Press.

Dovey, K. (2010). *Becoming places, urbanism/architecture/identity/power*. Routledge.

Dufrenne, M. (1973). *The phenomenology of aesthetic experience*. Northwestern University Press.

Durisch, T., & Zumthor, P. (Eds.). (2014). *Peter Zumthor: Buildings and projects, 1985-2013*. Scheidegger and Spiess.

Ekkekakis, P. (2012). Affect, mood and emotion. In G. Tenenbaum, R. C. Eklund, & A. Kamata (Eds.), Measurment in sports and exercise psychology (pp. 321-332). Human Kinetics.

Ferreira, M. I. (2018). *Cognitive architectures*. Springer.

Gandy, M. (2017). Urban atmospheres. *Cultural Geographies*, *24*(3), 353–374. doi:10.1177/1474474017712995 PMID:29278257

Griffero, T. (2016). Atmospheres and felt-bodily resonances. *Studi di estetica, 49*(4), 1-49.

Gullberg, J. (2016). Voids and bodies: August Schmarsow, Bruno Zevi and space as a historiographical theme. *Journal of Art Historiography*, 1-20.

Guthey, G. T., Whiteman, G., & Elmes, M. (2014). Place and Sense of Place: Implications for Organizational Studies of Sustainability. *Journal of Management Inquiry*, *23*(3), 254–265. doi:10.1177/1056492613517511

Hadot, P. (1995). *Philosophy as a way of life: Spiritual exercises from Socrates to Foucault*. Wiley-Blackwell.

Hasse, J. (2014). Atmosphere as expressions of medial power, Understanding atmospheres in urban governance and under self-guidance. *Lebenswelt*, *1*(4), 214–229.

Hay, R. B. (1998). Sense of place in a developmental context. *Journal of Environmental Psychology*, *18*(1), 5–29. doi:10.1006/jevp.1997.0060

Herzfeld, M., & Lenhart, M. D. (1982). *Semiotiqs 1980*. Plenum Press. doi:10.1007/978-1-4684-9137-1

Hough, M. (1990). *Out of place: Restoring identity to the regional landscape*. Yale University Press.

Ionescu, V. (2016). Architectural symbolism: Body and space in Heinrich Wölfflin and Wilhelm Worringer. *Architectural Histories*, *4*(1), 1–9. doi:10.5334/ah.213

Jackson, J. B. (1994). *A sense of place, a sense of time*. Walker Art Center.

Jencks, C. (2013). *Architecture becomes music*. Retrieved from The architectural review: https://www.architectural-review.com/archive/viewpoints/architecture-becomes-music/8647050.article

Jorgensen, B., & Stedman, R. (2001). Sense of place as an attitude: Lakeshore owners' attitudes toward their properties. *Journal of Environmental Psychology*, 233–248. doi:10.1006/jevp.2001.0226

Kahn, L. (1957). Architecture is the thoughtful making of spaces. *Perspecta*, (4): 2–3.

Kazig, R. (2016, April 27). *Presentation of Hermann Schmitz' paper, "Atmospheric spaces"*. Retrieved April 30, 2017, from Ambiances [En ligne]: http://ambiances.revues.org/709

Krohn, C. (2014). *Mies Van der Rohe the built work*. Birkhäuser Verlag GmbH. doi:10.1515/9783038212874

Kung, J. (1977). Aristotle on essence and explanation. *Philosophical studies. An International Journal for Philosophy in the Analytic Tradition, 31*(6), 361–383. doi:10.1007/BF01857029

LeCorbusier. (1948). *New world of space*. Reynal & Hitchcock.

Lukermann, F. (1964). Geography as a formal intellectual discipline and the way it contributes to human knowledge. *The Canadian Geographer. Geographe Canadien*, *8*(4), 167–172. doi:10.1111/j.1541-0064.1964.tb00605.x

Lynch, K. (1972). *What Time is This Place?* The MIT Press.

Merleau-Ponty, M. (2002). *Phenomenology of perception*. Psychology Press. doi:10.4324/9780203994610

Muminovic, M. (2015). Places as Assemblages: Paradigm Shift or Fashionable Nonsense? *Athens Journal of Architecture*, 295-309. doi:10.30958/aja.1-4-3

Norberg-Schulz, C. (1971). *Existence, space and architecture*. Studio Vista Limited.

Norberg-Schulz, C. (1980). *Genius Loci: Towards a phenomenology of architecture*. Academy Editions.

Pallasmaa, J. (2011). An architecture of the seven senses. In L. Weinthal (Ed.), *Theory, toward a New Interior: An anthology of interior design* (pp. 40–49). Princeton Architectural Press.

Pallasmaa, J. (2014). Space, place, and atmosphere: Peripheral perception in existential experience. In C. Borch (Ed.), *Architectural atmospheres: On the experience and politics of architecture* (pp. 230–245). Birkhauser. doi:10.1515/9783038211785.18

Relph, E. (1976). *Place and placelessness*. Pion.

Rendell, J. (2006). *Art and architecture: A place between*. I. B. Tauris.

Saitta, L., & Zucker, J.-D. (2013). *Abstraction in artificial intelligence and complex systems*. Springer Science & Business Media. doi:10.1007/978-1-4614-7052-6

Sartre, J.-P. (1993). *The Emotions: An Outline of a Theory* (B. Frechtman, Trans.). Carol Publishing Group.

Sauchelli, A. (2012). On Architecture as a Spatial Art. *The Nordic Journal of Aesthetics*, (43), 53–64. doi:10.7146/nja.v23i43.7497

Schmarsow, A. (1994). The essence of architectural creation. In H. F. Mallgrave & E. Ikonomou (Eds.), *Empathy, Form and Space: Problem in German aesthetics 1873-1893* (pp. 281–297). The Getty Center for the History of Art and the Humanities.

Schmitz, H., Owen, R., & Slaby, M. (2011). Emotions outside the box—The new phenomenology of feeling and corporeality. *Phenomenology and the Cognitive Sciences*, *10*(2), 241–259. doi:10.100711097-011-9195-1

Schonhammer, R. (2018). Atmosphere-the life of a place, the psychology of environment and Design. In J. Weidinger (Ed.), *Designing atmospheres* (pp. 141–179). Universitätsverlag der TU Berlin.

Schwarzer, M. W., & Schmarsow, A. (1991). The emergence of architectural space: August Schmarsow's theory of "raumgestaltung". *Assemblage*, *15*(15), 48–61. doi:10.2307/3171125

Scott, G. (1954). *Architecture of humanism*. Doubleday Anchore Book.

Shakibamanesh, A., & Ghorbani, M. (2017). Toward time-based design: Creating an applied time evaluation checklist for urban design research. *Frontiers of Architectural Research*, *6*(3), 290–307. doi:10.1016/j.foar.2017.05.004

Stanek, L. (2012). Architecture as space, again? Notes on the 'spatial turn'. *Spéciale'Z (Paris)*, *4*, 48–53.

Steele, D., Luka, N., & Guastavino, C. (2012). Constructing ideal soundscapes: a practical study on closing the gaps between soundscape studies and urban design. *Proceedings of the Acoustics 2012 Nantes conference*, 2163-2168. 10.1121/1.4709108

Steele, F. (1981). *Sense of place*. CBI Publishing Company, Inc.

Stefansdottir, H. (2018). The role of urban atmosphere for non-work activity locations. *Journal of Urban Design*, *23*(3), 319–335. doi:10.1080/13574809.2017.1383150

Taylor, C. (1975). *Hegel*. Cambridge University Press. doi:10.1017/CBO9781139171465

Terranova, C. N., & Tromble, M. (Eds.). (2016). *The Routledge companion to biology in art and architecture*. Routledge: Taylor & Francis. doi:10.4324/9781315687896

Thrift, N. (2004). Intensities of feeling: Towards a spatial politics of affect. *Geografiska Annaler. Series B, Human Geography*, *86*(1), 57–78. doi:10.1111/j.0435-3684.2004.00154.x

Tuan, Y.-F. (1979). Space and Place: Humanistic perspectives. In S. Gale & G. Olsson (Eds.), *Philosophy in Geography* (pp. 387–427). Springer. doi:10.1007/978-94-009-9394-5_19

Zumthor, P. (1998). *Thinking architecture*. Lars Müller.

Zumthor, P. (2006). *Atmospheres: Architectural environments-Surrounding objects*. Birkhauser.

ENDNOTES

1 The "mood" is a harmonious affective disposition that lasts for a long time. Its indicators appear through the feeling and the movement of the body, such as the symmetry, proportion, and vertical position (Ekkekakis, 2012).

2 Karin Dahlberg (2006) describes the difference between the characteristics of the essence and things by giving an example of the relationship between the table and the chair as two different phenomena. In her perception, each has something different from the other, each has its independent characteristics, and, consequently, its independent essence. Thus, Dahlberg believes it is not enough to consider the characteristics to know the essence. For example, it is possible to argue about the table having four legs, being made of wood, and being painted in blue, but without ever grasping its essence. Even if the observer made a shape from the table, added chairs to it, and put it in the dining room, all of this would not make the essence of the table.

3 Norberg-Schulz's point of view about the linkage between the spirit of place and its essence, determines the characteristics of a place and seems merely an abstract description. This description has nothing to do with the truth of Karen Dalberg's findings, according to which the determination of the characteristics of something does not mean the knowledge of its essence.

4 In this, Zumthor applied this notion in designing his buildings, particularly in the religious structures, such as the St. Benedict Chapel in the village of Sumvitg (1988). He designed also a museum built on the ruins of a church: the Kolumba Art Museum in Cologne, Germany (1997-2007) (Durisch & Zumthor, 2014).

5 Zumthor designed material-led projects, such as the Therme Vals thermal baths in Switzerland (1996) and the Kunsthaus Bregenz gallery in Austria (1997). Besides, he searched about how to achieve special atmosphere of intimacy and warmth by using materials, especially "'wood" in Haus Luzi, Jenaz, Switzerland (2002). He had displayed the notion of a presence in architecture to find an architecture that is free from symbolism and all about [the] experience. Also, he depends on the idea of avoiding the premature meaning in his work Steilneset Memorial, in Vardø, Norway (2011) (Durisch & Zumthor, 2014).

6 For more details, see endnote number 4, as well as Dahlberg's work (2006).

Chapter 4
Potentials of ICT and IoT in the Design of Smart Public Spaces

Rafee Ibrahim Hakky
Al Rasheed University, Damascus, Syria

ABSTRACT

This study aims to show that information and communication technologies (ICT) and the internet of things (IoT) can be used in order to improve the atmosphere of a space. This chapter is based on the assumption that if ICT and IoT are used to better the physical environment of the space, its atmosphere will also be enhanced. A summary of the design of public open spaces and how current ICT and IoT can help in this regard is presented. This theoretical background was the base for a proposal that uses current technology to allow constant changes to the physical layout of a space to accommodate the changing needs of users and improve the space's atmosphere. The proposal advanced here is intentionally simple in terms of its idea and its execution. It is also possible to be tested at any scale a designer or a researcher would prefer. The different and economically conscious experiments will lead to continuous improvements to the proposal.

INTRODUCTION

Open public spaces continued to act as social arenas for citizens all over the world for decades. In fact, thanks to the efforts of a large number of thinkers, designers, and writers, their social role has been increasingly emphasized by researchers since the sixties of the last century. The works of pioneers, such as William Whyte, Jane Jacobs, Paul Zucker, Rob and Leon Krier, and Jan Gehl, to mention only a few, dealt with open spaces within the realm of urban design from two main perspectives; namely, their sociability and their form. Knowledge gained from such studies has been reiterated, expanded, and reexamined during the past few decades, and thus it was focused, detailed, and substantiated.

However, questions related to ways of dealing with urban spaces within contemporary cities have surfaced again in light of the present possibilities and opportunities that Information and Communication Technologies (ICT) can deliver today to enhance the quality of these spaces (Aurigi, 2017). At the

DOI: 10.4018/978-1-7998-3856-2.ch004

same time, their form and detailing have also been opened to discussion at least in connection with the atmosphere which they may promote intentionally or unintentionally (Gandy, 2016).

Based on these two pressing issues, an interesting question emerges: can new information and communication technologies be used to improve urban public spaces in terms of their two main issues: sociability and atmosphere? As for sociability, the issue of concern is how to bring people to spaces, while atmosphere is, in simplified terms, the feel of the place. One possibility of addressing the above question is by finding ways to utilize ICT in order to improve the two issues, sociability and atmosphere. However, a more comprehensive view of the problem suggests a look at the three components (ICT, sociability, and atmosphere) as a threesome with a causal effect relationship.

The premise here, or at least the assumption, is that if it is possible to improve the sociability of the open public space through the use of ICT, more people would be interested in stepping into the space and spending time there. Having more people in the space will be reflected positively on its atmosphere. The realization of this assumption requires two consecutive stages: the first is theoretical and the second is practical. The first stage consists of two steps: the first is a study that aims to define physical and nonphysical elements that can support the creation of a positive atmosphere. The second step is a proposition of appropriate means to introduce the defined elements to urban spaces benefiting from the opportunities delivered by ICT. The practical stage would be an actual application of the proposed means leading to the materialization of the suggested elements.

This chapter focuses on the theoretical side of this endeavor only. It aims to identify elements that could positively affect atmosphere in open public spaces, and then propose possible applications of these elements through the use of information and communication technologies. Identification of the crucial elements will be done through the study of related literature, while their application on the design of urban spaces will be suggested based on the understanding of the potentials of ICTs.

ATMOSPHERE AND THE DESIGN OF PUBLIC OPEN SPACES

In order to identify elements that may contribute to the formation of an atmosphere in open urban spaces, three areas will be addressed: the concept of atmosphere and what constitutes it, the design of public spaces in connection with the question of atmosphere, and the uses of public spaces.

Atmosphere

'Atmosphere' as a term began to acquire two distinctive meanings as early as the beginning of the Nineteenth Century. The first meaning was the simple and commonly known one: the layer of gases that covers Earth; or in a more simple term, the known 'air.' The second meaning, however, "refer(s) to the prevailing mood of a place, situation, or cultural representation such as the feeling evoked by a film or a novel." (Gandy, 2016). It is worth mentioning here, though, that this definition of atmosphere is perhaps an oversimplification of a much more complex concept. Readings such as Gandy's (2016) and Albertsen's (1999) go through lengthy discussions about the different aspects of the term as seen by scholars from a variety of disciplines.

Albertsen (1999) relies heavily on the work of Böhme (1995) in his discussion of atmosphere. He explains that atmosphere can have many kinds, "serene, serious, terrifying, oppressive," (Albertsen, 1999). He distinguishes between the effect of an element and its reality; a red color, paraphrasing Böhme, radiates a warm atmosphere, yet it does not physically generate heat. Further, atmosphere, according to Böhme, "is something we perceive with all senses, something that affects us emotionally." (Albertsen, 1999).

As for architecture and atmosphere, Albertsen quotes Zumtor (1998): "Forget the form," because "(i) t is the light, the mood, atmosphere, and bodily state of being that matter in his (Zumtor) architecture." Albertsen also sees that atmosphere was definitely in Wright's mind who said: "Whether people are fully conscious of this or not, they actually derive countenance and sustenance from the 'atmosphere' of the things they live in or with" (Albertsen, 1999).

Albertsen (1999) later on in his informative article declares that three elements generate atmosphere in cities, namely; smell, sound, and form of life. He thinks that each city has its own smell or odor that is a result of its specific life style or environment. It does not matter whether the odor is pleasant or not, what matters is its existence which gives a place its uniqueness and thus its atmosphere. Sound has a similar effect in giving a place a distinct atmosphere. A city like Venice, as an example he explains, acquires a distinct atmosphere because of the sounds of footsteps and water splashes that are free from the noise of cars' engines. As for city life, Albertsen notes that people's behavior is reflected on its atmosphere. This is sensed through vendors' shouting out the names of their merchandizes, or drivers honking their horns. Apparently, one or all of the three can create an atmosphere that would be distinct and memorable.

In a similar view, Gandy (2016) quotes the geographer Ben Anderson who explains what is termed 'affective atmosphere' as "a class of experience that occur *before* and *alongside* the formation of subjectivity, *across* human and non-human materialities, and *in-between* subject/object distinctions." Gandy goes into a lengthy exploration of the meaning of atmosphere as a concept from a number of different but interrelated fields such as politics, phenomenology, and cultural geography. But what is of particular importance to this study is that he too dwells on the role of social life and sound in particular in the creation of atmosphere.

Atmosphere has been investigated by researchers from two perspectives: either by looking into the physical qualities of a space or by understanding how people experience that space (Michels, 2015). Böhme and Farrell (1992) suggest that "the experience of atmospheres is this experience of co-presence, where subject and object meet via 'aistheton, that is, what is perceivable'" (Michels, 2015). In simpler words, atmosphere is felt, not comprehended, not by either one of the two, but by both, the physical qualities of the space and state of mind of the observer. Thus, atmosphere for a space may not be constant or fixed, because both the physical qualities as well as the perceiver are constantly changing.

Atmosphere, perhaps, can be seen as a result of the physical elements in a space, but experienced through the personal state of mind of the beholder. The physical elements are all what make a space, such as light, sounds, symbols, texts, and other material objects which if worked into a space "atmospheres are 'enhanced', 'transformed', 'intensified', 'shaped'…" (Böhme, 2006). Regardless of the complexity of 'atmosphere' as a concept, it is a reality attached to any space. In the case of this study, it is the urban open space which acquires an atmosphere as a result of the composition of its physical and nonphysical components.

Visitors of such a space will experience this atmosphere. Although there is an overall atmosphere that will be felt by all visitors, there will be individual sensing of that atmosphere by every individual based on his or her personality and state of mind. It is not possible to influence the personal reaction of the visitor; nonetheless, it is in the hands of designers to create the overall atmosphere of the place by manipulating its physical and nonphysical components.

Design of Public Open Spaces

How spaces are to be designed has been a subject of study for decades. The famous volume written by Camillo Sitte *City Planning According to Artistic Principles* (originally published in1889) paved the way to a long and informative body of knowledge regarding the form of urban public spaces. His principles are still echoed in many contemporary studies on the subject. Krier's typological study of urban spaces (1979) was another landmark in the study of urban spaces' forms. However, it seems that contemporary work in connection with the design of open public spaces concentrates on what makes them more social and attractive to passersby. This is an expansion and continuation of the work of Whyte, *Social Life of Small Open Spaces* (1980) and Gehl's *Life between Buildings* (1971).

Based on the writings of Francis (1987), Korkmaz (2001) and Southworth (1989), Çakmak and Topçu (2018) argue that the following topics should be addressed in relation to design criteria of urban open spaces: function, diversity, accessibility, form, livability and identity. Spaces should provide diversity in functions to accommodate all users equally. They should also provide diversity when it comes to different times of the day or different seasons. Accessibility was always a crucial criterion which was stressed by Whyte and Gehl among others. The ability to easily access the space seems to be a very important ingredient to its success.

Çakmak and Topçu explain that form of the space is related to how users perceive it. This perception is affected by three issues: the users' characteristics, their behavior in the space, and the space characteristics. This echoes the issue of atmosphere discussed earlier. Atmosphere is a feel that is based on the form of the space. The form is perceived based on its qualities and the qualities of the person who perceives it. Here too, urban designers deal with the characteristics of the space which include its dimensions, elements, materials, form, among others. As for livability, the space should be comfortable and safe. And finally, identity is established through the legibility of the space and the possible meaning that it evokes.

Khakpour et al. (2017) adopt similar criteria for a successful public space: access and linkage, comfort and image, use and activities, and sociability. Their study concluded with a large number of effective factors that draw people to urban spaces. They categorized these factors under four headings: social-cultural, physical-environmental, management, and performance. Factors in the first two categories include existence of dynamic social-cultural relations, comfort, security, presence of various classes of people, accessibility, space attraction, flexible design, and low environmental pollution.

Pezzica et al. (2017) claim that many of the problems associated with urban spaces can be better dealt with if what they call 'efficiency targets' are properly addressed. Efficiency targets include centrality, connectivity, enclosure, thermal comfort, security, social equity and interaction. Overlap between these targets and effective factors identified by Khakpour et al. (2017) and topics introduced by Çakmak and Topçu (2018) is clear. Along the same line, Ja'afar et al. (2012) propose that street's character can be improved through physical elements (manmade and natural elements), activities (uses allowed because of available physical elements), and meaning.

Studies which go into discussing the physical elements to be in open spaces are many. Ja'afar et al. (2012) mention with details the importance of adequate seating, lighting, and garbage cans. They also emphasize the importance of introducing nature to public space particularly through trees. Pezzica et al. (2017) mention a number of practical solutions they applied to the open space which was under their investigation in order to improve its functionality. They talked about seating and lighting; they also mentioned the relocation of a water feature from the center of the space to one side of it in order to free the space for public activities and to benefit from the sound of the water to mask noises and provide some sort of privacy.

Similarly, Królikowski et al. (2019) point out the importance of lighting and urban furniture. Giddings et al. (2011) went back to earlier considerations of the form of open spaces starting with Sitte in his study of the appropriate size of squares, their dimensions, proportions, and openings; the researchers confirmed earlier findings. They also validate the importance of having appropriate temperature, wind speed, and noise control in open spaces. They further shared other studies in looking at the detailing of the spaces in terms of seating, public art, lighting, and shading devices.

Mehta and Bosson (2018) carried out a study about streets as public open spaces and not just as corridors for movement. They also saw the importance of making spaces meaningful to people and comfortably usable by them. Complementing other studies on the topic, they found that open spaces would be livelier if different types of seating are available and more activities around the space are found. Siramkaya (2019) talks about improving urban identity through the design of better urban spaces which guarantee three points: easy access, variety of surrounding activities in order to promote sociability, and provision of suitable green areas, water elements, and furniture in accordance with usage, climate and the character of the city itself.

Research done by Anderson et al. (2017) brings about similar results in principle. The group examined three types of behaviors a user of an open space can be involved in: to engage in some sort of social interaction (connecting), to participate in any kind of physical activity, and to observe the environment around him or her (taking notice). The research showed that low cost interventions to open spaces can increase the number of users and their period of stay. The intervention also resulted in substantial increases of the three-way behaviors (connecting, engaging in physical activities, and taking notice). What is crucial to notice here is that interventions are actually simple and low cost such as adding a few benches in critical locations and installing exhibition boards. In fact, these two items, it was found, are the most influential in making a successful space.

Perovic and Folic (2012) talked about many attributes of a successful open space, some of which are being "poly functional, dynamic, associative, homogeneous, inclusive, (and) accessible." These qualities resonate with findings of other studies mentioned earlier. Ferdous (2013) similarly reports that good level of enclosure of the space, presence of water features, sculptures, and vegetation are crucial elements in creating successful open space aesthetically.

In summary, open spaces need to have proper dimensions so that a feeling of openness at the same time a level of enclosure is clearly achieved. Spaces should have a variety of functions to accommodate different visitors, they should be easily accessible, and enjoy clear identity. Types of furniture in open spaces are also crucial for its success; seating in particular has been found as the most important item to improve the performance of a space. Vegetation and water are also found in many studies to be of special importance. Among other issues to be observed in order to improve open spaces' performance as a gathering place are flexible design, low environmental pollution, visual attraction, thermal comfort, lighting, and security. It was found in a study of a number of urban spaces in Europe that those which are not well taken care of received three times less users than those which are well maintained (Djukic et al. 2017).

Users of Public Open Spaces

Interest in public open spaces is a result of change in mentality, it seems. It is a result of a new view as well as vision of urban life and the rights of citizens of cities. This new vision is definitely the fruit of a lengthy struggle on the part of researchers in many fields but in particular in the fields of sociology and urban life. The story perhaps starts with the works of Jane Jacobs, Lewis Mumford, Kevin Lynch, and William Whyte, and then all those who continued working along the same line of study for a better urban life. And perhaps one can argue that social life in open urban spaces is part of social sustainability. Because "(p)ublic spaces are open to all, and as such represent a democratic forum for citizens and society.... Lack of public space is an insidious expression of a lack of democracy." (Giddings et al. 2011). Social sustainability is based on a democratic way of thinking, and open spaces for all citizens is an expression and a manifestation of people's right of having decent spaces to congregate for social and cultural activities.

Thus, public open spaces are made for people. They are the public's 'living room', they provide opportunities for people to sit, relax, and get away from their daily problems and worries to be part of a larger social entity. Therefore, it is expected that this 'living room' should meet the expectations of its users. And it goes without saying that the best way to grasp these expectations would be to talk to the users and get them involved in the design process of these spaces. Public participation, thus, has become, one dares to say, a must for a sincere effort to create responsive public spaces.

In this spirit, more designers are convinced that talking to people, asking them about their needs and views in relation to public spaces, and involving them in design decisions are practices that will guarantee better designs. "More dialogues do not mean less expertise; it brings dynamics in integrating the participants into design process of a certain place." (Bašová and Štefancová, 2017). According to the same authors, smart public spaces are the result of collaboration between those who administer and those who use the spaces. This is the secret of smart public spaces; smart devices are just tools, they cannot do much on their own.

Open public spaces are expected to meet users' needs. Al-Bishawi and Ghadban (2011) developed an informative model in which they suggested that three components work together when designers are addressing these needs. These components are use, rules, and design. Use is what activities and functions are expected to take place in the designed space. Rules are what direct, orient, or guide design such as regulations, codes, social norms, and the like. Design is the proposed physical solution that meets the needs. The authors explained that the designer should be aware of both use and rules; use informs the designer in terms of what needs to be met and rules tell the designer how to meet these needs. This simple and direct model can be of a great value when Internet of Things (IoT) and Information and Communication Technology (ICT) will be utilized for the betterment of open public spaces.

Another point which is critical to keep in mind when addressing the use of open public spaces is the fact that they are not a must-use places. They are used only when visitors feel like using them. Gehl (2011) explains that there are three types of outdoor activities: necessary activities, those which we need to practice such as going to work or school; optional activities, which are practiced when conditions of space are inviting and convenient such as taking a walk or sitting in a park; and social activities (or resultant activities) which result from the optional activities such as getting engaged in a conversation with a stranger in the park or watching an event in a public space. Optional activities, which are the cause of life in public open spaces, take place only when quality of the public space is good, and thus social life in urban spaces can only be generated when these spaces are pleasant, inviting, and meet the expectations of potential users.

It is mentioned earlier that spaces should be inclusive (Perovic and Folic, 2012), research has concluded that not all types of people are seriously considered when designing open spaces, in particular youth and children (Derr and Tarantini, 2016). Getting them involved in the design process is crucial to meet their needs and to encourage them to be part of the life of open spaces. Other important constituents of urban open space are people with disabilities. Many open spaces do not include sufficient or appropriate details to facilitate their use by this group of visitors (Oliveira Neto, 2018).

On the other hand, it has been observed by Hampton et al. (2015) that women's use of public open spaces has increased in recent years and that more men and women are spending time together in these spaces. It was also observed by the same group of researchers that the rate of use of mobile phones is small in public spaces. Based on their collected data (Banerjee, 2001), they also observed that people are using public spaces despite their extensive use of mobile phones. People also tend to spend more time in groups while in public. In other words, mobile phones did not cause an increase in social isolation in public spaces nor did they cause a decrease of use of such spaces as it was conveyed in earlier research (Banerjee, 2001). These are encouraging findings that should influence the way designers address open public spaces in cities.

Thus, public open spaces should accommodate all types of users, including youth, children, and people with disabilities. Women are using these spaces more often and they are enjoying them along with men. Therefore, in order to make spaces truly inclusive, it is vital that needs of all possible users are met. This is especially critical since these spaces are used only if they are inviting enough. They can be so, it is believed through research, if more public participation is allowed in their planning and design. Their participation will lead to a better understanding of their needs and how it can be accommodated through design that carefully considers expected uses and enforced rules.

Technology and Public Open Spaces

It has been evident that computer technology and social media are capable of providing researchers with information related to use of public spaces. For example, Cheliotis (2016) was able to prove that it was possible through the use of these technologies to document people's use of open spaces and correlate their use with weather and temporal conditions. Based on the reached results, he was able also to predict future daily life cycles for the studied space. The use of Information and communication technologies (ICT) is intend to help cities become more social and inclusive and not just more high-tech as Duarte and Mateus put it (2017).

It is also believed that ICT can be a tool to attract people to open spaces through social media; people tend to share their experiences and express their opinions and desires using their phones. It is obvious that by doing so, they invite more visitors (Menezes and Costa, 2017). Having free public Wi-Fi in open spaces will increase the chances of communication, and consequently will encourage people to use them. In fact, free internet connection will bring more people from coffee shops and other similar facilities to open spaces (Warbis, 2018). Of course, all needed gadgets to ensure proper use of the service should be well functioning. These gadgets should be satisfactory in terms of usage, comfort, quantity, accessibility, arrangement and aesthetics. They also should be protected from environmental factors and go through regular maintenance (Abdel-Aziz et al. 2016).

An informative study that supports the above listed characteristics of satisfactory gadgets in the context of Wi-Fi connections is done by Suchocka et al. (2017, a). These researchers categorized users under four groups: focused on work, focused on entertainment, transit pedestrians, and tourists and indigent users. Their expectations of hotspots slightly differ; however, they can be generalized in the following points:

- Ensure fast and stable internet connection
- Allow charging a mobile device
- Located in a comfortable place to work
- Secure
- Location is quiet
- Location is sheltered from weather conditions
- Location is aesthetically pleasing
- Location is close to restaurants or cafes

In another study, Suchocka et al. (2017, b) expanded on the subject of designing the hotspot by introducing detailed description for it and its surroundings. What is of interest in relation to the existence of hotspots, however, is their effect on the use of the locations where they were introduced. The study found that over 25% of hotspots users did not come to the location before knowing of the existence of a hotspot in it. 70% said that they are going more often to the location after internet became available in it. Moreover, the study found that 75% of respondents declared that when they had the choice of selecting a location with internet connection and another one without, they opted to go to the one with internet facility. These results are very encouraging.

Opportunities of the technology are many, one of which is augmented reality which can be used as an educational and entertaining tool. Experimentations in the use of augmented reality has been underway and they have been proven successful and promising (Duarte and Mateus, 2017). It is going to be an exciting challenge to benefit from this new method of communicating information and presenting parallel reality in an effort to make open spaces more inviting and pleasant for visitors. Furthermore, experimentations are now in progress in connection with the use of information technology as an educational tool in the context of smart cities (Bonanno et al. 2017). All types of information about places can now be communicated through smart phones including history, functions, activities, economics, and anything else. The combination of augmented reality and information dissemination via ICT will bring about very exciting results which can definitely be applied in the context of open urban spaces.

It is established that ICT and IoT are helpful in studies related to climate in open public spaces or in the analysis of density of use in spaces according to location and time. However, other uses geared towards enhancing social performance of spaces are what designers hope that these new technologies can provide. One example is ways of benefiting from ICT in studies related to sound in open spaces. Sound is typically not fully utilized as a design element. ICT's provide a unique opportunity to better utilize sound to enhance the overall sensory experience of open spaces' users. Three types of soundscapes are at the disposal of sound designer (Botteldooren, 2017):

- **Background Soundscape:** Sound is meant to be the least noticed in the experience of a visitor to a space; thus, it does not contribute significantly in the overall composition of the space. This type of soundscape is used the most ignoring other more influential possibilities. An example would be elevator music.

- **Supportive Soundscape:** Sound supports the sensory experience of the space without being its primary focus. This is the case of sound, whether music or words, in museums for instance.
- **Focused Soundscape:** Sound is the main reason for being in the space; it is what a visitor looks for. This sound is experienced in theaters and cinemas. In such places, acoustics are also crucial.

How ICT can utilize these three types of soundscapes for the benefit of any design is an exciting question.

To conclude this section, it is well established that ICT and in extension IoT can provide a very rich technical support system that can enhance the designer's understanding of the site's conditions which he intend to design. ICT provides information related to climate and location and movement of users. Yet, it can be used also to enhance the sensory experience of users such as in the case of sound management. How can ICT and IoT enhance the visual experience is an interesting question; perhaps it is the main question. An increasing number of studies show that ICT can be used as an educational and informative tool. Studies also found that availability of ICT and IoT in the disposal of users of open spaces even in its simplest form, i.e., Wi-Fi connections, can increase public use of outdoor spaces. One requirement for successful use of the technology is the provision of functionally supportive and aesthetically pleasing environment to house it. The important question here is how to maximize the use of available as well as future technologies for the benefit of socially oriented urban spaces?

ICT AND THE 'ATMOSPHERE' OF PUBLIC OPEN SPACES

As mentioned earlier, the objective of this study is to observe possibilities of utilizing ICT and IoT for the creation of better atmosphere in public open spaces. This is to be achieved through the improvement of these spaces at two levels: physically (spaces' visual appearance) and socially (spaces' usability). To this point, the study explored four points: the meaning of 'atmosphere', design of pubic open spaces, issues related to users of public open spaces, and information and communication technologies in public open spaces. This section will put results of the four points together in order to form a holisitc picture. It will then explore points where information and communication technologies can help improve spaces physically and socially in order to bring about a special atmosphere.

A Holistic Picture

Studies related to atmosphere have become relatively complicated and extensive. For the purpose of this study, a simple description will suffice. Atmosphere is the feel of the space, its vibe. It is sensed, not comprehended and it is a result of the interaction of the physical characteristics of a space with the state of mind of the person who is experiencing the space. Physical characteristics include everything that compose the space be it its shape, size, enclosing elements, materials, colors, furniture, light, shade, and sound.

Designers' main concern is the shape of the open space which can accommodate both aesthetic and social requirements. As mentioned earlier, it is assumed that a successful open space in connection with these two aspects will generate a pleasant feel in the mind of the user, and thus the space will enjoy a positive atmosphere. Studies indicate the importance of the space's dimensions, proportions, shape, and degree of enclosure. Also, accessibility, clear identity, flexibility, furniture, vegetation, and water elements are found to be of great influence in generating both aesthetically pleasing and socially inviting space.

As for the social dimension, it is advised that public open spaces should accommodate the needs of all types of people including youngsters, children, women, and people with disabilities. The only practical way to be able to accommodate the needs of all these types of expected users is to rely on public participation in the design decision making process. Needs of users are met as a result of the interaction of a trio: uses (the activities that are requested), rules (all kinds of guidelines, standards, and regulations that dictate how things should be done), and design (the physical manifestation of meeting the needs within the limitations of the rules).

Information and Communication Technologies (ICT) as well as Internet of Things (IoT) provide designers with valuable information related to conditions of a site, whether climatic or locations of users. ICT and IoT, however, have much more capabilities; they should be able to assist designers in many other aspects. For instance, they are used now for educational and informative purposes. The premise of this study is to explore other potentials of the new technologies in order to satisfy the needs of users and bring about more pleasing spaces that are capable of generating exciting atmospheres. From a practical point of view, the use of technology should be observed in a manner that enhances the spaces' functionality and attractiveness.

How Can Communication Technology Help?

Again, the main line of thought in this work stems from the assumption that if a public space can satisfy the needs of its users including the aesthetic dimension; it is possible that a positive atmosphere in that space will result. The question is, then, how can technology help? In order to address this question, issues that are summarized in the last section are examined in terms of whether they can be treated with the help of ICT. It is understandable that more issues related to the physical composition of open spaces and the atmosphere generated by them exist and are influential. But for the sake of controlling the study, only those presented and summarized earlier will be addressed. Issues can be categorized in order to help formulate an overall framework. This framework can be modified in accordance with the case under investigation. The process of examining the issues of concern is basically twofold; taking notice of the particular issue and then reacting to it. This twofold process is detailed in eight steps as presented in table 1. The table shows these eight steps with an application as an example.

Solutions to issues would come from two sources:

1. Programmed solutions: All involved devices are to include a set of possible solutions to expected issues programmed in their software based on the particulars of the open public space where these devices are installed.
2. Public opinion: Constant collection of public opinions in connection with the space in terms of its use and design is done regularly. Results of public opinions are filed for future reference and use.

Data to be collected and filed, whether from programmed solutions or collected ones from public opinions, can be based on the issues which are summarized in the previous section. Tables 2 and 3 present a way of categorizing the most crucial issues. Table 2 shows physical issues while table 3 is for social issues.

Table 1. The eight-step process of addressing an issue by ICT and IoT using size of space as an example

	The step	Explanation	Application
1	Sensing the issue	ICT and IoT gadgets will sense a concern	• Size of space is large • Activity for 4 days requires smaller space
2	Inquiring about the issue	Issue may need further investigation in order to understand its magnitude	• Sense of scale is important because participants are mostly children (programmed: if users are children re-evaluate scale of space) • Activity is during the evenings
3	Evaluating the issue	Assigned device assesses the actual need to respond to the issue	• Space may be very unfriendly to children • Being evening activity, lighting can be the solution (programmed: control of lighting in order to visually sub-divide the space).
4	Responding to the issue	Assigned device will propose solutions and rank them in accordance with already fed set of criteria	• Provide lighting in the area of the activity only • Use partitions (Programmed: if lighting can meet the need, use it; if not, use partitions.)
5	Communicating the issue and the possible responses	Assigned device will send all information related to the issue to the Person in Charge	• Space too large for the activity • Sense of scale is crucial because activity is for children • Activity at night • Best solution: adjust lighting to make space feel smaller, or • Second solution: Use partitions (not necessary in this case).
6	Receiving order of action	Person in Charge makes his/her selection	Person in Charge selects use of lighting
7	Executing the selected response	Responsible device will take action	• Lighting is used in the area of the activity • Lighting is dimed in the rest of the space
8	Observing the performance	Devices will monitor use of space and the success of solutions in handling the need	• Report of action is generated and sent to Person in Charge • Space is monitored to ensure lighting provided the required feel of the space (people are concentrated in the lit area only).

Based on the above proposal, there would be two types of devices to be installed in a smart open space:

1. Data collection devices: These devices collect information from the space via audio-visual cameras and sensors. Information is filed according to their categories as specified in the system. Data will include climatic information and use of space (moving people and stationary people. Information includes numbers, type of activities, and duration of activities). Devices need not to be seen by people.
2. Public opinion collection devices: These devices require interaction with people; thus, they have to be designed in a way that they are easily accessible, user friendly, and resist vandalism. Questions asked by these devices should be short, easy to comprehend, and easy to answer.

Table 2. Partial example of filing of issues to be treated with the support of ICT and IoT in connection with physical issues

Main Area	Issues addressed	Source of data
Design of public open space	Form of space	Programmed
	Length of space	Programmed
	Width of space	Programmed
	Proportion of length to width	Programmed
	Possible solutions to modify length to width proportion	Programmed Public opinion
	Proportion of length to height	Programmed
	Possible solutions to modify length to height proportion	Programmed Public opinion
	Openings (Northern elevation)	Programmed
	Openings (Eastern elevation)	Programmed
	Openings (Southern elevation)	Programmed
	Openings (Western elevation)	Programmed
	Sense of enclosure	Public opinion
	Possibilities of modifying sense of enclosure	Programmed Public opinion
	Etc.	
Accessibility	Northern Elevation: Entrance 1: • Width • Type (steps, ramp, flat, etc.) • Features (gate, flooring, etc.) • Etc. Entrance 2: Etc.	Programmed
	Eastern Elevation: Entrance 1: • Width • Type (steps, ramp, flat, etc.) • Features (gate, flooring, etc.) • Etc. Entrance 2: Etc.	Programmed
	Etc.	
Views from outside	Northern elevation: • Location of exposure • Degree of exposure • Width of view	Programmed
	Eastern elevation	
	Etc.	
Etc.		

Table 3. Partial example of filing of issues to be treated with the support of ICT and IoT in connection with social issues

Main Area	Issues addressed	Source of data
Seating	Location 1: • Number of seats • Design of seats • Conditions of seats • Orientation of seats • Etc.	Programmed
	Adequacy of seats	Programmed Public opinion
	Location 2: • Number of seats • Design of seats • Conditions of seats • Orientation of seats Etc.	Programmed
	Adequacy of seats	Programmed Public opinion
	Etc.	
Lighting	Location 1: • Number of lighting devices • Type of lighting devices • Conditions of lighting devices	Programmed
	Adequacy of lighting devices	Programmed Public opinion
	Location 2: • Number of lighting devices • Type of lighting devices • Conditions of lighting devices	Programmed
	Adequacy of lighting devices	Programmed Public opinion
	Etc.	
Planting beds	Location 1: • Number of planting beds • Type of planting beds • Conditions of planting beds	Programmed
	Adequacy of planting beds	Programmed Public opinion
	Location 2: • Number of planting beds • Type of planting beds • Conditions of planting beds	Programmed
	Adequacy of planting beds	Programmed Public opinion
Etc.		

The Other Side of the Coin

The above discussion dealt with the technical side which addresses collecting information, communicating it, and generating responses to it. In order for these responses to be of use, they should be able to communicate with the open space. Communication can be either manual or automated. Manual response is when recommendations approved by the Person in Charge are communicated to workers on the ground to execute them. So, if the recommendation were, for instance, that more seats to be introduced to the space because of an increased number of users in one occasion, workers on site would bring out from storage moveable seats and distribute them.

Automated responses, on the other hand, would be more in line with the idea of a "smarter" open space which would be equipped with different kinds of furniture pieces that allow needed modifications to take place. This would include moveable and foldable partitions, seats, planting beds, shading devices, and lighting fixtures. These pieces would be connected with the main system, and thus become part of it. They would be designed and programmed so that they move, recline, fold, and unfold in accordance with orders controlled by the ICT's running the place.

Design of open spaces will have to be able to accommodate the requirements of moveable elements. The design may have to use a specific grid for instance or certain materials to ensure the smooth operation of these elements. If responses are to be automated, the design and detailing of the space will have to accommodate all programed options the ICT's and IoT's may have. In a simplified way, the public space would act as if it were a 3d plotter executing the orders of these technologies at a one-to-one scale.

Table 4 lists the elements that recent researchers have indicated important for the improvement of the usability and sociability of open urban spaces as discussed in this chapter.

Table 4. Elements that can contribute to the betterment of open urban spaces' usability and sociability as collected from recent literature in the field

Subject	Element
Sensory	views – smell – sound
Meaning	symbols – texts – images
Furniture	Seats – lighting – planting beds – trash bins – information signs – shading devices – pieces of sculpture – water elements
Design criteria	function – diversity – accessibility – form – livability – identity – linkage – security – attraction – flexible design – inclusiveness comfort – low environmental pollution – connection with nature – connectivity to ICT –
Form criteria	dimensions – proportions – shape – enclosure – openings – materials – colors
Thermal comfort	sun – shading – wind protection – breezes – reflection
Activities	passive activities – active activities – individual activities – group activities – social activities – observation – use of augmented reality
Design processes	public participation

A Theoretical Scenario

Table 1 showed the eight-step process of operation. One more scenario is exhibited here in order to present a typical routine of operation in response to a particular circumstance. The scenario is a case when the number of users increases in a beautiful sunny day.

1. Established preparations in case of an increase in number of users in a sunny day:
   a. The space is provided with the following gadgets:
      i. Planting beds' edges can expand in their width to become 50 cm wide.
      ii. Short dividing elements in the space have folded boards that can be unfolded to be used as seats.
      iii. Shading devices can unfold to cover specific sunny areas according to the time of the day.
      iv. Lids of extra trash bins located in the space can be opened to accommodate need.
      v. Shutters of extra vending machines can be opened as needed.
      vi. Extra screens for information and advertisements are located in strategic locations in the space.
      vii. Street venders are on calling lists updated through apps with live visitor numbers to be informed about level of use on a daily/hourly basis.
      viii. Etc.
2. The Information and Communication Technology system provided for the space is programmed as follows (a sample list):
   a. If number of users goes over (x) users, then expand (y) meters of the planting beds' edges and unfold (z) meters of the dividing elements. If number of users goes over (x2) users, then expand (y2) meters of the planting beds' edges and unfold (z2) meters of the dividing elements, and so on.
   b. If seats in area (1) are unfolded, then open shading area (1), if seats in area (2) are unfolded, then open shading area (2), and so on.
   c. If seats in area (1) are unfolded, then open lids of trash bins in area (1), and so on.
   d. If seats in area (1) are unfolded, operate screens (1-a, 1-b), if seats in area (2) are unfolded, operate screens (2-a, 2-b), and so on.
   e. If seats in area (1) are unfolded, open shutters of vending machines (1 and 2), if seats in area (2) are unfolded, then open shutters of vending machines (3 and 4), and so on.
   f. If number of users goes over (x) users, inform list (1) of street vendors, if number of users goes over (x2), inform list (2) of street vendors, and so on.
   g. Etc.
   i. Response to an increase number of users in a sunny day: Table 5 shows a step by step process in response to an increase of number of users in the space.

Table 5. A process responding to an increase of number of users in the open space

	Steps	Action	Obtained data
1	Sensing the issue	• ICT and IoT gadgets sense an increase in number of users • ICT and IoT gadgets sense increased area of space exposed to sun	• Number of users is calculated • Locations of sunny areas are documented
2	Inquiring about the issue	• ICT and IoT gadgets locate areas of increased activities and movement and calculate their numbers • ICT and IoT gadgets calculate areas of sunny spots, temperature, and wind speed to determine level of discomfort	• Location of users • Number of users • Temperature • Wind speed • Areas of sunny spots
3	Evaluating the issue	• ICT and IoT gadgets determine the need for more seating facilities • ICT and IoT gadgets determine the need for shading • ICT and IoT gadgets determine the need for other supporting services (trash bins, vending machines, advertisement boards, screens, etc.)	• Location and length of needed seating • Location and area of needed shading devices • Location and number of other needed facilities (trash bins, vending machines, etc.)
4	Responding to the issue	• ICT and IoT gadgets work out possible scenarios that determine number of seats to put in operation and their locations • ICT and IoT gadgets determine numbers and locations of other services (shading devices, trash bins, etc.) • lined for shading • ICT and IoT gadgets determine the list of street venders to be informed. This is done based on an estimate of number of users in the following hours of the day.	• Seating devices to be operated is determined • Shading devices to be operated is determined • All other services to be operated are determined Numbers, locations, and exact pieces of equipment are determined in accordance with the constantly changing climatic and use of the space.
5	Communicating the issue and the possible responses	• ICT and IoT gadgets convey to the Person in Charge a concise report showing all the information and the proposed responses	• A comprehensive and concise report is delivered to the Person in Charge
6	Receiving order of action	Person in Charge makes his/her selection	• Order from Person in Charge to the system
7	Executing the selected response	Designated devices will take action	• Designated seating devices are operated • Designated shading devices are operated • All other designated services are operated
8	Observing the performance	Devices will monitor use of space and the success of solutions in handling the need	• Report outlining actions is generated and delivered to Person in Charge • Space is monitored to ensure all services are operating properly • Action is taken in response to any change in use following the eight-step process

BACK TO ATMOSPHERE

It was explained earlier that atmosphere is the feel of a place based on its physical and non-physical elements gained through the senses of a visitor and filtered by his or her state of mind. Predicting or controlling a person's state of mind is beyond the scope of designers' work, but looking at a space's vibrating feel is part of any designer's job. However, it is to be noticed that a well-orchestrated atmosphere can and may influence the state of mind of people, both ways, negatively and positively. One way that may help create a positive atmosphere in a public space is to make it responsive to the needs of its users.

This was the premise of the study presented here, and accordingly the proposal was all about finding ways to accommodate the changing needs of users.

The proposal addressed physical modifications in response to physical changes in the use of the space. Yet, ICT and IoT can help create a positive atmosphere in one more way. That is by sensing the mood of people and working with it. It is known that water and its sound can calm people down, and that music will change people's mood, and color can give different hidden messages, and light can attract people's eyes.

Using these abilities and the same eight-step process explained earlier, the available technology will not only detect weather and use, but also people's emotions in the place. Through cameras and sensors, it will be possible to detect the kind of people who are in the space. A simplified example, speed of passersby and facial recognition can indicate if the majority of the users are young or old. Accordingly, ICT and IoT can suggest a certain music or lighting. It can also suggest a certain play of water fountains, or perhaps the amount of shading. These suggestions will calm the elderly or excite the youngsters.

The possibilities are limitless and the applications are not complicated in terms of the needed technology. These possibilities allow continuous informative experimentations that can inspire designers in their pursuit of creating exciting and positive atmosphere in their designs. It goes without saying that any change in the form or details of a space for the sake of accommodating a physical requirement will create a new mood in that space. In other words, the atmosphere of any space is constantly changing when changes are applied to accommodate a physical need. Designers should only exercise more caution when suggesting any change so as to improve the atmosphere of a space while answering for a need.

THE NEXT STEP

Any further exploration of the proposed concept and process will be too detailed or redundant. The next useful step is to take this theoretical proposal to a practical experimentation. This will take place through two consecutive steps:

1. Development of an appropriate software (or the repurposing of existing software and gadgets) as to respond to the two segments of the process:
    - Sensing and evaluating the circumstance; and
    - Responding to it.

    This is a task to be done by a computer engineer in collaboration with an interested designer.

2. Testing the program through a practical application. This can take place on a limited scale first, and then it can be a comprehensive application. A candidate space will have to be prepared for such an experiment. One element of the space's furniture can be designed to be able to perform two actions:
    - Move or change form in accordance with the designer's vision (planting beds' edge can expand, partitions can unfold, etc.).
    - Communicate with and respond to ICT and IoT which control the space. The element should be able to receive orders (to be folded or unfolded), send information (it is used as intended, it is not used anymore), and respond physically (execute the orders).

A small scale experiment will provide researchers with very valuable information about the practicality of the proposal so as they can bring it to a realistic and manageable platform. The encouraging point about this proposal is that it is simple in principle and relies on presently existing and not very complicated technologies, yet it can be very useful in practice. This step needs industrial designers, furniture manufacturers, and landscape architects among others.

CONCLUSION

This chapter is based on the premise that atmosphere of an open space can be improved when physical and non-physical needs of users in that space are met. It is also believed that ICT and IoT can help meet these constantly changing needs of users. Therefore, the study looked in some detail into the design of public spaces from two perspectives: their forms and their usability. It then discussed ways of how available technology can be part of the efforts of improving the livability of public spaces. A proposal was introduced that, with the help of current technologies, can improve the performance of public open spaces functionally and can enhance its atmosphere.

Although the proposal is very theoretical at this stage, it is not unrealistic. Existing current technologies can meet the technical requirements of the proposal. In fact, what is required are a number of different kinds of sensors and automatically operated mechanical apparatuses with very basic tasks such as opening and closing, folding and unfolding, and expanding and contracting. Another positive point to this proposal is that its testing can be scaled as desired, i.e., it is possible to test the logic on one edge of a planting bed or on the whole space.

ACKNOWLEDGMENT

This research received no specific grant from any funding agency in the public, commercial, or not-for-profit sectors.

REFERENCES

Abdel-Aziz, A., Abdel-Salam, H., & El-Sayad, Z. (2015). The Role of ICTs in Creating the New Social Public Place of the Digital Era. *Alexandria Engineering Journal*, *55*(1), 487–493. doi:10.1016/j.aej.2015.12.019

Abdulkarim, D., & Nasar, J. (2014). Do Seats, Food Vendors, and Sculptures Improve Plaza Visitability? *Environment and Behavior*, *46*(7), 805–825. doi:10.1177/0013916512475299

Al-Bishawi, M., & Ghadban, S. (2011). A Methodological Approach for Reading Urban Open Space. *Archnet-IJAR*, *5*(1), 73–85.

Albertsen, N. (1999). Urban Atmospheres. *Sociologi I Dag*, *29*(4), 5-29.

Anderson, J., Ruggeri, K., Steemers, K., & Huppert, F. (2017). Lively Social Space, Well-Being Activity, and Urban Design: Findings from a Low-Cost Community-Led Public Space Intervention. *Environment and Behavior*, *49*(6), 685–716. doi:10.1177/0013916516659108

Aurigi, A. (2017). Space Is Not a Platform – Foregrounding Place in Smart Urban Design. In *Enhancing Places Through Technology* (pp. 69–81). Edições Universitárias Lusófonas.

Banerjee, T. (2001). The Future of Public Space. *Journal of the American Planning Association*, *67*(1), 9–24. doi:10.1080/01944360108976352

Bašová, S., & Štefancová, L. (2017). Creative and Smart Public Spaces. *International Journal of Liberal Arts and Social Science*, *5*(1), 17–33.

Böhme, G. (1995). *Atmosphare. Essays zur Neuen Asthetik*. Frankfurt a. M: Suhrkamp Verlag.

Böhme, G. (2006). Atmosphere as the subject matter of architecture. In P. Ursprung (Ed.), *Herzog and Meuron: Natural History* (pp. 398–407). Lars Muller Publishers.

Böhme, G., & Farrell, J. (1992). An Aesthetic Theory of Nature: An Interim Report. *Thesis Elev., 32*, 90-102, Doi:10.1177/072551369203200107

Bonanno, P., Pierdicca, R., Franco, F., Martinez, A., Marcheggiani, E., & Malinverni, E. (2017). A Connectivist Approach to Smart City Learning: Valletta City Case-Study. In *Enhancing Places Through Technology* (pp. 69–81). Edições Universitárias Lusófonas.

Botteldooren, D. (2017). Cyber Enhancing the Urban Soundscape. In *Enhancing Places Through Technology* (pp. 181–187). Edições Universitárias Lusófonas.

Çakmak, B., & Topçu, M. (2018). An Evaluation of Urban Open Spaces in Historical City Center of Konya in the Context of Pedestrian Mobility. *International Journal of Human Sciences*, *15*(4), 1827–1846. doi:10.14687/jhs.v15i4.5506

Cheliotis, K. (2016). Capturing Real-Time Public Space Activity Using Publicly Available Digital Traces. In *The Workshops of the Tenth International AAAI Conference on Web and Social Media City-Lab, Technical Report WS-16-16* (pp. 8-13). Association for the Advancement of Artificial Intelligence.

Derr, V., & Tarantini, E. (2016). 'Because We Are All People': Outcomes and Reflections from Young People's Participation. *Local Environment: international Journal of Sustainability and Justice, 21*(12), 1534-1556.

Djukic, A., Vukmirovic, M., Jokovic, J., & Dinkic, N. (2017). Tweeting in Open Public space – Case Study Belgrade. In *Enhancing Places Through Technology* (pp. 191–203). Edições Universitárias Lusófonas.

Duarte, T., & Mateus, D. (2017). The Contribution of ICT in Planning of Public Open Spaces Reflections on the City of Lisbon. In *Enhancing Places Through Technology* (pp. 41–56). Edições Universitárias Lusófonas.

Ferdous, F. (2013). Examining the Relationship between Key Visual Characteristics of Urban Plazas and Aesthetic Response. *SAGE Open*, *3*(2), 1–10. doi:10.1177/2158244013485581

Gandy, M. (2016). Urban Atmospheres. *Cultural Geographies*, *24*(3), 353–374. doi:10.1177/1474474017712995 PMID:29278257

Gehl, J. (2011). *Life between Buildings*. Island Press.

Giddings, B., Charlton, J., & Horne, M. (2011). Public Squares in European City Centres. *URBAN DESIGN International*, *16*(3), 202–212. doi:10.1057/udi.2011.6

Hampton, K., Goulet, L., & Albanesius, G. (2015). Change in the Social Life of Urban Public Spaces: The Rise of Mobile Phones and Women, and the Decline of Aloneness over 30 Years. *Urban Studies (Edinburgh, Scotland)*, *52*(8), 1489–1504. doi:10.1177/0042098014534905

Ja'afar, N., Sulaiman, A., & Shamsuddin, S. (2012). Article. *Social and Behavioral Sciences*, *50*, 643–656.

Jacobs, J. (1961). *The Death and Life of Great American Cities*. Random House.

Karimi, K. (2012). A Configurational Approach to Analytical Urban Design: 'Space Syntax' Methodology. *Urban Design International*, *0*(0), 1–22. doi:10.1057/udi.2012.19

Khakpour, A., Sabouri, S., & Harirchian, M. (2017). Analysis of Effective Factors on Presence of Citizens in Urban Spaces, Case Study: Towhid Square in Tehran. *Journal of Sustainable Development*, *10*(3), 35–45. doi:10.5539/jsd.v10n3p35

Krier, R. (1979). *Urban Space*. Academy Editions.

Krier, R., & Krier, L. (1979). *Typological and Morphological Elements of the Concept of Urban Space*. AD and Acroshaw Ltd.

Królikowski, J., Rybak-Niedziólka, K., Skibinska, M., & Wlazlo-Malinowska, K. (2019). Indentification of the Genius Loci Category in the Process of Recognition of the City Landscape Design Issue. *Technical Transactions*, *2*, 29–40.

Mehta, V., & Bosson, J. (2018). Revisiting Lively Streets: Social Interactions in Public Space. *Journal of Planning Education and Research*, 1–13. doi:10.1177/0739456X18781453

Menezes, M., & Costa, C. People, Public Space, Digital Technology and Social Practice: an Ethnographic Approach. In *Enhancing Places Through Technology* (pp. 167–180). Edições Universitárias Lusófonas.

Michels, C. (2015). Researching Affective Atmospheres. *Geographica Helvetica*, *70*(4), 255–263. doi:10.5194/gh-70-255-2015

Oliveira Neto, J. (2018). *Inclusive Smart Cities: Theory and Tools to Improve the Experience of People with Disabilities in Urban Spaces* (Unpublished doctoral dissertation). Université Paris-Saclay, Universidade de São Paulo, Brésil.

Perovic, S., & Folic, N. (2012). Visual Perception of Public Open Spaces in Niksic. *Procedia: Social and Behavioral Sciences*, *68*, 921–933. doi:10.1016/j.sbspro.2012.12.277

Pezzica, C., Lopes, J., & Rebelo Paio, A. (2017). Informed Design Decision-Making: From Digital Analysis to Urban Design. *Gestão & Tecnologia de Projetos*, *12*(3), 9–26. doi:10.11606/gtp.v12i3.133936

Pierdicca, R., Malinverni, E., Khromova, A., Marcheggiani, E., Bonanno, P., Franco, F., & Martinez, A. (2017). The Integration of an Augmented Reality Module within the Way-Cyberparks App. The Case Study of Valletta City. In *Enhancing Places Through Technology* (pp. 57–68). Edições Universitárias Lusófonas.

Siramkaya, S. (2019). The Effect of Change in Urban Squares on Urban Identity: The Case of Konya. *Architecture Research*, *9*(1), 7–15.

Sitte, C. (1965). *City Planning According to Artistic Principles*. Random House.

Suchocka, M., Kimic, K., Maksymiuk, G., & Kolodynska, N. (2017). Outdoor Hotspots as a Tool for Enhancing Healthy Lifestyles of ICT Users Design and Development Principles. In *Enhancing Places Through Technology* (pp. 153–165). Edições Universitárias Lusófonas.

Suchocka, M., Maksymiuk, G., Kimic, K., & Kolodynska, N. (2017). Behaviour, Expectations and Preferences of 'Digital Natives' in Regard to the Design of Urban Public Spaces. In *Enhancing Places Through Technology* (pp. 117–128). Edições Universitárias Lusófonas.

Warbis, M. (2018). *Making the Spaces of Cities Smarter: Why Cities (and Their Citizens) Should Embrace the Digital Public Realm*. Retrieved from https://medium.com/@warbismichelle/making-the-spaces-of-cities-smarter-why-cities-should-embrace-the-digital-public-realm-c12afc810aaa

Whyte, W. (1980). *The Social Life of Small Urban Spaces*. Conservation Foundation.

Zucker, P. (1970). *Town and Square: From the Agora to the Village Green*. MIT Press.

Zumtor, P. (1998). Questioning Images. Interview with Peter Zumtor. *Daidalos*, *68*, 90–101.

ADDITIONAL READING

Anderson, B. (2009). Affective Atmospheres. *Emotion, Space and Society*, *2*(2), 77–81. doi:10.1016/j.emospa.2009.08.005

Djukic, A., Vukmirovic, M., Jokovic, J., & Dinkic, N. (2017). Tweeting in Open Public Space. In *Enhancing Places Through Technology* (pp. 191–203). Edições Universitárias Lusófonas.

Doorley, R., Noyman, A., Xiong, Z., Alonso, L., Grignard, A., & Larson, K. (2019). RevUrb: Understanding Urban Activity and Human Dynamics through Point Process Modelling of Telecoms Data. *Smart Cities Symposium Prague*. DOI: 10.1109/SCSP.2019.8805683

Karimi, K. (2012). A Reflection on 'Order and Structure in Urban Design'. *The Journal of Space Syntax*, *3*(1), 38–48.

Klichowski, M., & Patricio, C. (2017). Does the Human Brain Really Like ICT Tools and Being Outdoors? A Brief Overview of the Cognitive Neuroscience Perspective of the CyberParks Concept. In *Enhancing Places Through Technology* (pp. 223–239). Edições Universitárias Lusófonas.

Knöll, M., Neuheuser, K., Cleff, T., & Rodolph-Cleff, A. (2018). A Tool to Predict Perceived Urban Stress in Open Public Spaces. *Environment and Planning B. Urban Analytics and City Science*, *45*(4), 797–813.

Krier, R. (1988). *Urban Space*. Rizzoli.

Mitrache, G. (2012). Architecture, Art, Public Space. *Procedia: Social and Behavioral Sciences*, *51*, 562–566. doi:10.1016/j.sbspro.2012.08.206

Nouri, A., Costa, J., & Matzarakis, A. (2017). Examining default urban-aspect-ratios and sky-view-factors to identify priorities for thermal-sensitive public space design in hot-summer Mediterranean climates: The Lisbon case. *Building and Environment*, *126*, 442–456. doi:10.1016/j.buildenv.2017.10.027

Chapter 5
Transformation of Urban Space by Smart Technologies:
Evidence and Traces From Istanbul Historical Peninsula

Burak Belli
Istanbul Technical University, Turkey

Turgay Kerem Koramaz
Istanbul Technical University, Turkey

ABSTRACT

This chapter evaluates the Istanbul Historical Peninsula, which has been the historical, cultural, and financial core of the city, in its quest for becoming smart(er) with reference to the five principles that are put forth by Mitchell, in his 1999 book e-topia. The chapter aims to define the role of ICTs in urban planning and management of Istanbul Historical Peninsula by depicting its fields of use in redesigning the societal, financial, and cultural aspects of urban life in the area. Additionally, the transformative feature of these technologies, used in urban planning, conservation, and heritage management, will be validated whether they serve for a vision and a strategic framework of transitioning the community into an information society.

INTRODUCTION

Technological advances, today, which enable new methods of big data and smart technologies, initiate major opportunities on the knowledge convergence regarding entire urban planning and management policies, as well as re-shaping of our urban environments. Thus, information and communication technologies (ICTs) can be acclaimed as a set of innovative tools, which support entire sets of these policies, including preservation of authentic quality of cultural heritage and possible interventions to meet contemporary needs of related quarters. Around the World and in Turkey, many kinds of researches reveal

DOI: 10.4018/978-1-7998-3856-2.ch005

the increasing need for progressive technological solutions in different stages of urban planning and management in historical sites. These studies mainly focus on the reflection and the transfer of cultural heritage value through the utilization of ICTs. Comprising digital museum studies, interactive platforms, and digital documentation and archiving of movable cultural assets, the concept of e-Heritage constitutes a common research field in the World. Internet and mobile applications, crowdsourcing communication channels and social media, digital interactive interfaces, and virtual and augmented reality, present the benefits of employment of ICTs in collaborative urban planning processes. Briefly, researches about the use of ICT in heritage site management mostly praise the operational functions and aim to facilitate the transfer and sharing of the conservation value of cultural heritage through these technologies. Besides, smart technologies, as claimed in this chapter, should be referred with transformative attitudes towards becoming an information society through all segments of the community, encouraging embeddedness in varying urban planning programs and participation in site management.

In the chapter, the aforementioned technologies and functions will be investigated through a procedure of urban planning and site management of the case area registered by UNESCO as a World Heritage Site, Istanbul Historical Peninsula, Turkey. The site is the core historic city of Istanbul, which is the leading city for public and private technology enterprises and conservation efforts on an imperial capital of former civilizations. The chapter aims to define the role of ICT in urban planning and management through the capacity of improving planning and conservation skills and enhancing promotional values in destination management. Additionally, the transformative feature of these technologies, used in urban planning, conservation and heritage management, will be validated whether they serve for a vision and strategic framework of transitioning community into the information society.

The case study in the chapter uses descriptive methodology with a narrative investigation from the plan documents, reports, and online resources of related institutions in Istanbul, Turkey. This investigation is conducted to undercover Istanbul case with an integrated look at experiences under the principles of future cities. Despite the existence of limited experiences, addressing transformative vision and/or strategies for an information society, actual technologies will be presented for querying Istanbul case in detailed perspective. These investigations will be conducted through five principles of "lean and green future cities" reported in William J. Mitchell's (1999) book, titled, "e-topia". From the in-depth investigation of both governmental and non-governmental bodies and institutions will be interpreted with the use of smart technologies and digitization of entire commodities, services.

THEORETICAL BACKGROUND SMART TECHNOLOGIES

Investigations in this chapter, as mentioned above, will be conducted through five principles of "lean and green future cities" reported in William J. Mitchell's (1999) book, titled, "e-topia".

Mitchell's "e-topia" is emphasized as a key concept on the evolution of technology and its impact on our social and spatial environment (Maeng and Nedović-Budić, 2008). This concept is associated with the third wave revolution of our civilization - information age- which started with the "electronic cottage" concept put forward by Toffler (1981), after the completion of agriculture and industrial revolutions. Among other theories, Maeng and Nedović-Budić (2008) report that Mitchell's idea of "e-topia" can be presented in a perspective that reveals the "new urban form" spatial structuring. In the search for "new urban form", Toffler's (1981) "electronic cottage" concept will be expressed as a pioneering concept in highlighting households with technologically strengthened and mixed activities. "Ubiquitous city", as one of the recent concepts in the last decade aims to optimize the use of information in the infrastructure and services of urban environments (Lee, et. al, 2008, pp.). Maeng and Nedović-Budić (2008) refer to Mitchell's theory with the necessity of a new economy, which tackles the uncontrollable growth of cities, which work smarter, not harder. Audirac and Fitzgerald (2003, pp. 483), sharing the common views above, highlighted the fact in Mitchell's "e-topia" that ICTs support urban planning and design professionals to overcome challenges of the physical construction of smart buildings and communities.

Mitchell's five principles of "lean and green future cities", with the exception of one, can be associated with the embedding new infrastructure, networks and tools into the existing cities. When applying smart technologies in urban sites, public spaces are prominently equipped with technologies under these four principles, which are dematerialization, demobilization, mass customization, and intelligent operation. For instance, conversion of conventional promotional services and exhibitions in museums into digital collections via virtual and augmented reality represents the dematerialization principle, whereas the emergence of online collections facilitates the opportunities of demobilization that is constructed onto telecommunication network. Such improvements in public spaces of heritage sites can be created with efficient utilization of resources via mass customization and intelligent operation in heritage management processes (e.g. smart ticket systems and online reservation). These technologies may overcome the challenges of corrosion and depreciation, significantly occurred by the overuse and crowdedness in heritage sites.

Addition to these four principles, one of the preeminent principles, which is called as soft transformation, is more distinctive than the rest. In heritage sites, transformation of existing building stock through smarter applications and tools may be harder than the installation of the aforementioned tools and technologies into the newly constructed quarters. Soft transformation principle is about the regeneration of existing infrastructure, which meets the requirements of smart technologies in living environments of the future. This final principle of "lean and green future cities" highlights the significance of small-scale and adaptive interventions in historic buildings, quarters, and cities, which can be referred as urban conservation and renewal in digital age and will be discussed under this term.

In his first book, including "city" in title -*City of Bits: Space, Place, and the Infobahn*- Mitchell (1995) extended the technology uses in architecture and urbanism in a discourse of saying it was a revolution manifestation. *E-topia: Urban Life, Jim—but not as We Know It* was his second forthcoming book after City of Bits. Mitchell (1999), in his book of "e-topia", cited global digital network, more powerfully how future cities would be shaped its apparent and explicit built structures such as pedestrian circulation, mechanized transportation systems, neighbourhoods, and community life, etc [1]. When talking about technology, as in science fiction films, the utopian and / or futurist theorists are strongly expected to answer questions about the future implications of technology in our environment (Mattern, 2017). In this context, in the structure of this chapter, e-topia book and five principles of "lean and green future cities" have been chosen in order to highlight the Istanbul case.

Five principles of "lean and green future cities" by Mitchell (1999) are contributing to the creation of synergy between social and economic digital strategies and microscale urban planning and design strategies. Although five principles of "e-topia" have been relatively less referred to in smart and sustainable technologies related literature, the book itself and Mitchell's literature has been widely used as a significant reference. Since 2005, in Scopus indexed literature, there are 18 valuable publications related to Mitchell's "e-topia" and with the keywords of "smart" and "sustainable" (h-index is 10; and 43.7 citation per publication). Among these publications, one of the researches which is led by the researcher team Anna Kramers and Mattias Höjer, has a differing significant reference of "lean and green future cities" developed by Mitchell (1999). In their research, they put forward an analytical framework combining ICTs' opportunities titled with these five principles (Mitchell, 1999) and energy use typology of household functions (Höjer et. al, 2011), derived from an original case study for Greater Stockholm Area (Kramers et. al, 2014). This study has let these five principles to verify that ICTs were appropriate for promoting urban energy sustainability (Haarstad and Wathne, 2019), under the domains of household functions, such as housing services, durable goods, transport services, semi-durable goods, amount of purchased energy and other services. Apart from this critical work, it is accepted that these five principles can be evaluated by the professions related to industrial product design, architecture, urban planning and design as an additional statement of Mitchell (Hendarti, el. al, 2016; and Maté et. al, 2016).

Deakin (2011; and 2014) states that we need digitally inclusive regeneration plans for our future cities, also highlighted by both of Mitchell's books, which have become valuable assets in managing global strategies. Mitchell's literature is significant because it refers not only to the high-profile intelligence of network cities that have been criticized as a conventional engineering empowerment strategy but also to the discourse of a global strategy on good governance. Similarly, this discourse, which can be seen as a major criticism of 'smart city' manifestation, supports the idea of horizontal empowerment of citizenship, inserted into social equality and ecological integrity, instead of favouring privileged minority groups vertically strengthened in the purchase and use of technology. Otherwise, embedded technologies in our environment would fail because of not being capable of guiding in eliminating social divisions, environmental degradation and pollution (Graham and Marvin, 1999; Deakin, 2011).

Graham and Marvin (1999) warned against exaggerating the power of telecommunications and information technologies while requiring integrated policymaking in contemporary urban strategies. Just before the beginning of the 2000s, social exclusion in digital technologies and telecommunication was explicit enough in theoretical arguments that restricted zones for information elites and media consumption space have been already associated with high-income privileged groups. These argument overlaps with an emerging term of "digital divide" which can be briefly determined as gap in access to, use of and participate via information technologies and digital services among social groups and geographies (Dragulanescu, 2002). Empirical studies in the World (Korupp, 2006; Wilson 2004) and in Turkey (Koramaz, et al, 2019) revealed the multidimensionality of digital divide issues, but concentrated on the use of and access to computers and the internet, the capacity of local inhabitants when becoming online and institutions when transmitting knowledge.

In addition to contemporary urban strategies, cultural policies and heritage management strategies also need an emphasis on inquiry, dialogue, and collaboration in the whole society. Darmawan (2011) also emphasizes the need to address cultural heritage information and its value in a broader context with digital technology. Both the discourse on heritage information and its epistemological organization must envisage the operational use of technology in the inventory, documentation, and transfer of heritage knowledge, as well as the establishment of communication and collaboration functions between various heritage stakeholders, including institutions and individuals (Koramaz, 2018a). Such an improvement will extend the capabilities of smart mobility of society in the destination management process; increase the awareness of new practices and the meaning of heritage sites (Leite, 2009).

Then this chapter will discuss the entire digital services and information in Istanbul's Historic Peninsula, where has been the focal point with its central business district and managerial activities throughout history. Due to its rich and multi-layered culture, which includes significant historic sites from the Roman Empire to the Ottomans, UNESCO registered the site in 1985 as a World Heritage Site within the title of Istanbul's Historic Areas (WHC-UNESCO, 1985). Although there have been limited efforts on urban planning and conservation in Istanbul Historical Peninsula, concerns on policies and strategies could not overcome the challenges of physical deterioration, social decay, gentrification through a harsh forcing to eviction process of low income and disadvantaged groups by area-based programs (Koramaz, 2018b). From the very early urban conservation project efforts, strategies on appreciating cultural heritage site in Istanbul Historical Peninsula for future generations has been always a strong intention (Gulersoy et al, 2008). Strategies on becoming an information society might at least tackle challenges of social development in the heritage site.

IS ISTANBUL HISTORICAL PENINSULA LEAN AND GREEN?

In this chapter, the smart applications that are available in Istanbul Historical Peninsula are sorted out according to their compatibility with Mitchell's principles which are dematerialization, demobilization, mass customization, intelligent operation and, soft transformation. The smart applications that are deployed in Historical Peninsula have been specified within the scope of Belli's master's thesis titled "Dimensional Evaluation of Retrofitting Smart Cities: The Historical Peninsula Case" (2019), through using wide-ranging sources of information, which include national and local strategy documents, academic research papers, official websites and mobile applications of municipal companies, news articles, recent interviews of local government officials and social media accounts of local institutions.

Dematerialization

Dematerialization is the first of the five points proposed by Mitchell (1999), which would make cities operate smarter and consume lesser resources so that the cities are sustained for future generations. The principle suggests that if a procedure or a service that operates in a physical setting is completely or partially transferred to virtual space; it will increase the efficiency by achieving the same results without needing as much physical material for structures and for the operation itself. Meaning that, instead of physical structures and outdated procedures, employing physically downsized virtual alternatives in provision of urban services, decreases material usage, energy consumption and waste production, thus helps to create greener cities. In light of this, among all of the smart applications, rationale of e-Government Gateway, e-Pulse, Provision of Broadband Networks and Istanbulkart can be attributed to dematerialization. As e-Government and e-Pulse moves material-consuming physical operations into online platforms that save time and resources, Provision of Broadband Networks and Istanbulkart provide much more efficient services in their field of operation with reduced need for material, space and construction.

E-Government Gateway

In line with its Information Society Strategy, Turkey has established the e-Government Gateway in 2008, though the earliest attempts were initiated in 2000, with the National Information Infrastructure Plan (TUENA) published in 1999 and e-Turkey Initiative Action Plan in 2000. After TUENA and e-Turkey Action Plan both could not be put into practice, 2006-2010 Information Society Strategy and Action Plan was confirmed as the guiding policy paper that would direct Turkey towards an innovative way of administration of public services (Information Society Department, 2015).

With the help of ICT, e-Government Gateway enables citizens to access state services from different institutions and organizations from a single point (Department of e-Government Services, 2016). By going to the website, citizens, businesses and government entities can access services from hundreds of institutions in a fast and efficient manner. Official institutions related to every public sector including justice, environment, agriculture, state and legislation, education security, health, telecommunication, traffic and transportation, tax, fees and penalties etc., offers services such as payment transaction, e-services, information services or updates and announcements. According to United Nation's survey (2018), Turkey ranks 53rd among 193 countries in terms of e-Government Development Index which is calculated according to the countries' human capital and performance on provision of online services and telecommunication infrastructure. Considering that, the country was in the 68th place in the 2016 survey, it is indisputable that Turkey is taking substantial steps towards increasing its e-Government performance and realizing the vision of becoming an Information Society.

There are also a number of municipal services provided by Fatih Municipality at e-Government Gateway, including document verification and inquiries related to land values, certifications and payments (Digital Transformation Office, 2019). Fatih Municipality also provides these e-services through its own website and mobile application (Fatih Municipality, 2019a).

E-Pulse

Developed by Turkish Republic Ministry of Health, e-Pulse is an online public service that collects healthcare data of individuals and share them with citizens, physicians and professionals from public hospitals and subsidiary healthcare facilities. The system stores personal healthcare information of individuals regardless of which hospital they were treated and allow them to access and manage their records via single database. E-Pulse is available through its online website and mobile application. Developers of the system claim that among all of world's healthcare information services, e-Pulse is the most comprehensive one, which diminishes the time spent for diagnosis and enables a high-quality treatment process (Turkish Republic Ministry of Health, 2018).

Provision of Broadband Networks

The municipal company Istanbul Electronic Communications and Infrastructure Services Industry and Trade Co. (ISTTELKOM) has signed a protocol with operator companies TURKCELL and VODAFONE, which ensures that the entities will work in cooperation and build common communications infrastructure services. With the protocol planned to make a significant contribution to IMM's smart city vision, ISTTELKOM will build common broadband (fiber optic) infrastructure for the operator companies. The aim of this protocol is to prevent the waste of resources and to provide wide-bandwidth internet to more households faster and more economically. General Director of ISTTELKOM states that this service model which is already implemented in Stockholm, creates a sustainable solution to provision of fiber infrastructure by eliminating the possibility of repeated excavations by separate operators, thus decreasing the construction costs, which actually constitutes the 68% of the total cost.

There is also another project carried out by the Turkish Republic Ministry of Transport and Infrastructure called Electronic Communication Infrastructure Information System (EHABS) that aims to gather the electronic communication infrastructure information of the different operators in a single database, and use the collected data in strategy and policy making processes. After the introduction of EHABS, the information about the length (km), capacity (GB) and route (address or geographic location) of the fiber optic infrastructure (km), will be available for any operator from any location.

According to the latest statistical report prepared by Information and Communication Technologies Authority (BTK) (2018), there are 68,869,578 broadband subscriptions (both mobile and fixed) in Turkey by the end of 2017, 25.15% of which is from Istanbul with 17,322,496 subscriptions. Among these broadband subscriptions, 3.314.055 (19.1%) of them are fixed broadband subscriptions, while 14,008,441 (80.9%) of them are mobile broadband subscriptions. Also, penetration rate of fixed broadband subscriptions is 14.75% in Turkey, whereas the same rate is 22.05% in Istanbul and around 29% in OECD member countries (Organization for Economic Co-operation and Development) (ISTTELKOM, 2017).

Istanbulkart

The digital public transit payment system of Istanbul is called Istanbulkart. The smart card which is developed by BELBIM A.S (Istanbul Electronic Currency and Payment Systems Co.) is used in mainly used for public transportation. Other than public transport, Istanbulkart can be used in ISPARK car parks, iTaksi (e-hailing service), Isbike (smart bike sharing), beaches and public toilets. However, BELBIM A.S states that in a short while, the smart card can be used for grocery shopping and payment of bills (BELBIM, 2019b).

Istanbulkart also has a mobile application with the same name. The application that is developed with Financial Technology (FinTech) infrastructure, allows users to monitor their balance in real-time, track their activities from last 3 months, locate nearest top-up points on the map, and form instructions to load money to their smart cards. Currently, the system only loads money when users form an instruction on the mobile application and tap their cards on a top-up point. However, according to BELBIM, citizens soon be able to charge their smart cards simply by tapping them on the backside of their NFC-enabled (Near Field Communication) smart phones without having to find a top-up point (BELBIM, 2019a).

Demobilization

Since modes of transportation and travel times have a direct impact on the everyday life, mobility is considered to be the most problematic area in cities, which deals with timeless urban problems such as traffic congestion, traffic accidents or lacking parking spaces (Ning et al., 2017). Among these problems, congestion is the most severe one compared to others, especially in the metropolitan areas. Inadequacy of parking spaces or overpricing of off-street parking are major problems contributing to congestion. According to the survey conducted by Shoup (2006), although the rate differs from city to city, an average of 30% of the traffic in cities are resulted from cruising for parking. That is, among all of the vehicles circulating in traffic, 30% of them are searching for convenient parking spaces. For example, people tend to look for curb parking (roadside parking), when other options are limited or more expensive.

Congestion has impacts on almost every aspect of life including environmental, financial and social aspects. Spending extra time in traffic evidently wastes more fuel, thus causes extreme carbon emissions. A study conducted in 15-block business district in Los Angeles, shows that looking for curb parking in that area created approximately 950,000 miles of extra travel distance in a year -which is also "equivalent to 38 trips around the earth, or four trips to the moon"-, and ultimately led to 47,000 gallons of wasted fuel and 730 tons of extra carbon dioxide emission (Shoup, 2007). Considering this example is only from a small business district in Los Angeles, consequences for metropolitan areas are undoubtedly more damaging. Other than wasting quality time and polluting the air, congestion also cost citizens tremendous amounts of money because of excess fuel usage, which is estimated in 2015 report by Frost & Sullivan to be 600 million USD per year for an average city.

Regarding this, Mitchell (1999) argues that ICT's can be solutions for congestion in cities in two way. First one is eliminating the need for transport with the help of telecommunication such as through enabling citizens to work from home or providing government services via online platforms so that they do not have to travel to public institutions. The second is reducing the volume of congestion by decreasing the length of daily commutes with compact and mixed-use neighbourhoods, by redirecting citizens away from peak hours and congested zones and by promoting eco-friendlier options of transport. Accordingly, the bike sharing platform Isbike, IBB YolGosteren which is a mobile application for real time navigation and multimodal public transport information, Smart Parking System of Transport Management Centre in Istanbul and Mobile Application of ISPARK, can be linked to demobilization.

Isbike

Isbike is the smart bike sharing system of Istanbul that is established recently by the ISPARK, the municipal company responsible for parking spaces of transport vehicles across the city. By being supported with the technological database and being easily integrated to the urban transport network of the city, the system is an alternative way of commute which is also highly sustainable. Isbike aims to eliminate the need for motor vehicles for the 3 to 5 km travels, thus allowing citizens to have the opportunity to use a healthier and more environmentally conscious mode of transportation that has zero carbon footprint and also decreases the pressure on the public transport (ISPARK, 2019a).

Isbike currently has 380 bikes in 38 smart stations in the Anatolian side and 1120 bikes in 102 stations in the European side of Istanbul, among which, 134 bikes in 15 stations of them are located in Historical Peninsula (Figure 1). The system operates in two different which are credit card use and subscriptions. In the first method, citizens use the kiosks located in the smart bike sharing stations with their credit cards, which give them a password to unlock the bikes from the stations. While, in the second method, they use the Isbike website or Isbike mobile application to register for an online subscription (ISPARK, 2019b).

IBB YolGosteren

IBB YolGosteren is the mobile navigation application that is developed by Istanbul Metropolitan Municipality and can be used in all of the cities in Marmara Region. The application uses the live traffic information provided by Transport Management Centre and navigates users to reach their destination in routes that are most convenient. By simply searching their destination in the application by typing or voice command, users are able to create alternative routes with different distances and arrival times. After selection of the most suitable route, the application vocally navigates users to their destination. In case of a traffic accident, road maintenance work on the route or false exit from the route, the application generates an alternative one automatically. IBB YolGosteren is also integrated with public transport option of the city including buses, subways, Metrobus line, tram lines, minibuses, shared taxis (dolmush), ferry lines, funicular line, rail and cableways, thus it can guide users with the most sustainable way of transport as well. Similar to "IBB Cep Trafik" (IMM Mobile Traffic), IBB YolGosteren also provides the location of available parking spaces, real-time streams from traffic cameras and location of open

Figure 1. Isbike smart bike sharing stations in Historical Peninsula
Source: (ISPARK, 2019a)

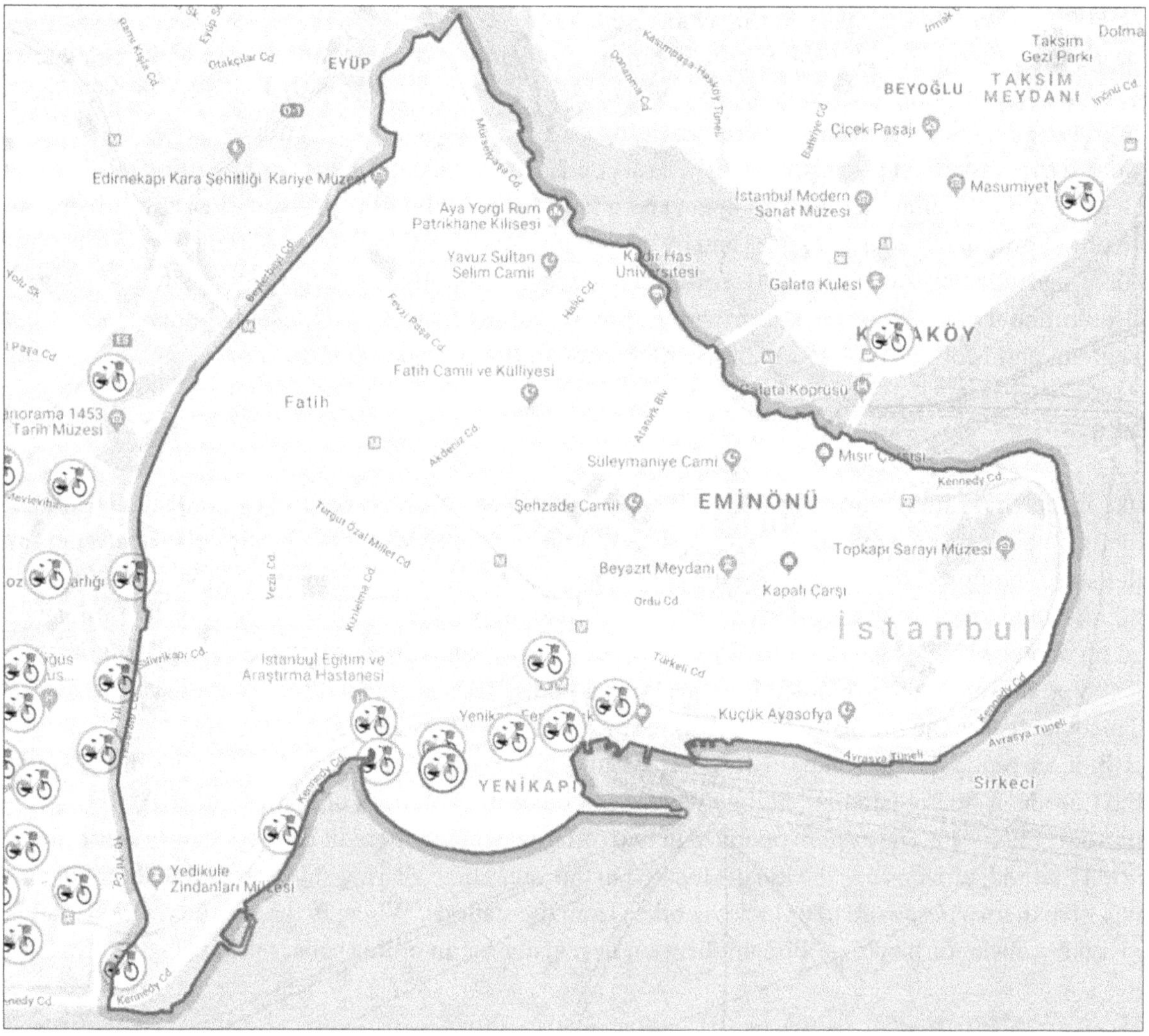

pharmacies. Users also can share information about roadworks, traffic jam, accidents or vehicle breakdowns from their positions in the traffic (Figure 2) (Istanbul Metropolitan Municipality, 2017).

Developed by Istanbul Metropolitan Municipality Department of Transportation, Smart Parking System informs drivers on the roads with nearby available parking spaces. The system uses LED screens placed on traffic signs to display the information about real-time occupancy of nearest parking areas. By receiving live data from the Transport Management Centre, Smart Parking System directs drivers to the available parking lots without them wasting their valuable time. This way, the system eliminates the stress caused by cruising for parking in congestion, saves fuel and decreases congestion in the traffic. Drivers also can access the location an occupancy information about parking spaces via traffic density map of TMC or its mobile application IBB Cep Trafik (Figure 3) (Transport Management Centre, 2017).

Figure 2. User interfaces of IBB YolGosteren Smart Parking System of TMC
Source: IBB YolGosteren Mobile Application

Mobile Application of ISPARK

ISPARK which is the municipal company that is responsible for parking lots of IMM, uses its mobile application with the same name to share real-time information about the occupancy rates of parking spaces in Istanbul. Contrary to smart parking system of TMC, ISBAK only shows information about its own parking spaces in 600 locations in Istanbul. Users are able to access location and price information about nearby parking spaces of their preferences among open parking lots, parking garages or curb parking. Also, beside paying with cash and credit cards, citizens can also pay their parking fees with Istanbulkart which is the smart card that is mainly used for public transportation (ISPARK, 2017). Moreover, ISPARK too takes the advantage of LED-enabled variable message signs (VMS) provided by Transport Management Centre, displaying the live occupancy information about the nearest parking spaces in Istanbul's traffic.

Mass Customization

According to Mitchell (1999), the technologies that have the ability to customize their products to individual preferences allow users to achieve more sophisticated services based on their needs and enable providers to operate more intelligently without redundant repetition. Among Mitchell's (1999) five principles, mass customization is the one with the most indirect outcomes, however, the convenience coming with the automated individualization is nothing to be overlooked. Accordingly, Adaptive Traffic Management System (ATAK) and iTaksi can be given as examples to mass customization in Historical Peninsula.

Figure 3. Smart parking system showing the parking spaces in Historical Peninsula
Source: (Transport Management Centre, 2019a)

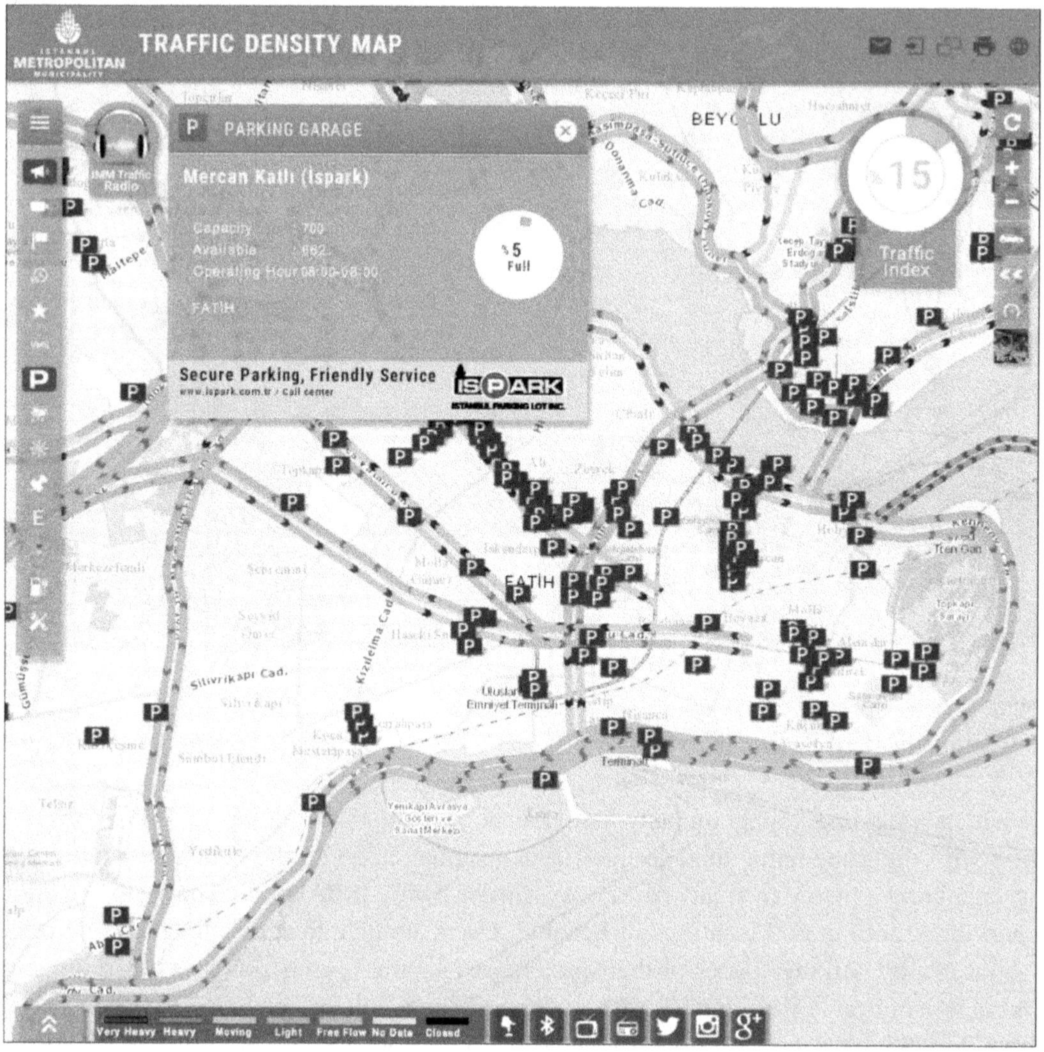

Adaptive Traffic Management System

Developed by ISBAK A.S. (Istanbul IT and Smart City Technologies Co.), ATAK is a real-time traffic management system that collects data about the number of vehicles at the intersections with the help of magnetic sensors embedded in the asphalt surface. Using a domestic software developed by local engineers, the sensors instantly and continuously transmit the collected data to the traffic signal controller at the spot which later transfers it to Adaptive Traffic Management System operational at IMM's Transport Management Centre (TMC). After algorithm-enabled calculations of the density at the intersections, the system optimizes the durations of green lights for each direction. By doing so, ATAK reduces the density of vehicles at problematic intersections simultaneously (Figure 4) (Transport Management Centre, 2018).

ATAK system was first piloted in 80 problematic intersections in late 2018, including maintenance in crowded quarters of the Historic Peninsula, especially with narrow roads in organic pattern (such as Ayvansaray, Aksaray Kucuk Langa, Edirnekapi, Koska and 10. Yil Street intersections). Moreover, it is planned to increase the ATAK-operational intersection number to 200 in a short-term period and ultimately deploy the system at all 2250 intersections in Istanbul thanks to the development processes at domestic and national facilities that will produce the same operational equipment for nearly half price (Istanbul Metropolitan Municipality, 2018; Haberturk, 2018).

Figure 4. ATAK's operation at Koska Intersection in Historical Peninsula
Source: (TMC, 2018)

By reducing delays, congestion and improving traffic flow, ATAK claims to have several benefits in social, financial and environmental regards. People wouldn't waste their valuable time in congestion, thus have more quality time for their social and financial endeavours. Less time spent in traffic directly affects the fuel consumption. According to model study of Koska Intersection (given above), the decrease in the congestion equals about 312,000 litres of fuel saved, 227,760 hours of valuable time not wasted and 4680 kg of carbon not emitted per year. All in all, according to the results from 56 ATAK-operational intersections, the system decreases travel time 20%, carbon emission by 18%, fuel consumption by 15%, and average delay by 30%, which ultimately creates 1,700,000 TL in productivity and 700,000 TL in fuel savings per intersection per year. Considering the system costs about 120,000 TL for each intersection, it can be asserted that ATAK is highly profitable smart application that has many outcomes that are beneficial for cities and citizens (TMC, 2018; Haberturk, 2018).

iTaksi

iTaksi is a private e-hailing application for taxis in Istanbul's traffic. The application is developed by ISBAK A.S. (Istanbul IT and Smart City Technologies Co. as one of the semi-public enterprises) for IMM and is operated by ISPARK. The system is available only via its mobile application which allows users to call for a taxi of their preference among yellow taxi, turquoise taxi, black taxi or sea taxi (Figure 5). There are also several payment options including cash, credit card or Istanbulkart. Other benefits of iTaksi include safe travelling, driver rating, route tracking and estimated fare calculation (ISBAK, 2019). As of late 2018, among 17,395 yellow taxis, 4085 of them are registered to iTaksi. There are also 39 turquoise taxis (upper-middle segment) and 22 black taxis (upper segment) (Hurriyet, 2018).

Figure 5. User interfaces of iTaksi
Source: iTaksi Mobile Application

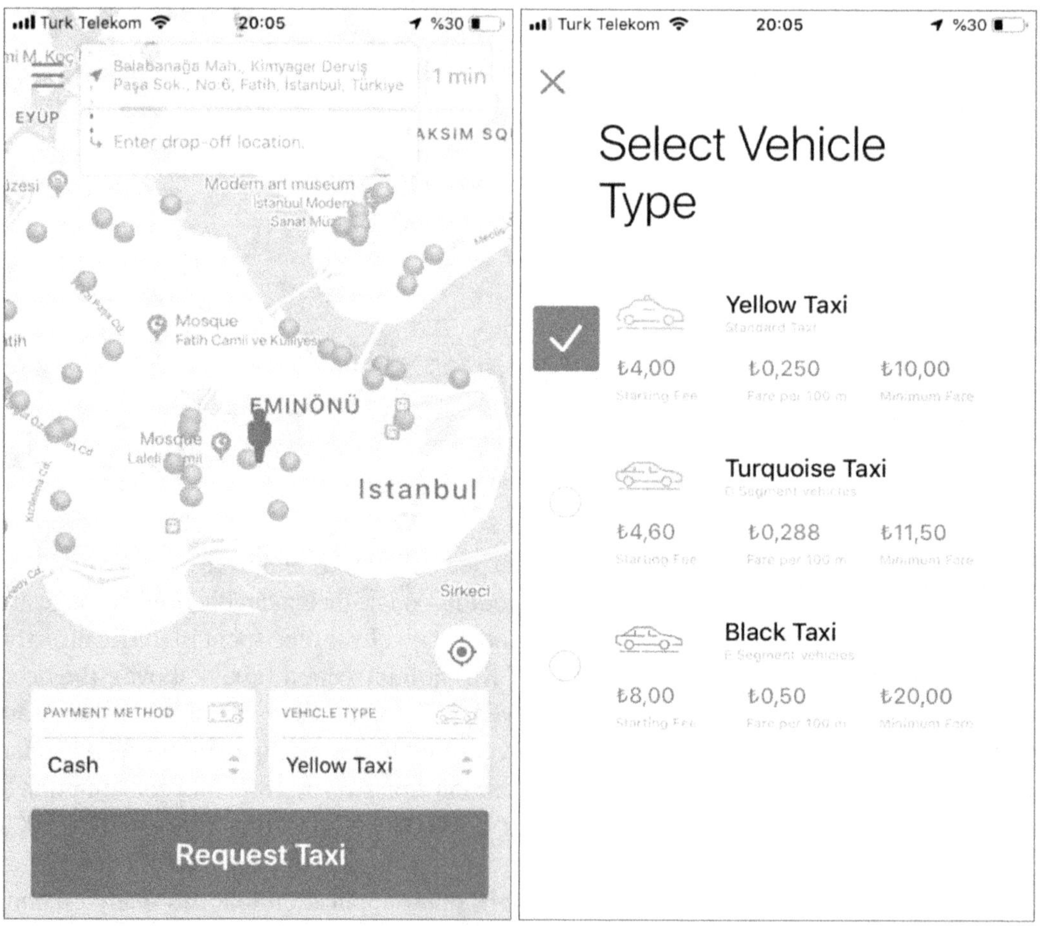

Intelligent Operation

At the centre of intelligent operation lies smart technologies which have been beneficial tools for local governments to augment urban governance, increase liveability and tackle timeless urban problems such as traffic, crime, waste management, disaster mitigation and congestion. Cities embedded software-enabled technologies to every aspect of everyday life in order to be able to remotely monitor, govern and operate provision of public services more intelligently with the help of real-time sensing technologies and ubiquitous computing (Townsend, 2013; Kitchin 2014). Enhancing the ways that urban areas are managed in, ICT-enabled technologies have become instruments to achieve intelligent operation and to increase the quality of life in cities with the aim of creating a lively community that will eventually foster a vibrant economy.

Transport Management Centre (TMC)

It is no surprise that Istanbul's traffic management is a challenging process as the city spans over two continents with over 15 million inhabitants. Having a complex and interconnected variety of transport options provided both in European and Asian Side, Istanbul's traffic generates unprecedented amounts of data every second (Table 1).

Table 1. Statistics about transportation in Istanbul (Source: IMM, 2015)

Population	14.38 M
Daily Trips	30 M (1.5 M over bridges)
Registered Vehicles	3.38 M (2.25 M in traffic)
New Vehicles Added to Traffic	503
Length of Total Road Network	30.000 Km
Average Travel Time	45 minutes
Car Ownership	158/1000 persons

Established by Istanbul Metropolitan Municipality Directorate of Traffic in 1997, Transport Management Centre monitors and controls the traffic from its operation building in Bakirkoy, Istanbul (Figure 6). The centre uses different types of sensors and cameras to gather information about the traffic and utilize that information with smart transport software to optimize traffic flow, to create real-time traffic density map and to inform drivers, passengers and pedestrians through mobile applications, variable message signs (VMS) and traffic signals (VTS). Created instantly with the data received from sensors, traffic density map provides information about, real-time traffic conditions, spots of traffic accidents and road maintenance works, predicted traffic, estimated travel duration, available car parks, announcements about traffic via IMM Traffic Radio and so on (Figure 7). The map gives the locations of traffic cameras, variable message signs and traffic signals, and by clicking on a particular location, users can display camera views and signal information directly. Traffic density map and above-mentioned information are available at both online website and mobile application called "IBB Cep Trafik" (IMM Mobile Traffic) (Transport Management Centre, n.d.b).

Figure 6. Transport Management Centre of Istanbul
Source: (TMC, n.d.b)

Traffic Monitoring Cameras

There are 46 traffic monitoring cameras located in Historical Peninsula. Placed at the top of 20 m poles, the cameras are operational 24/7 (Figure 8). The cameras provide a wide range of visibility as they can move around 360-degree horizontally and 130-degree vertically. Beside traffic monitoring, camera streams are also being used by safety and security related issues. Police departments uses these cameras in criminal cases, whereas ISKI (Istanbul Water and Sewerage Administration) use them for flood control. Disaster Coordination Center (AKOM) has access to the streams in case of an emergency. In fact, in case the TMC is out of operation, DCC would still have its independent monitoring thanks to taken measures (Transport Management Centre, n.d.c).

The camera streams are available via TMC Website, IBB Cep Trafik, IMM TV, IBB YolGosteren, smart televisions, Turksatanalog Broadcast, Digital Broadcast, Digiturk, screens at airports and malls etc. (TMC, n.d.c). There are also other surveillance camera streams provided by IMM that includes, touristic cameras, animal shelter cameras, city cameras, construction site cameras and Beykoz summer scout camp cameras. Among these streams only touristic cameras are available for Historical Peninsula from locations such as Beyazit Tower, Beyazit Square, Spice Bazaar, Eminonu, Grand Bazaar, Sarachane, and Blue Mosque (Figure 9-10) (Istanbul Metropolitan Municipality, 2019d).

Mobile Electronic Violation Detection Systems (EDS)

Mobile Electronic Violation Detection Systems (EDS) is the system that is developed by IMM to detect the drivers who breach the traffic rules such as red-light violation, emergency lane violation, reverse direction violation, preferred bus lane violation, tram line violation and parking violation. Contrary to standard EDS technology that depends upon the immobile smart sensors placed in poles across the

Figure 7. Traffic Density Map of Istanbul
Source: (Transport Management Centre, 2019b)

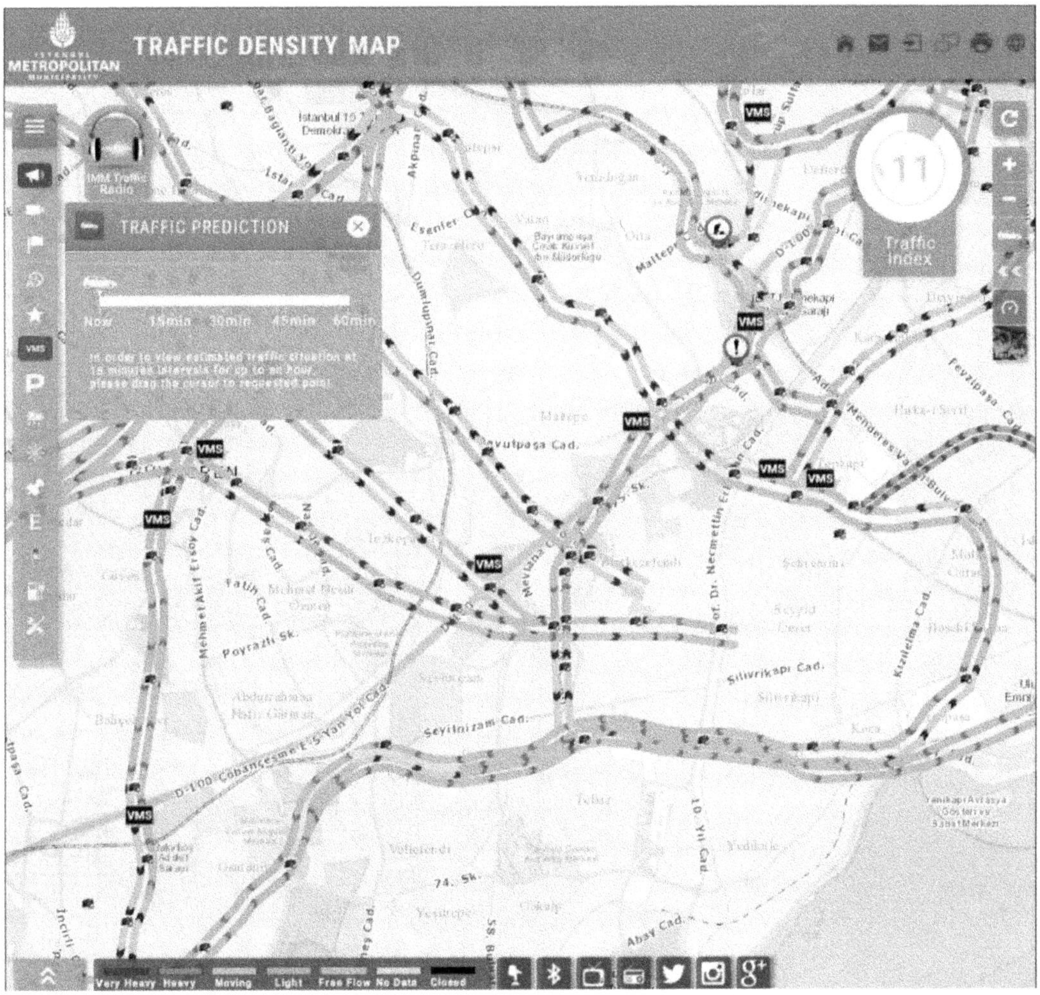

transport networks, Mobile EDS operates with sensors placed on moving vehicles that are circulating in the traffic. Even though system has many benefits and detects many more violations that standard EDS couldn't catch, there are some violations happening in the historical neighborhoods with narrow streets such as Historical Peninsula that vehicle cannot enter to have clear shot of the situation. For this kind of situations, operators use smart glasses to take images of the violations and transfer them to Transport Management Center (Figure 11) (Transport Management Centre, n.d.a).

Istanbul Fire Department Automation Project

GIS Academy of IMM ensured that location information of the calls to 110 (Fire Department Call Center) are immediately detected and are displayed instantly on the map. Beside the address information of the relevant location, the information of the fire department that is responsible for the that area is too shown on the map. This way, the operator at the Fire Department who responds to the emergency call, starts the

Figure 8. Traffic monitoring cameras in Historical Peninsula
Source: (TMC, 2019b)

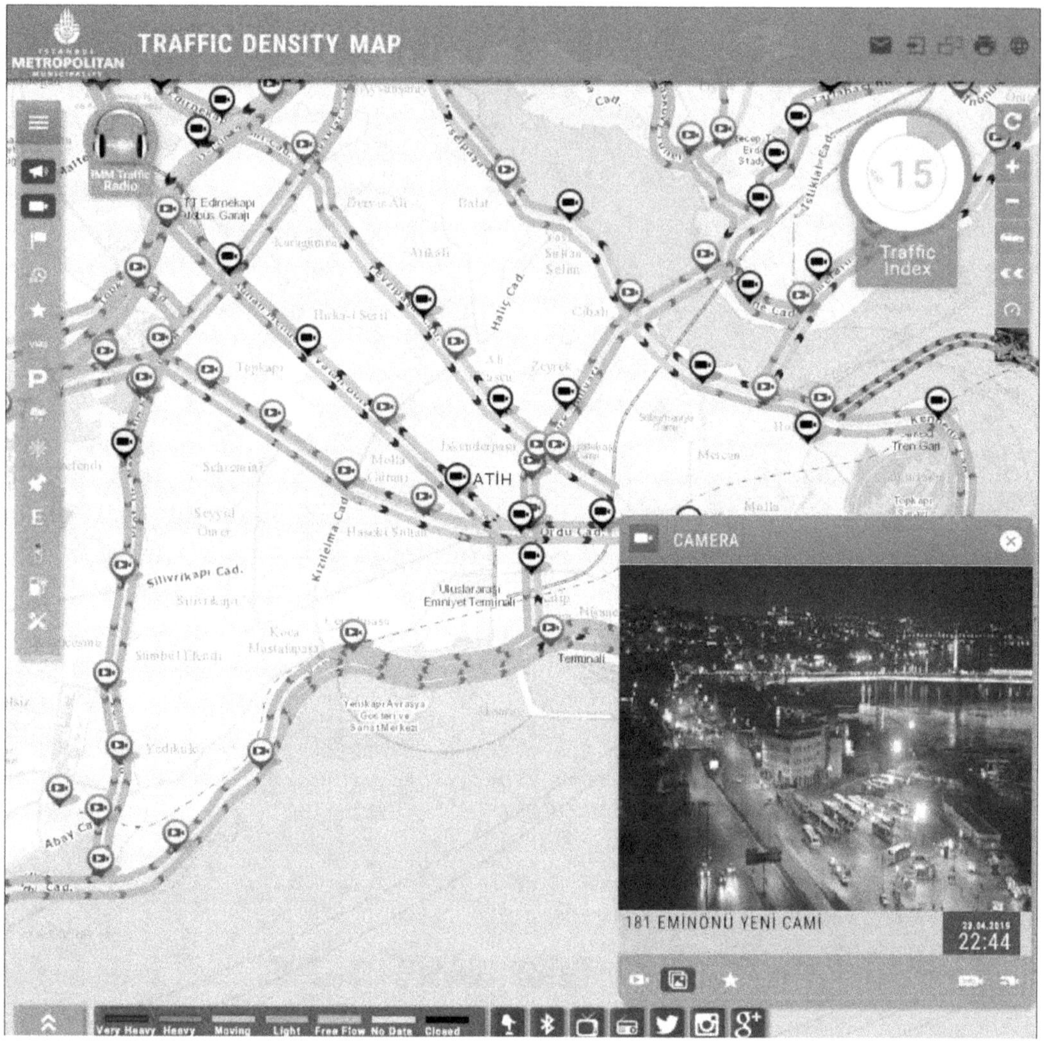

talk with crucial information obtained beforehand. Within three seconds, the event that is entered on the map, also gets processed in to the entire application. Anyone using the application called GEOERP in the Istanbul Fire Department can instantly see the information about type of the incident (fire-vehicle accident-animal rescue), the exit time, the stages of the incident, the responsible stations, the number of vehicles and personnel intervened (Figure 12) (GIS Academy, n.d.).

With this project, streets of Istanbul which are impossible or difficult to be reached, were marked on the map and alternative routes were determined. Pictures of these streets can also be taken and integrated into the system together with the problem types. By this way, before fire trucks arrive at the scene, they can be guided to alternative routes and prevent time loss (GIS Academy, n.d.).

Figure 9. Live touristic camera stream from Blue Mosque
Source: (IMM, 2019d)

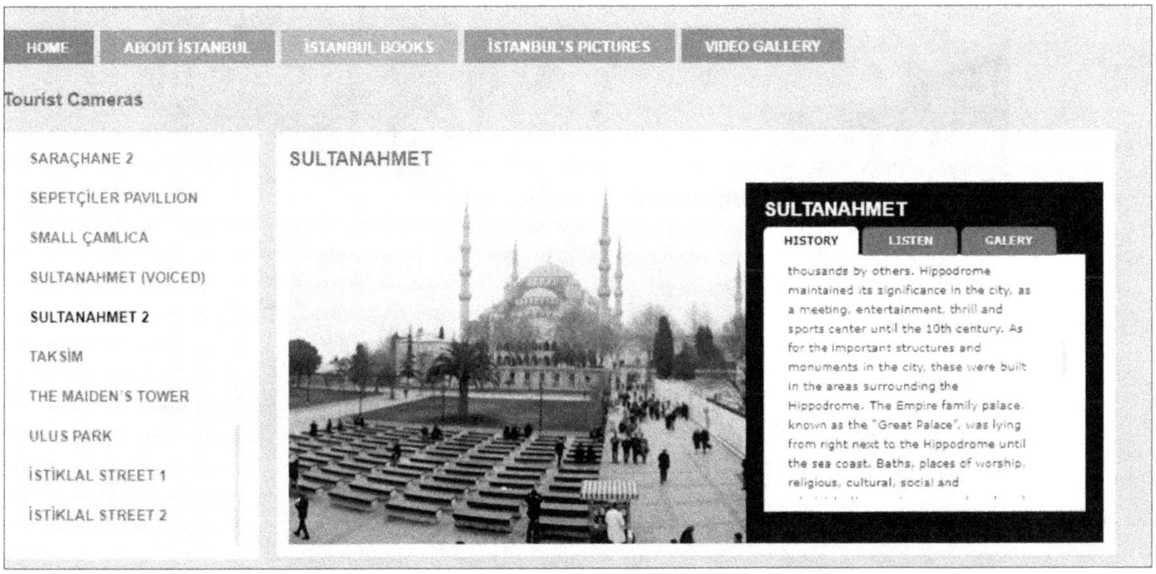

Figure 10. Live touristic camera stream from Grand Bazaar
Source: (IMM, 2019d)

Online Healthcare Scheduling: CDAS

Centralized Doctor Appointment System (CDAS) that is developed by Turkish Republic Ministry of Health in 2010, helps citizens to search for medical care and schedule their appointments with doctors of their preferences from public hospitals and subsidiary healthcare facilities. The system is available with

Figure 11. Operator wearing smart glasses to take images of parking violations
Source: (TMC, n.d.a)

a call to 182 or via the online website or mobile application. Centralized Doctor Appointment System (CDAS) is among the 20 civil services that are adopted in European Union. The system is accessible to the 99.6% of Turkey's population including residents of Historical Peninsula through 182 call center, online website and mobile application (Turkish Republic Ministry of Health, 2016).

Istanbul Air Quality Monitoring Centre

Istanbul's Air Quality Monitoring Centre, which started with 2 mobile air quality measurement vehicles in 1995, currently serves the people of Istanbul with approximately 40 air quality monitoring stations provided by the Istanbul Metropolitan Municipality and the Turkish Republic Ministry of Environment and Urbanization. Air quality measurement values are published online in real-time in order for citizens to better understand the air quality of the environment they live in and possible negative effects on their health. There are 2 air quality measurement stations located in Historical Peninsula, one being in Aksaray and the other in Catladikapi (Figure 13). There is also another project that is being developed by ISBAK called IoT Taxi Caps that will measure temperature, humidity, noise and air quality with the IoT sensors on the taxi caps (Turkish Republic Ministry of Environment and Urbanization, 2019).

SCADA Technology

ISKI which is the administration that is responsible for Istanbul's water and sewerage systems, uses SCADA (Supervisory Control and Data Acquisition) system to manage and monitor the water distribution systems through satellite communication. SCADA allows ISKI to measure water levels in instantly

Figure 12. Istanbul fire department's emergency map
Source: (GIS Academy, n.d.)

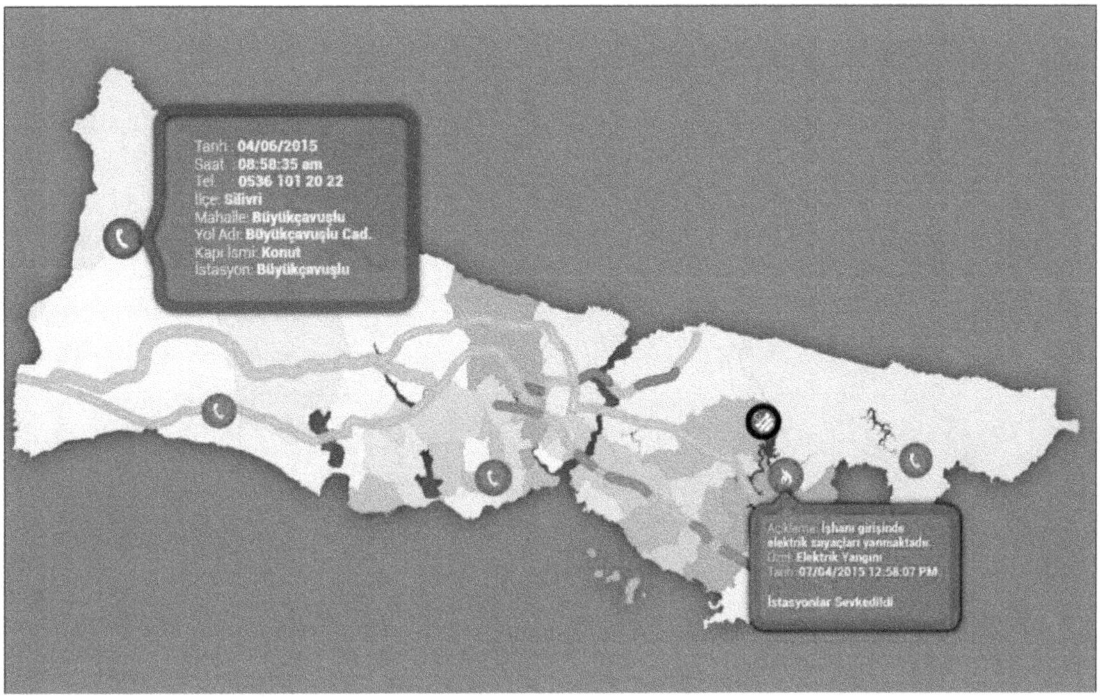

during or after a rainfall and prevent water loss by predicting and interfere in any possible breakdown in the distribution system. Aiming to establish a smart network, the system is able to detect leakage or illegal water use in the system by monitoring the amount of water that is supplied to the system and the amount of water that is billed to citizens. From its centre in Kagithane, SCADA also monitors data related to water levels, water pressure, water flow and water quality in the dams, distribution lines, pumping stations and water tanks that belong to ISKI. The system also provides live stream camera views from the stations and enables access to the data in case of an emergency (ISKI, n.d.a; ISKI, 2016).

Smart Water Meters

Smart water meters are developed by ISKI to serve the administration's vision of a smart water network. Smart meters are installed free of charge by ISKI, and operates upon a prepaid system of water consumption. Citizens are given a smart card that that they use for buying water from payment points and later for transferring that prepaid water amount information to the smart water meters (ISKI, n.d.b). Smart meters enable real-time monitoring of water consumption both for ISKI and its subscribers. The meters are connected to ISKI web services and provide the administration with remote data in one-hour interval or less. The water consumption information received from smart meter are instantly embedded in a single database where it is shared with consumers through online website or mobile application. Being more aware of their water consumption, users are expected to be more inclined to save water. Currently operates with 4021 smart meters across Istanbul (Sonmez, n.d.), the system also reduces labour costs and increase accuracy and efficiency, because it eliminates the need for manpower to read water meters manually (ISKI, 2017).

Figure 13. Istanbul air quality monitoring centre's online service
Source: (Istanbul Metropolitan Municipality, 2019a)

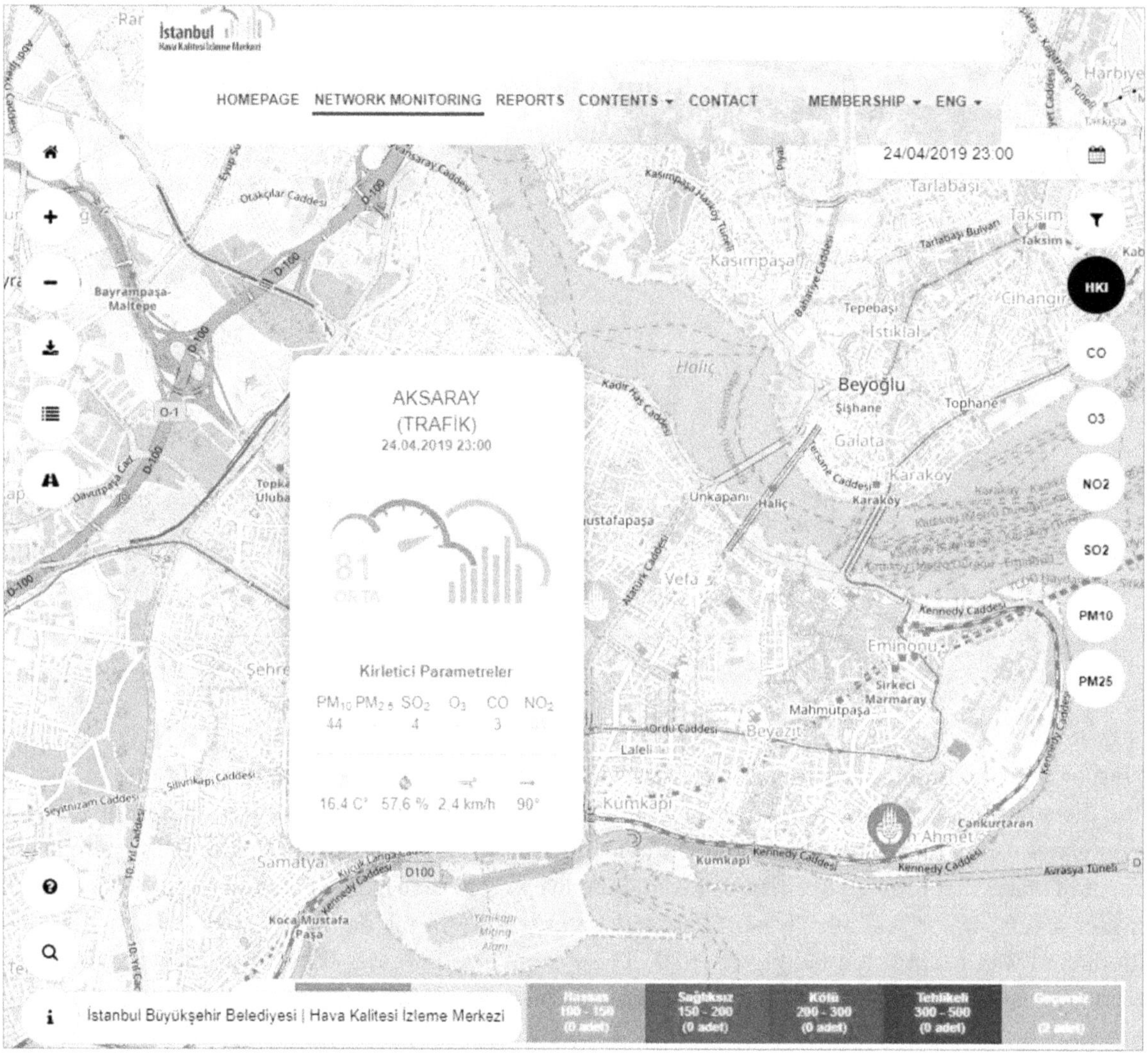

IMM White Desk (IBB Beyaz Masa)

Managed by IMM Directorate of Public Relations, White Desk is a municipal platform where citizens can submit their complaints, requests and opinions. The platform is operational with contact points enabling face-to-face contact in several locations and also available via 153 call centre and its mobile application. White desk which started its operations in 1994, aims to better public services and bring solutions to problems of citizens by taking their opinions and requests into consideration. The platform receives applications through several procedures including phone call, social media, mobile application, online website, text message, fax and face to face interview. There are 6 stages of operation in White Desk. The process starts with the engagement with the citizen where the citizen is listened and the source of the problem is specified. After the problem is identified, an application is formally received which later is forwarded to the related directorate. This marks the beginning of the solution process where the responsible directorate creates a solution to the problem. Lastly, the citizen is informed with the solution after the solution is approved by the directorate (Istanbul Metropolitan Municipality, 2016; Istanbul Metropolitan Municipality, 2019b; Turkish Republic Ministry of Environment and Urbanization, 2019).

Soft Transformation

There are also physical constraints in the design of retrofitting smart cities which sometimes is resulted from the errors in former urban planning activities or as in the Historical Peninsula case, they are due to the old building stocks and narrow road networks from early ages of the city. For example, autonomous vehicles depend on well-defined traffic lanes; and for a touristic and traditional district such as Historical Peninsula which has narrow roads and inconsistent road widths, and whose traffic constantly gets interrupted by tramlines and pedestrians, rollout of autonomous vehicle would be a challenging process. On the other hand, it seen that Historical Peninsula lags behind in terms of deployment of smart energy applications. Because, when looking at the energy efficient strategies, it is seen that they include energy efficient infrastructure, efficiency of energy distribution systems and use of renewable sources of energy on neighborhood scale; and on building scale, they include natural lighting, energy efficient facades and roofs, demand controlled ventilation, passive cooling systems and water systems where rainwater and gray water (from lavatories and showers) are collected, treated and reused in flushing and irrigation (Whitman et al., 2008). It is almost impossible to implement these strategies in Historical Peninsula because of its deep-rooted physical infrastructure and mostly listed historical buildings that are neither suitable for building automation systems nor compatible for solar panels.

LEAN AND GREEN SOLUTIONS FOR HERITAGE MANAGEMENT IN ISTANBUL HISTORICAL PENINSULA

When embedding technologies to heritage management, it should be recognized that heritage knowledge is transformed into 'e-heritage' in a new concept that includes digital museum work and interactive platforms, digital documentation and archiving of portable cultural assets. While such applications configure technology with operational functionality, another function is to transmit this information through new communication tools and interfaces such as internet and mobile applications, mass communication channels and social media. It is evident that these technologies enhance the capacity of conservation skills (at all stages of planning and management), increase the promotional value of conservation sites, and strengthen the destination management of cultural tourism. The purpose of this chapter is to evaluate Mitchell's (1999) five "lean and green future cities" principles in which the use of smart technologies can be assessed in the heritage management and urban planning procedures of the Historic Peninsula of Istanbul. Thus, the essential strategies for the digitization of entire heritage assets and related cultural services can be validated by these five principles (Koramaz, 2017).

One of the crucial strategies on dematerialization is to digitalize and complete current inventory studies regarding the Historic Peninsula and to broadcast the results of related researches performed through digital media mediums. Istanbul Metropolitan Municipality and responsible directorates have been working on digital archiving works, as 3D Surface Models of the Historic Peninsula utilizing laser technology (LIDAR) as a basis to allow more detailed examination of the impact of new projects (Kersten et al, 2009; and Buyuksalih, et al., 2018), and Web- Based 3D Guide Application established with panoramic images embedded to city guide portal (Figure 14) It can be said that these efforts endeavor to realize approved strategies in the site management and planning program of the Istanbul Historic Peninsula with a knowledge infrastructure that aims to develop and document cultural heritage. At the same time, in the last decade, the Istanbul city guide portal reveals that it has contribute into the dematerialization principle with the objective of establishing easily understandable and accessible maps, documents, publications and web sites that promote the distinctive features and aesthetic quality.

The key strategy supporting the demobilization principle is to share the findings of heritage research with the public on the website. Unfortunately, all government agencies, such as the site management directorate, the ministry of culture and tourism, and other cultural institutions, do not have interactive functions on their websites, and only have e-mail addresses without hot links to the relevant e-services for communication (http://www.alanbaskanligi.gov.tr; http://www.kultur.gov.tr; https://korumakurullari.ktb.gov.tr). As the authorized administration of Istanbul Historical Peninsula, Fatih Municipality, has an online information science learning program since 2007. This is a certified online training program that is organized in order to increase the use of data processing technologies inside and outside the institution. On the other hand, both the district and metropolitan municipalities have the project of Mobile Municipality and Transition to Mobile Signature Application either, like other governmental offices in Turkey. As one of the critical strategies for demobilization strategy, it is suggested to develop interactive museum works for world-famous historical sites such as Istanbul Archaeology Park, Topkapi Palace, Museum of Turkish and Islamic Art.

Mass customization can be called as a challenging principle in order to create an intelligent adaptation, personalization and fitting individual preferences (in other words demand management) in heritage management. For instance, Bath Site Management Plan prepared for the period of 2016-2022, aims to support outreach work to help people (especially the young), engage in heritage issues, working with local societies & interest groups (https://www.bathworldheritage.org.uk/management-plan). While addressing the preferences of different user groups and residents, social media and smartphone applications can be an effective tool for different stakeholder groups with changing preferences to access and incorporate information on heritage programs. Both Istanbul Metropolitan Municipality and Fatih District Municipality have mobile applications and social media profiles but have limited applications on the aforementioned information and program.

The principle of intelligent operation in heritage management, seeking effective operational resources development and monitoring and feedback capabilities can be emphasized primarily through ticketing systems for museums and heritage sites. Turkish Republic Ministry of Culture and Tourism has started an integrated museum pass system since 2008 (https://muze.gov.tr). In the process of convergence with the e-municipality, Fatih District Municipality has established a city information system based on geographic information system technologies where the residents can get information about the planning and conservation decisions in the local plan details (Figure 15).

In assessing conservation and planning challenges in the Istanbul Historical Peninsula, criticisms are directed at the preservation of authenticity and uniqueness in a limited understanding, rather than the role of technology and innovation in future urban and architectural change. In 2017, the UNESCO-WHC Committee report indicated that there was a threat of the loss of vernacular timber houses and the outstanding universal value of the entire Peninsula. The report recommends developing a long-term conservation strategy for the Ottoman timber buildings instead these houses have been demolished and replaced with new ones. (Conclusion by ICOMOS Advisory Bodies in 2017, ICOMOS, 2017). In order

Figure 14. Istanbul city guide portal
Source: (Istanbul Metropolitan Municipality, 2019c)

Figure 15. Fatih municipality city information system
Source: (Fatih Municipality, 2019b)

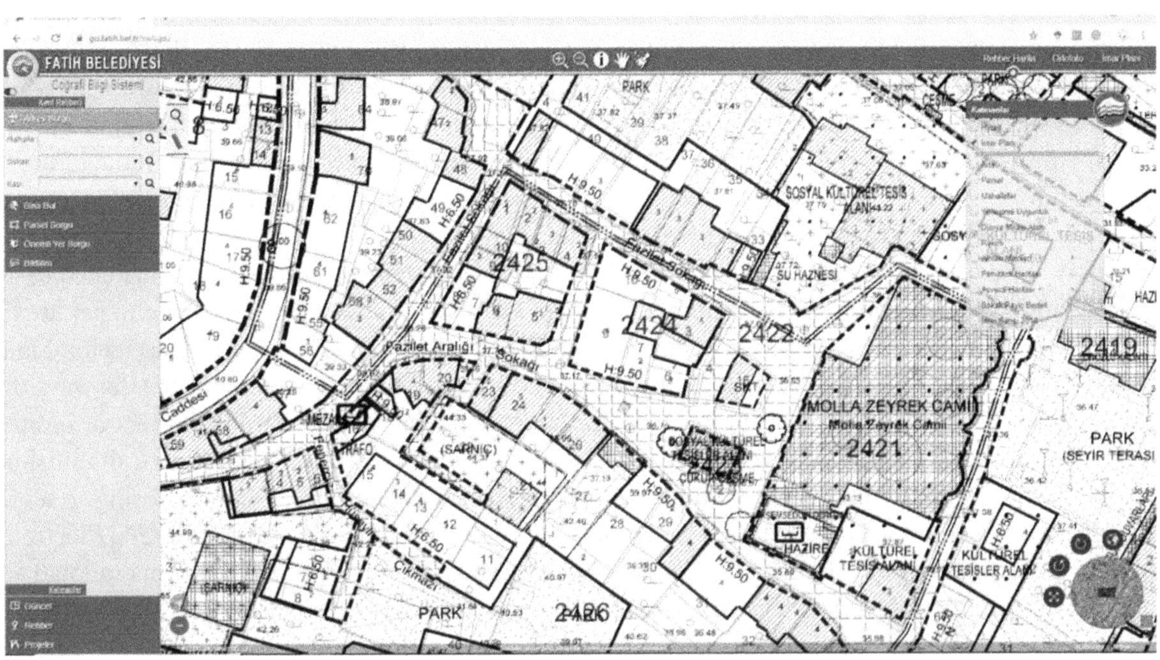

to respond to these recommendations, Istanbul and Historical Peninsula management should consider that soft transformation of the site could meet requirements of information paradigm, small-scale and adaptive interventions in buildings, neighbourhoods, and cities where technology and digital age meet.

DISCUSSION

In this study, the role of ICT-enabled technologies in the urban planning and heritage management processes in Historical Peninsula have been sorted out in accordance with Mitchell's (1999) principles, with the aim of uncovering where the historical, cultural and financial hub of the city stands in the age of rapid digital advancement. The rationale behind the choice of methodology is that Mitchell (1999) is among the pioneers in the field who approached the digital revolution with an architectural and urban designer point of view. Considering that the role of urban planning and policy making were underestimated by the promoters of then experimental idea -now vision- of "information society" or "cyberculture" (Graham and Marvin, 1999), the visionary principles of the Mitchell suggest invaluable contribution to the redesigning of the urban systems. Concluding remarks indicate that these principles can meet the requirements of the new era of technology and the new practices of living when adjusting our cities and societies.

Surely, Istanbul has taken notable steps towards smartness, and shared and realized Turkey's determined vision of becoming an information society. The deployed technologies in Historical Peninsula justify the city's efforts and comply with Mitchell's (1999) principles. However, among these smart applications, some of them have more impact. For instance, e-Government Gateway and e-Pulse are significant examples of dematerialization, which are being widely embraced by people, simply because they are easy to use user-oriented interfaces that allow citizens to save time and resources while accessing state and healthcare services. Similarly, there are smart applications geared towards demobilization that have also reached broad utilization. The success of these applications can be linked to the advantages which they pose especially for the public, whether by reducing the need for transport via online platforms for government services, work or education as in dematerialization; or by decreasing the volume of traffic via real-time navigation systems (IBB YolGosteren), smart parking systems (Transport Management Centre and ISPARK), and bike sharing systems (Isbike).

In this regard, it is safe to say that, whatever the extent of deployment, the services attributed to dematerialization and demobilization receive wider-spread adoption or attention, since the citizens are the main beneficiaries. Whereas, for the applications related to intelligent operation, the smart technologies themselves take the centre stage as they are used by authorities as instruments to enhance the ways that urban areas are managed in. In this approach, even though citizens can benefit from direct or indirect outcomes, the motivation of local governments mostly is to increase efficiency through diminishing the operational costs, eliminating the need for manual labour or conducting easier monitoring. Accordingly, smart applications in Historical Peninsula such as Traffic Monitoring Cameras, Mobile Electronic Violation Detection Systems (EDS), SCADA Technology, Smart Water Meters provide operational and economic advantages, while services like Istanbul Fire Department Automation Project, Centralized Doctor Appointment System (CDAS), Istanbul Air Quality Monitoring Centre, IMM White Desk deliver more observable benefits for the welfare of the citizens.

Another critical component of the smart development of Historical Peninsula is the collection of the big data and availability of open source data portals. The case study shows that majority of Istanbul's big data is related to mobility and is captured by autonomous real-time sensors and traffic cameras across the transport network. Although the captured data is processed and used in different platforms (such as Adaptive Traffic Management System - ATAK, IBB YolGosteren - IMM Navigation, smart parking system of Istanbul and smart surveillance), these procedures are all carried out by Traffic Management Centre, which does not share any raw data and let other actors act upon it and make further studies. In addition, there are no open data platforms related to water quality and water consumption habits, and none regarding the affairs of citizenry or cultural heritage in Istanbul. Even though, open data platforms are induced in the national strategy documents in Turkey and there are smart applications about water management (by ISKI); civic engagement (by IMM White Desk); and documentation of cultural heritage (by Fatih Municipality), the data is not available for the public or the other actors whom actually can make use of it. It is claimed that this is due to business-led approach of the local governments that perceive citizens as the customers of these smart services rather than the natural beneficiaries (Hollands, 2008; Batty, 2017).

Adoption can be referred as the most crucial component of smartness as it is the most challenging one to achieve. Compared to financial and administrative topics, citizen engagement is generally overlooked by municipalities, governments and global tech-companies in aspiring smart cities. Rather than focusing on the potential impact of smart technologies on citizen behaviour, the above-mentioned (mostly non-academic) actors give attention to comprehensive embedding, integration and operation of these marketable technologies that are developed around main themes such as mobility and energy (Batty, 2017). In the case study conducted in Historical Peninsula, there is no discussion related to citizen engagement to the deployed smart applications, came across from the survey. Depicted applications show little to no participatory features like crowdsourcing interfaces that would allow citizens to share their opinions, give feedback and be a part of the development processes of these smart applications such as online care search and scheduling or digital administrative citizen services.

CONCLUSION

With the 2023 vision of becoming an information society, in early 2000s Turkey specified science and technology as the strategic tools to achieve this vision where the infrastructure of information and communication technologies is retrofitted and Turkey has the ability to produce knowledge, to transform it into economic and social benefit (TUBITAK, 2004). Surely, Istanbul has been the natural habitat of this vision of pursuing smartness as the city has been experiencing urban and environmental problems on a much greater level than any other city in Turkey. But most importantly, the potential economic results are the most promising for Istanbul, since the city produces almost 1/3 (31.2% in 2017) of the total GDP of Turkey (TURKSTAT, 2017).

Thus, Istanbul started to implement smart solutions in variety of domains related to economic or daily activities. However, findings of the case study show that, the smart application rollout in Historical Peninsula is more improved in the principles that have direct financial outcomes such as intelligent operation, dematerialization and demobilization, while mass customization that has subtler effects seems to be overlooked in the area. On the other hand, soft transformation is arguably the most challenging principle to follow for Historical Peninsula, since it is crucial to transform the district's old building stocks, urban spaces and narrow road networks which can be dated back to Byzantine Period, to the smart technologies in a way that is not detrimental to its cultural, architectural and socio-economic value. Also, for the citizen engagement part of the future smart developments, it is recommended for local authorities not to just put the technology out there but to put people first, to monitor the outcomes and to ensure that the advantages are intended but not coincidental. Because the opposite would legitimize the claims of the critics who suggest that local governments use smart technologies to promote their locations in the global competitiveness while tech-companies aim to accumulate more capital through sales of software and infrastructure.

Concluding remarks of this chapter has a strong motivation for deliberating how Istanbul Historic Peninsula Site can be managed with revised policies and programs, already introduced in a recently approved policy document, titled as of Information Society Strategy and Action Plan (2015-2018), prepared by Government of Turkey. Creation of public services, personalized and fit into individual preferences of varying user groups, is significantly missing in the heritage management and urban planning process, to overcome inadequacies with user-oriented and effective public services. Social media and crowd-sourcing applications can be improved extensively in public involvement in heritage site management. Additionally, the establishment of living labs, aiming inclusive smart city programs, can also produce such participation and collaboration interfaces on digital content for cultural heritage between technology and creative narratives. Remarks from the case study will guide decision-makers and policy developers for re-interpretation of comprehensive and integrated frameworks, improving cultural heritage perception and awareness with smart technologies and transforming public space by means of heritage site management. This guidance is nothing to be overlooked especially since the recent turn of events reminded us the significance and urgency of redesigning the societal, financial and cultural aspects of urban life in accordance with the necessities of the new era, given that all of these aspects are now paralyzed with the effects of the Covid-19 pandemic around the world. Dematerialization and demobilization in particular prove once again that Mitchell's (1999) principles are invaluable guidelines for future studies and developments.

ACKNOWLEDGMENT

This chapter is based on Burak Belli's master's thesis titled "Dimensional Evaluation of Retrofitting Smart Cities: The Historical Peninsula Case" which was submitted in 2019 to Istanbul Technical University, under the supervision of Assoc. Prof. Dr. T. Kerem Koramaz.

REFERENCES

Audirac, I., & Fitzgerald, J. (2003). Information technology (IT) and urban form: An annotated bibliography of the urban deconcentration and economic restructuring literatures. *Journal of Planning Literature*, *17*(4), 480–511. doi:10.1177/0885412203017004002

Batty, M. (2017). *The Age of the Smart City*. London: UCL Working Paper Series.

BELBIM. (2019a). *Istanbulkart Mobile*. Retrieved from https://www.istanbulkart.istanbul/istanbulkart-mobil

BELBIM. (2019b). *Our smartcards.* Retrieved from https://www.istanbulkart.istanbul/kartlarimiz

BTK. (2018). *Annual statistics bulletin on provincial basis for the electronic communication sector* [Elektronik haberleşme sektörüne ilişkin il bazında yıllık istatistik bülteni]. BTK.

Büyüksalih, G., Bayburt, S., & Jacobsen, K. (2018). Analysis of height models based on KOMPSAT-3 images. *International Archives of the Photogrammetry, Remote Sensing and Spatial Information Sciences-ISPRS Archives*, *42*(3W4), 115-119.

Darmawan, R. (2011). Web 2.0 and idiosyncrasy of cultural heritage: a perspective from indonesia. In G. Styliaras, D. Koukopoulos, & F. Lazaniris (Eds.), *Handbook of Research on Technologies and Cultural Heritage: Applications and Environments* (pp. 481–494). IGI Global. doi:10.4018/978-1-60960-044-0.ch024

Deakin, M. (2011). From the City of Bits to E-Topia: Space, Citizenship and Community as Global Strategy in the Governance of the Digitally-Inclusive Regeneration Thesis. In D. Piaggesi, K. J. Sund, & W. Castelnovo (Eds), *Global strategy and practice of e-governance: examples from around the world* (pp. 124-141). Hershey, PA: IGI Global.

Deakin, M. (2014). From the city of bits to e-topia: Space, citizenship and community as global strategy. *International Journal of E-Adoption*, *6*(1), 16–33. doi:10.4018/ijea.2014010102

Department of e-Government Services. (2016). *2016-2019 National e-Government Strategy and Action Plan.* Author.

Digital Transformation Office. (2019). *e-Government Gateway.* Retrieved from https://www.turkiye.gov.tr

Dragulanescu, N. G. (2002). Social impact of the "digital divide" in a Central–Eastern European country. *The International Information & Library Review*, *34*(2), 139–151. doi:10.1080/10572317.2002.10762570

Fatih Municipality. (2019a). *e-Municipality.* Retrieved from http://www.fatih.bel.tr/ebelediye/

Fatih Municipality. (2019b). *Fatih Municipality city information system.* Retrieved from https://gis.fatih.bel.tr/webgis/

Frost & Sullivan. (2015). *Strategic Analysis of Smart Parking Market in Europe and North America.* Author.

GIS Academy. (n.d.). *Istanbul Fire Department automation project.* Retrieved from https://cbsakademi.ibb.istanbul/proje/istanbul-itfaiyesi-otomasyon-projesi/

Graham, S., & Marvin, S. (1999). Planning cybercities: Integrating telecommunications into urban planning. *The Town Planning Review*, *70*(1), 89–114. doi:10.3828/tpr.70.1.w34454x3475g2858

Gulersoy-Zeren, N., Tezer, A., Yigiter, R., Koramaz, T. K., & Gunay, Z. (2008). Istanbul Project: Istanbul Historic Peninsula Conservation Study; Zeyrek, Süleymaniye and Yenikapı Historic Districts: Vol. 1. *Conservation of Cultural Assets in Turkey.* Istanbul Technical University.

Haarstad, H., & Wathne, M. W. (2019). Are smart city projects catalyzing urban energy sustainability? *Energy Policy*, *129*, 918–925. doi:10.1016/j.enpol.2019.03.001

Haberturk. (2018). *"ATAK" model for Istanbul traffic: Green light will be on according to intensity.* Retrieved from https://www.haberturk.com/tv/gundem/video/son-dakika-haberi-istanbul-trafigine-atak-modeli-yogunluga-gore-yesil-isik-yanacak/505204

Hendarti, R., Tomasowa, R., & Katarina, W. (2016, December). Study on Smart City Concept of Energy Source and Distribution for A Special Economic Zone in Jakarta. In *2016 IEEE 18th International Conference on High Performance Computing and Communications; IEEE 14th International Conference on Smart City; IEEE 2nd International Conference on Data Science and Systems (HPCC/SmartCity/DSS)* (pp. 1312-1316). IEEE. 10.1109/HPCC-SmartCity-DSS.2016.0186

Höjer, M., Gullberg, A., & Pettersson, R. (2011). *Images of the Future City. Time and Space for Sustainable Development.* Springer. doi:10.1007/978-94-007-0653-8

Hollands, R. G. (2008). Will the real smart city please stand up? Intelligent, progressive or entrepreneurial? *City*, *12*(3), 303–320. doi:10.1080/13604810802479126

Hurriyet. (2018). *Here are the new taxis of Istanbul! Turquoise, black...* Retrieved from https://www.hurriyet.com.tr/gundem/iste-istanbulun-yeni-taksileri-turkuaz-siyah-40902239

ICOMOS, (2017). *UNESCO-WHC Committee report and Conclusion by ICOMOS Advisory Bodies for Historic Areas of Istanbul (Turkey) No 356 Bis.* Retrieved from file:///C:/Users/ITU/Downloads/356bis-ICOMOS-2255-en.pdf

Information Society Department. (2018). *2015-2018 Information Society Strategy and Action Plan.* Author.

ISBAK. (2019). *iTaksi.* Retrieved from https://itaksi.com/en/#Welcome

ISKI. (2016). *ISKI 2017 budget approved.* Retrieved from https://www.iski.istanbul/web/tr-TR/kurumsal/haberler1/haberler-detay/iski-2017-butcesi-onaylandi

ISKI. (2017). *2017 Annual Report.* ISKI.

ISKI. (n.d.a). *Scada technology.* Retrieved from https://www.iski.istanbul/web/en-US/kurumsal/iski-hakkinda/scada-technology

ISKI. (n.d.b). *Smart meter application.* Retrieved from https://www.iski.istanbul/web/tr-TR/musteri-hizmetleri/kartli-saya-uygulamasi

ISPARK. (2017). *Smart parking systems.* Retrieved from https://ispark.istanbul/ispark-ile-akilli-otopark-sistemleri/

ISPARK. (2019a). *Isbike.* Retrieved from https://www.isbike.istanbul

ISPARK. (2019b). *Isbike: How does it work?* Retrieved from https://www.isbike.istanbul/nasilcalisir

Istanbul Metropolitan Municipality. (2015). *Traffic Management for a Smarter City.* Author.

Istanbul Metropolitan Municipality. (2016). *2016 Annual Report.* Author.

Istanbul Metropolitan Municipality. (2017). *IBB YolGosteren: Your New Guide on the Road.* Author.

Istanbul Metropolitan Municipality. (2018). *"ATAK" model from IMM to Istanbul traffic*. Retrieved from https://www.ibb.istanbul/News/Detail/34842

Istanbul Metropolitan Municipality. (2019a). *Air quality monitoring centre*. Retrieved from https://havakalitesi.ibb.istanbul/Pages/AirQuality

Istanbul Metropolitan Municipality. (2019b). *IMM White Desk*. Retrieved from http://alo153.ibb.gov.tr/

Istanbul Metropolitan Municipality. (2019c). *Istanbul city guide portal*. Retrieved from https://sehirharitasi.ibb.gov.tr/

Istanbul Metropolitan Municipality. (2019d). *Watch Istanbul: Touristic cameras*. Retrieved from http://www.istanbuluseyret.com/TuristikKameralar/

ISTTELKOM. (2017). *Common communications infrastructure is coming to Istanbul*. Retrieved from https://isttelkom.istanbul/en/common-communications-infrastructure-is-coming-to-istanbul/

Kersten, T. P., Büyüksalih, G., Baz, I., & Jacobsen, K. (2009). Documentation of Istanbul historic peninsula by kinematic terrestrial laser scanning. *The Photogrammetric Record*, *24*(126), 122–138. doi:10.1111/j.1477-9730.2009.00528.x

Kitchin, R. (2014). The real-time city? Big data and smart urbanism. *GeoJournal*, *79*(1), 1–14. doi:10.100710708-013-9516-8

Koramaz, T. K. (2017) Information - Communication Technologies and Management Plans for Heritage Sites. In *Abstract Book of 30th Annual AESOP 2017 Congress, SPACES OF DIALOG FOR PLACES OF DIGNITY: Fostering the European Dimension of Planning* (pp. 599-600). Lisbon, Portugal: Técnico Lisboa, The Institute of Geography and Spatial Planning, Faculty of Architecture.

Koramaz, T. K. (2018a). Information and Communication Technologies in Cultural Heritage Management. In L. Reynolds, A. Campelo, A. Lindgreen, & M. B. Beverland (Eds.), *Cultural Heritage: A Research Anthology* (pp. 155–165). Routledge. doi:10.4324/9781315107264-12

Koramaz, T. K. (2018b). Housing Renewal Sites and Spatial Features of Deterioration and Deprivation in Istanbul's Historic Peninsula. *Journal of Urban Planning and Development*, *144*(1), 1-10.

Koramaz, T. K., Nasrollahzade, S., & Ozdemir, Z. (2019) Spatial Pattern of Digital Divide in Turkey. In *Proceedings of REAL CORP 2019*. Karlsruhe, Germany: Karlsruher Institut für Technologie.

Korupp, S. E. (2006). Causes and trends of the digital divide: a European perspective. In *Proceedings of Deutsche Gesellschaft für Soziologie (DGS), Soziale Ungleichheit, kulturelle Unterschiede: Verhandlungen des 32. Kongresses der Deutschen Gesellschaft für Soziologie*. Campus Verl.

Kramers, A., Höjer, M., Lövehagen, N., & Wangel, J. (2014). Smart sustainable cities–Exploring ICT solutions for reduced energy use in cities. *Environmental Modelling & Software*, *56*, 52–62. doi:10.1016/j.envsoft.2013.12.019

Lee, S. H., Han, J. H., Leem, Y. T., & Yigitcanlar, T. (2008). Towards ubiquitous city: concept, planning, and experiences in the Republic of Korea. In *Knowledge-based urban development: Planning and applications in the information era* (pp. 148–170). IGI Global. doi:10.4018/978-1-59904-720-1.ch009

Leite, J. M. V. (2009). Cultural heritage and monument, a place in memory. *City and Time*, *4*(2), 23–33.

Maeng, D. M., & Nedović-Budić, Z. (2008). Urban form and planning in the information age: Lessons from literature. *Spatium (Belgrade)*, (17-18), 1–12. doi:10.2298/SPAT0818001M

Maté, A., Peral, J., Ferrández, A., Gil, D., & Trujillo, J. (2016). A hybrid integrated architecture for energy consumption prediction. *Future Generation Computer Systems*, *63*, 131–147. doi:10.1016/j.future.2016.03.020

Mattern, S. (2017). A City Is Not a Computer. *Places Journal*. doi:10.22269/170207

Mitchell, W. J. (1995). *City of bits: space, place, and the infobahn*. MIT Press.

Mitchell, W. J. (1999). *"e-topia:" Urban life, Jim—but not as we know it*. MIT Press. doi:10.7551/mitpress/2844.001.0001

Ning, Z., Xia, F., Ullah, N., Kong, X., & Hu, X. (2017). Vehicular social networks: Enabling smart mobility. *IEEE Communications Magazine*, *55*(5), 16–55. doi:10.1109/MCOM.2017.1600263

Shoup, D. (2007). *Gone parkin'*. Retrieved from https://www.nytimes.com/2007/03/29/opinion/29shoup.html

Shoup, D. C. (2006). Cruising for parking. *Transport Policy*, *13*(6), 479–486. doi:10.1016/j.tranpol.2006.05.005

Sönmez, M. S. (n.d.). *ISKI monitoring systems smart grid applications* [İSKİ izleme sistemleri akıllı şebeke uygulamaları]. Retrieved from https://www.tarimorman.gov.tr/SYGM/Belgeler/%C4%B0zleme%20H%C4%B0E%20SUnumlar/iski-mehmet%20sait.pdf

Toffler, A. (1981). *The Third Wave*. Bantam Books.

Townsend, A. M. (2013). *Smart cities: Big data, civic hackers, and the quest for a new utopia*. WW Norton & Company.

Transport Management Centre. (2017). *Smart Parking Systems*. Retrieved from https://uym.ibb.gov.tr/corporate/news-and-announcements/smart-parking-systems

Transport Management Centre. (2018). *Adaptive traffic management system (ATAK) in Istanbul*. Retrieved from https://uym.ibb.gov.tr/corporate/news-and-announcements/adapt%C4%B1ve-traff%C4%B1c-management-system-atak-%C4%B1n-istanbul

Transport Management Centre. (2019a). *Smart parking system of TMC*. Retrieved from https://uym.ibb.gov.tr/YHarita/Harita_EN.aspx

Transport Management Centre. (2019b). *Traffic Density Map of Istanbul*. Retrieved from https://uym.ibb.gov.tr/YHarita/Harita_EN.aspx

Transport Management Centre. (n.d.a). *Electronic violation detection systems*. Retrieved from https://uym.ibb.gov.tr/services/electronic-violation-detection-systems

Transport Management Centre. (n.d.b). *Traffic information systems*. Retrieved from https://uym.ibb.gov.tr/services/traffic-information-systems

Transport Management Centre. (n.d.c). *Traffic measuring and monitoring.* Retrieved from https://uym.ibb.gov.tr/services/traffic-measuringandmonitoring

TUBITAK. (2004). *National Science and Technology Policy 2003-2023 Strategy Document* [Ulusal Bilim ve Teknoloji Politikaları 2003-2023 Strateji Belgesi]. Retrieved from https://www.tubitak.gov.tr/tubitak_content_files/vizyon2023/Vizyon2023_Strateji_Belgesi.pdf

Turkish Republic Ministry of Environment and Urbanization. (2019). *Smart cities white paper* [Akıllı şehirler beyaz bülteni]. Author.

Turkish Republic Ministry of Health. (2016). *Centralized doctor appointment system.* Retrieved from https://www.mhrs.gov.tr/Vatandas/hakkimizda.xhtml

Turkish Republic Ministry of Health. (2018). E-Pulse v.2.0 User Manual. Author.

United Nations. (2018). *E-Government Survey 2018: Gearing E-Government to support transformation towards sustainable and resilient societies.* United Nations.

Vanden, B. G., Knapp, S., & Doe, J. (2001). *Role of reference elements in the selection of resources by psychology undergraduates.* Retrieved from http://jbr.org/articles.html

WHC-UNESCO. (1985). Inscription: Historic areas of Istanbul: Turkey. In *9th Session of the Committee.* Retrieved from http://whc.unesco.org/en/decisions/3864

Whitman, C. T., Reid, C., von Klemperer, J., Radoff, J., & Roy, A. (2008, March). New Songdo city– The making of a new green city. *Proceedings of the 8th World Congress Council on Tall Buildings and Urban Habitat (CTBUH'08).*

Wilson, E. J. (2004). *The information revolution and developing countries.* MIT Press.

ENDNOTE

1 For detailed information about William J. Mitchell's publications under MIT Press, https://mitpress.mit.edu/contributors/william-j-mitchell

Chapter 6
Towards a Smart Neighbourhoods Simulated Model

Nada M. Alhakkak
https://orcid.org/0000-0002-5435-0730
Baghdad College of Economic Sciences University, Iraq

ABSTRACT

The literature generally describes a conventional neighbourhood as an urban area that consists of houses, schools, shopping, religious buildings, and open spaces. This work focuses on the challenge of establishing smart neighbourhoods. The contribution here is in presenting a new model for a smart neighbourhood that contains two frameworks. The first framework is related to input specific parameters from the user and gets reports for timetable, map, and materials. The second framework uses the virtual reality aspect and compound of multiple layers. The concluding remarks figured out that the importance of using the model's work depends on reusing and ABC clustering algorithm.

INTRODUCTION

The world is changing fast, cities are growing, and the urban population is rising. More people than ever are living in cities, and by the end of this century, humans will be having urban issues as our cities grow, so do the challenges. The sustainable building today holds the technological key to the future. The term neighbourhood is used to describe part of the urban settlement; it is where a group of people share a place for living; the clustering of multiple neighbourhoods' forms towns, villages, and cities. A neighbourhood is a term that refers to various blocks of houses or buildings (Allaire, 1960). It's a place with its own identity and services, where people interact with others (Elshater, 2016). Several researchers describe the best neighbourhoods place is the one that has a walkability feature together with economic and social services (Elshater, 2012; Elshater, Towards Refined a New Concept of New Urbanism in Egypt, 2013; Kumar, 2015; Jarass & Heinrichs, 2014). Moreover, the neighbourhood represents the start point of city planning unit (Elshater, 2016). People culture accepts deal with this term in both social

DOI: 10.4018/978-1-7998-3856-2.ch006

and physical ways (Moulaert, Martinelli, Swyngedouw, & González, 2010). Neighbourhood's design is affected by multiple factors (e.g., support legal framework, analysis of the local society and economy, the infrastructure technology and capacity and institution's capacity) (Western Australian Planning Commission, 2007).

This chapter aims to design a model that takes its input from designers in order to build a smart, sustainable neighbourhood. To achieve this goal, some related background topics are discussing in the following section. To follow this aim, the current chapter is organised in four sections after the introduction: background, literature review, methods, data sets and discussion, and finial the conclusion that ends with future directions of research.

BACKGROUND

The last update of 'green' term is now called smart, and as green was and still necessary, the smart term will be always be relevant. Smart product means intelligent and able for connecting, sensing, processing, and react according to other's actions. However, there is no static definition for the term 'smart city', instead any city could be smart if it becomes more liveable with high response to new technology's challenges (Ahuja, 2016).

People like to live close in sections or blocks of area, those shared blocks so-called neighbourhood, and multiple neighbourhoods form clusters called cities. However, any living place should contain spaces inside it. Those Spaces include homes, buildings, and spaces between all those parties to get smart spaces. This connection via all devices should be managed by using new technologies like DS2O by Pahl, Niedermayer, Kinkelin, and Carle (2013) that handles all devices inside/outside houses and buildings.

Generally speaking, people like to live close in sections or blocks of the area. Those shared blocks called a neighbourhood, and multiple neighbourhoods form clusters called cities. Akl (2015) classified neighbourhoods into three main types of patterns, showing the main features together with advantages and disadvantages. This classification includes interconnected neighbourhood, neighbourhood as a cell, and neighbourhood in the linear township.

However, for any designer or engineer that intend to design or build a smart, sustainable neighbourhood, they should check their progress by going through a checklist. Akl (2015) reported in the "UN-Habit" five principles that used to measure neighbourhood analysis. Those five principles presented a simplified framework for managing the design process that composite of five levels (Table 1).

Moreover, the sustainable neighbourhood is composed of multiple systems' types (also called layers). It depends on how we look to it. Mainly, each system (layer) has different importance needs or requirements, and any neighbourhood should contain those layers, but it differs from one another. There are five main types of systems (Table 2).

In order to achieve the above five layers in our design, we should use clustering as a tool to group each layer component separately into its related layer. Clustering analysis depends on similarity to divide set of objects. It is described, in this work, like blocks or neighbourhoods. The clustering uses are portioning methods that divide the collection of objects into a set of groups. One of it is well-known methods so-called prototype-based clustering algorithms where the centre of the cluster is the backbone of each cluster. These algorithms depend on the K-means algorithm as its primary tool. Here, the chapter uses Bee Colony Optimization BCO as a clustering algorithm which is classified as a swarm-based data clustering technique (Mizooji, Haghighat, & Forsati, 2012).

Table 1. The design process for sustainable smart neighbourhood

Levels	Sub-levels
1- Getting started	a. Collect project objectives
	b. Determine the used strategy for achieving those objectives
	c. Resource requirements
	d. Funding and resource constraints
	e. Active participation encouragement
1- Appreciating the context	a. Data collection and analysis
	b. Asset's search
	c. Define constraints that affect the development process
	d. Data collection and analysis
2- Creating the urban structure and making connections	a. Combine the design principles with i. Community need ii. Economic availability iii. Engineering feasibility
	b. Develop design proposals using improvement ideas
	c. Option's evaluations
	d. Present the design to the team, and then to the community
3- Detailing the place	a. More research and studies required for the development
4- Following up	a. Agree on detailed design
	b. Confirm implementation

This work is specialised with connecting the old school of sustainable neighbourhood design with the new idea of smart cities and smart neighbourhood that uses smart electronic devices and sensors in its infrastructure. Hence, there is a severe need for connection and analysing the transformed data to enhance the performance of smart, sustainable neighbourhoods to achieve the performance goal we need to get benefit from cloud and fog computing services. However, Cloud computing is a new technology where computer resources like; hardware and software; are shared over the internet and could be purchased as services from any computer in the world (Sinha & Shekhar, 2015). Moreover, there are three main types of clouds; public, private, and hybrid. The last one it's the most popular. It has the advantage of a private cloud security and public cloud's open access (Zou, Deng, & Qiu, 2013). Moreover, the fog is a cloud close to earth ground devices; it deals with remote devices to achieve low latency, high reliability, and high efficiency for end-user applications. Fog computing is essential to transferee data to it from the internet of things IoT and gets a faster response, instead of sending to the cloud all the details. The fog can be in a middle way between cloud computing and IoT (Dubey, et al., 2015; Abusaada & Elshater, 2019).

Table 2. The systems of smart neighbourhood

System	Description
1- Natural systems	Land
	Water
	Climate
	Habitat
2- Land-use systems	Designation of uses
	Mixed-use development and social mix
	High-density development
	Planning and building legislation
	Neighbourhood services and facilities
3- Mobility systems	Roads network
	Walkable environment
	Transportation
	Roads types and their design
4- Energy's systems	Involving stakeholders
	Baseline survey
	Develop options
	Refining solutions and implementation
5- Environment management system	Water system: i. Reducing demand ii. Harvesting rainwater iii. Grey water treatment and reuse iv. Blackwater treatment
	Solid waste system

LITERATURE REVIEW

On the same vein, Al-Thani, Skelhorn, Amato, Koc, and Al-Ghamdi (2018) present another review of Doha's neighbourhood and then used Delhi's study for neighbourhood characteristics' investigation. El-khateeb (2019) works on Dubai City, as a case study of a sustainable neighbourhood (SN). All said researchers presented three features of SN and cities: compact, integrated and connected. Together with five principles, the concept of smart neighbourhood supports these features so-called UN-Habitat. Also, Akl (2015) presents them in his work as follow:

1. Create enough space for a street that is connected to an efficient network of streets.
2. High density or population's growth, it is the base of neighbourhood sustainability. This action refers to people and the activity of the concentration.
3. Mixed land-use focuses on neighbourhood and block-level. This action means that a part of the neighbourhood space should be allocated for economical use. The part also should be 40 per cent of the floor space. Which contains the following advantages:
    a. Local job availability

b. Local economy encouragement
c. Car independence
d. Encouragement of train and traffic cycle
e. Minimise fragment landscape
f. Public services reachability

4. Social mix provides houses with different prices for different level income accommodation.
5. Limitation of land-use is for preventing any single use of blocks or neighbourhoods.

Their study applied the UN-Habit five principles on the sustainable city of Dubai and found that not all the principles were applied to it. However, Dubai's case study was very close to this work because they used the UN-Habit in their work.

Other studies, classified neighbourhoods into three main types of patterns, showing the main features together with advantages and disadvantages, as mentioned previously in the background section (Akl, 2015). On the other side, Kumar (2015) and Abusaada & Elshater (2019) argue that using social media networks together with information and technology, would help in creating a better neighbourhood. They presented multiple case studies from Europe Middle East and Northern Africa and India. Their vision depends on Information and Communication Technology (ICT) that provides an interface for infrastructure improvement. They argued that to get smart neighbourhood, decision-makers do not need to invest in the infrastructure too much and spend money. You just need to use social media and networking. So, the investment would be on social Economic and Environmental sustainability by involving the community with social networks.

Park and Rogers (2015) propose a conceptual framework contained a new hierarchy of neighbourhood composed of four levels Regarding the face-blocks, residential neighbourhood, institutional neighbourhoods, and community; each level includes some other details. They used the basic idea of clustering on each level that presents the upper level starting from multiple face blocks on the first level reaching the final level that contains specific culture centres; without any particular details about the clustering method or how it works. They also argued that there are four main requirements affects neighbourhood those are: size, boundary, and core-use.

Shekhar & Tripathi (2015) present the parameters of the smart neighbourhood. These parameters are:

1. Density: higher density results netter urban
2. Infill Development: developing the whole neighbourhood with all its small details that serve the people in everything.
3. Mixed-use development: using a traditional model of the neighbourhood that design a vibrant, smart neighbourhood uses smart buildings connected with smart streets.
4. Street space design: streets are affected by traffic volume and green walking places.
5. Traffic calming: it composed of two main parts: reducing vehicle speed, and reducing traffic volume.
6. Parks and open spaces: focus on open natural spaces.
7. Improving neighbourhood equity: things that should be present at the neighbourhood level; equity, sustainability, ecological.

However, their work did not use those parameters on any model or case study. However, this work uses them as input parameters for the first framework (framework1). López, Moura, Custodio, and Moreno (2012) develop a model on the smart grid used multiple types of real-world scenarios such as short, medium, and long scenarios of a neighbourhood of the smart grid. They based their model on a collected data from ENERship project. However, their work focused on how to communicate using IEEE parameters in smart grids, but during the whole work they did not mention what the neighbourhood position in the proposed model was. They used ENERship architecture which deals with neighbourhood domain that contains M2M GW; as a network area. They focused on the way of connecting houses using electric distribution infrastructure.

Although there are many studies in a smart neighbourhood remains limited, more work is necessary. To fill this literature gap, this chapter presents a new model for modelling and simulating smart neighbourhood hybrid cloud computing environment that is connected with IoT devices via fog computing which reuses previous components in building the neighbourhood map.

METHODS

The concept of a smart city aims at building the neighbourhood of the future with sufficient and effective way. Building a smart neighbourhood is not impossible or complicated, but the method of building is what matter. The principal objective of this work is about saving time and efforts in building smart neighbourhoods by reusing some available part used previously in the neighbourhood buildings.

This work presents a smart neighbourhood model which contain two frameworks; framework1 called a physical smart neighbourhood. The second framework so-called a virtual smart neighbourhood (framework2) (Figure 1).

As discussed above, framework1 deals with stockholder where it requests inputs (i.e., parameters, neighbourhood map, building time table, and building features). This framework responds in reports that give the needed map, timetable, and materials list, at the same time the work is built within a hybrid cloud, as seen in Figure 2. It is significant to mention that the reusing feature is in framework1 because after receiving the request from the stockholder decision is made; either forwarding the request to framework2 or reusing existing reports. In other words, frameowork1 use in its work some saved components (i.e., smart home, smart light, smart building) for building the smart neighbourhood but if not exist then making these components for today users can save them for future reuse.

Going bottomless, framework2 deals with new requests where it contains four main layers; as follows:

- Layer1-simple neighbourhoods contain heterogeneous components of neighbourhoods; they represent the smallest units. Therefore, it cannot use to represent physical planning. To cluster the above components into groups with similar needs, similar design, same architectural style, similar housing values, and serve people with similar income. ABC algorithm is the clustering tool that used in current layer to get the next layer.
- Layer2-residential neighbourhoods contain groups of similar blocks represent residential neighbourhoods that are homogenous physically; these groups have been created using the ABC algorithm in the previous layer. Those groups or clusters have similar architecture contains central serve points like schools, stores, and other daily-serve needing points.
- Layer3-low level neighbourhoods are a result of using the same clustering algorithm but with different parameters related to simple services or functions to have service centres like shops, healthcare, schools, and other types of low-level services. Those centres often have their names or logos.

Figure 1. Smart neighbourhood model
source: The author of the present chapter

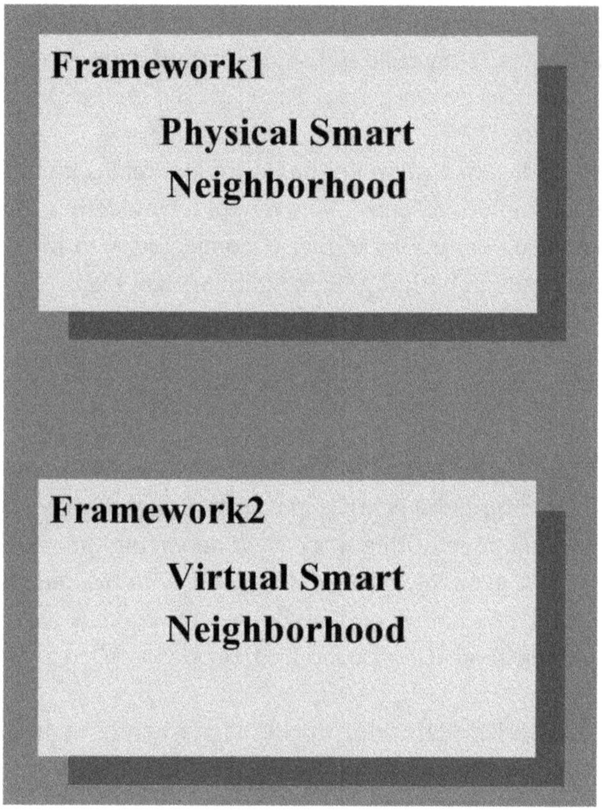

- Layer4-high level neighbourhoods represented by suburbs or townships, those are specific clusters or parts of the city. Those clusters have high-level features of neighbourhoods represented by universities as an example.

DISCUSSION, FUTURE DIRECTION AND CONCLUSION

The main idea of this work built upon the work of Park and Rogers (2015) that used four layers to build the model depending on the clustering process on each layer to get the next layer without any other obvious details or add to their work. However, the two frameworks interact with each other in the following way, to build their smart neighbourhood (SN):

1. The user interacts only with framework1 he/she use Graphical User Interface (GUI) for request/replay process.
    i. The user starts the request by input the details of the preferred (SN)'s. with used parameters in Shekhar and Tripathi (2015).
    ii. Framework1 process request by sending it to framework2 to build the desired (SN).

2. Framework2 gets its input from framework1 to build its four layers with specific details about each item on those layers.
3. Framework2 can deal with each item on each layer, wither build the things from scratch or reuse some previous item from old designs to save time and efforts.
4. Framework2 send the results to framework1 to build time table with the SN-map and other details, then send a replay message to the user (Figure 3).

Figure 2. Framework1-physical smart neighbourhood
source: The author of the present chapter

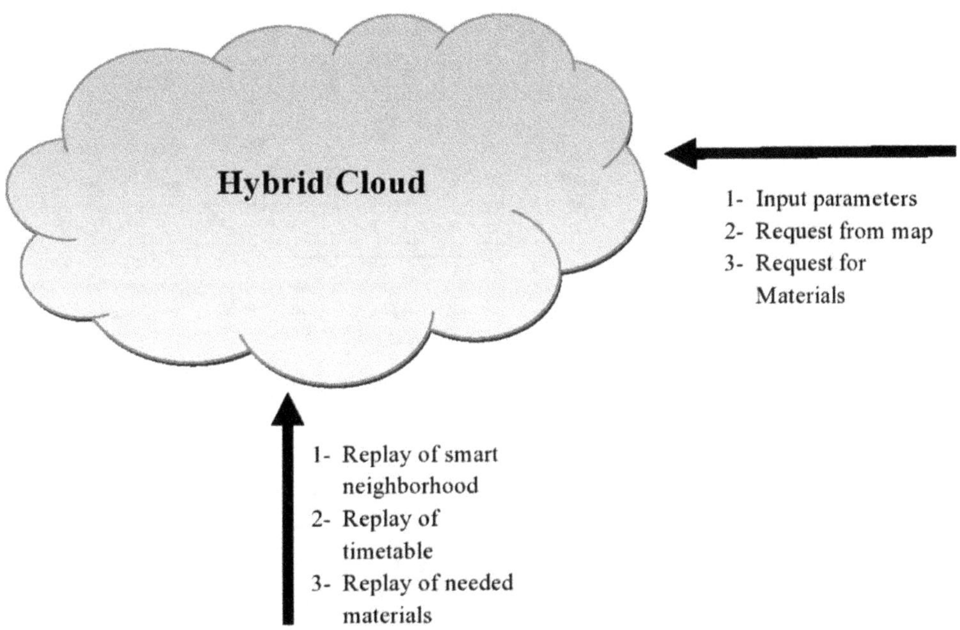

This work is built within cloud computing with the help of fog computing as a middleware between the IoT devices that can be used by end-users or stockholders to overcome the overhead on the cloud. The proposed model position in the pyramid of cloud-fog-IoT is on the top, as shown in Figure 4.

For optimization needs, should use different clustering types on each layer of framework2. Besides, the whole proposed model should be tested in a simulation environment like the work discussed by Akl (2015) who presented numerous types of simulations for urban and neighbourhood analysis, that helps in designing and constructing buildings and urban spaces, and may use in comparing between multiple designs. Examples of these simulators are: Autodesk Ecotect Analysis 2011, Revit, Autodesk CFD software, ENVI-met, and SOLARCHIVISON.

Finally, this research paper aimed to understand the term smart neighbourhood and its relation to smart cities with the important factors that affect designing smart neighbourhoods. Because of the lack of time and efforts, the author decided not to implement the proposed model to get further results. This aspect of research suggested a new model called 'smart neighbourhood model' that include two main frameworks; one for physical use and the other for virtual use. Future research is needed on simulating smart neighbourhoods that use new technologies like blockchain

Figure 3. Interaction between framework1and farmework2
source: The author of the present chapter

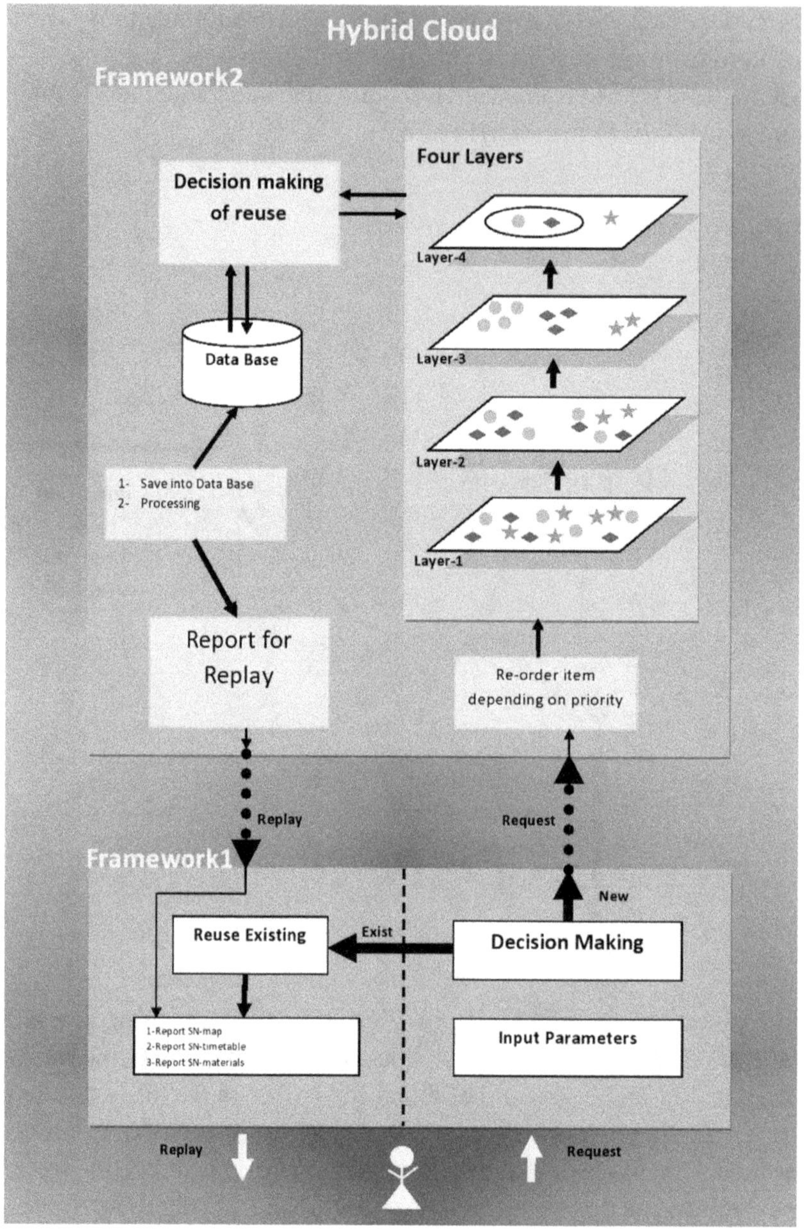

Figure 4. Smart neighbourhood model's position
source: The author of the present chapter

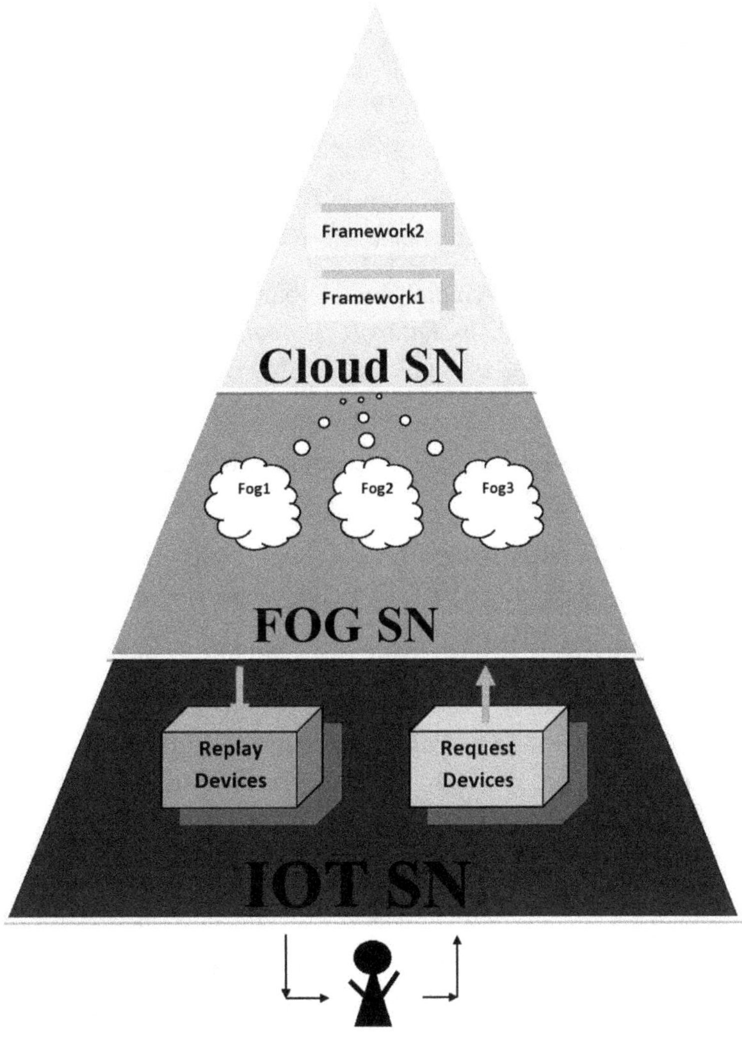

REFERENCES

Abusaada, H., & Elshater, A. (2019). Internet of Things Interpreting a Myth: Internally displaced persons transform cities of hardship into chaos cities. In H. Abusaada, C. Voulguth, & A. Elshater (Eds.), *Handbook of Research on Digital Research Methods and Architectural Tools in Urban Planning and Design* (pp. 47–81). IGI Global. doi:10.4018/978-1-5225-9238-9.ch004

Ahuja, A. (2016). Integration of Nature and Technology for Smart Cities (3rd ed.). Springer. doi:10.1007/978-3-319-25715-0

Akl, M. H. (2015). *Towards Sustainable Internet and ICT for Sustainability Egypt*. Mansoura University.

Al-Thani, S. K., Skelhorn, C. P., Amato, A., Koc, M., & Al-Ghamdi, S. G. (2018). Smart Technology Impact on Neighborhood Form for a Sustainable Doha. *Sustainability, 10*(4764), 17. doi:10.3390u10124764

Allaire, J. (1960). *Neighborhood Boundaries*. American Society of Planning Officials.

Dubey, H., Yang, J., Constant, N., Amiri, A. M., Yang, Q., & Makodiya, K. (2015). Fog Data: Enhancing Telehealth Big Data Through Fog Computing. In *Proceedings of the ASE bigdata & socialinformatics* (p. 6). ACM.

El-khateeb, S. M. (2019). Sustainable Neighborhoods in Arab world; a Path Towards Sustainability The Case of the Sustainable City in Dubai. *The Academic Research Community Publication, 3*(437), 14.

Elshater, A. (2012). The New Urbanism Principles vs. Urban Design Dimensions towards Behavioral Performance in Egyptian Neighborhood. In *Future Communities: Socio-Cultural & Environmental Challenges* (pp. 826-843). Giza: AicE-Bs- ASIA Pacific International Conference on Environment-Behavior Studies.

Elshater, A. (2013, March). Towards Refined a New Concept of New Urbanism in Egypt. *Asian Journal of Environment-Behavior Studies, 4*(12).

Elshater, A. (2016). The Ten-Minute Neighbourhood Is[not] a Basic Planning Unit for Happiness in Egypt. *Archnet-IJAR, 10*(1), 344–357. doi:10.26687/archnet-ijar.v10i1.878

Jarass, J., & Heinrichs, D. (2014). New Urban Living and Mobility. *Transportation Research Procedia, 1*(1), 142–153. doi:10.1016/j.trpro.2014.07.015

Kumar, P. (2015). Smart Neighbourhood to Enhance Socail Sustainability and Inclusive Planning in Smart Cities. In *National Conference*. Sonipat: Deenbandhu Chottu Ram University of Science and Technology Murthal.

López, G., Moura, P. S., Custodio, V., & Moreno, J. I. (2012). Modeling the Neighborhood Area Networks of the Smart Grid. In *Selected Areas in Communications Symposium* (pp. 3357-3361). IEEE.

Mizooji, K. K., Haghighat, A. T., & Forsati, R. (2012). Data clustering using bee colony optimization. In The Seventh International Multi-Conference on Computing in the Global Information Technology (pp. 189-194). IARIA.

Moulaert, F., Martinelli, F., Swyngedouw, E., & González, S. (2010). *Can Neighbourhoods Save the City?: Community development and social innovation*. Routledge. doi:10.4324/9780203849132

Pahl, M.-O., Niedermayer, H., Kinkelin, H., & Carle, G. (2013). Enabling Sustainable Smart Neighborhoods. *Sustainable Internet and ICT for Sustainability*, 6.

Park, Y., & Rogers, G. O. (2015). Neighborhood Planning Theory, Guidelines, and Research: Can area, population, and boundary guide conceptual framing? *Journal of Planning Literature*, *30*(1), 18–36. doi:10.1177/0885412214549422

Shekhar, S., & Tripathi, A. (2015). Smart Neighbourhood: A way to Sustainable Development. *Annual Conference Proceedings,* 17.

Sinha, U., & Shekhar, M. (2015). Comparison of Various Cloud Simulation tools available in Cloud Computing. *International Journal of Advanced Research in Computer and Communication Engineering*, *4*(3). Advance online publication. doi:10.17148/IJARCCE.2015.4342

Western Australian Planning Commission. (2007). *Liveable Neighbourhoods Community Design Code: A Western Australian Government Sustainable Cities Initiative*. Western Australian Planning Commission.

Zhang, Q., Yung, E. H., & Chan, E. H. (2018). Towards Sustainable Neighborhoods: Challenges and Opportunities for Neighborhood Planning in Transitional Urban China. *Sustainability*, *10*(406), 23. doi:10.3390u10020406

Zou, C., Deng, H., & Qiu, Q. (2013). Design and Implementation of Hybrid Cloud Computing Architecture Based on Cloud Bus. In *9th International Conference on Mobile Ad-hoc and Sensor Networks* (p. 5). IEEE. 10.1109/MSN.2013.72

Chapter 7
Impact of Urban Ambiance on the Productivity of the Built Environment Using Cluster Analysis:
Sustainable Built Environment

Saurav Dixit
https://orcid.org/0000-0002-6959-0008
RICS School of Built Environment, Amity University, Noida, India

ABSTRACT

Productivity is the key measure of success and efficiency of any process or activity, and it is dependent on a number of factors. The ambiance of cities and the use of latest technology and environment also affect the productivity of the residents working in the city. The objective of this chapter is to identify the attributes of urban ambiance from the literature review and to analyse the impact of satisfactory urban/city ambiance on the productivity of the built environment. The methodology adopted for the study is to collect the data using a structured questionnaire survey on the impact of urban ambiance on the productivity form the different cities of the world. The collected data shall be analysed using reliability analysis, relative importance index, exploratory factor analysis, and cluster analysis. The finding of the study shall provide an understanding of urban ambiance on the productivity of the built environment and the ways and means to improve the situation.

INTRODUCTION

The contemporary community is viewed in a multi-space sense with which individuals communicate on a daily and regular basis. These are not just the experiences with the shape that was created but are worlds which are metabolised and transformed in an ethos or "atmosphere" of perception: a vivid combination of time, space and place. In many cases, these perceptive atmospheres are not entirely pleasant

DOI: 10.4018/978-1-7998-3856-2.ch007

and result in negative or unhealthy experiences. This paper discusses the urban worlds with which we connect every day to encourage the development of healthy experiential cities. They should pay special attention to how they personalise our hostile and unsanitary conditions and how we cope with them. The experiential city and our modern urban landscapes are a mixture of changing circumstances, which we know through our livelihoods, working conditions and recreation. The densely populated urban grid of the city blocks and laneways offers an excellent playground for the personal physical and social intelligentsia.

The different scales of the design (form) visually target the economic successes of businesses and private enterprises. Such social environments vary greatly in a person's perception of healthy or unhealthy experiences. Many of our modern urban environments are institutional, sterile and even tough industrialised–they lack healthy spatial and material properties, otherwise available in the natural world. We will define factors that synthesise our balanced modern experiences in order to develop more fitting conditions to improve the quality of our modern experiences. The use of the notion of ambiance enables one to examine normal societies from a sensory perspective and to address social problems from an esthetical viewpoint. To this end, the standard structures of perception need to be extended. The urban life is multi-faceted and relies on a variety of sensory worlds and ways to experience them. This is analogous in all cases to unfamiliarity – breaking experiments – as a tool and as a way to move emphasis, such as field anthropologists in the into urban community, as it includes attempting to dissolute assumptions, an attempt to shake up existing definitions, a certain "sensorial transition" (Rimbaud). We need to underline the value of the creation of an approach that is considered to be a simple and obvious way of dealing with the world. In other words, dealing with urban cultures involves the ability to communicate with and among various forms of being in the world.

In the era of exhausting natural resources, developing worldwide environmental change, corporate social responsibility & demands for ecologically friendly solutions, several organisations are progressively dedicated to developing new ways of producing & operating the building. Talking about new ways of developing a building, the decision is to create & develop green building which is not only the need of the hour but also represents a timely business decision & a good investment. In addition to other things, green buildings are considered more productive & more aesthetic for the occupants. Nowadays, organisation focus on activities that ensure that the land and building asset base of an organisation is optimally structured in the best corporate interest of the organisation concerned, aiming to maximize human interaction with the organization. In the present era, the profession of corporate real estate and workplace encompasses business strategy for companies which is based on three important factors- that is,

1. How and where people work,
2. How it affects people, productivity, profit,
3. And increasingly important - the planet.

BACKGROUND

When we understand how we perceive our physical environment, we have to first and foremost appreciate our environment by primarily using our mind and body. Our social perspective is dependent on several key factors: at a specific time, we live in the world, we create a "sense of place" in our memories and imagination, and our external envelopes encapsulate them. This phenomenological interpretation of our eternal nature illustrates a "body-centred" style of our place in this world. Security is the key concern of every individual irrespective of gender, location or financial status. The sense of security builds dignity and confidence to do better in life.

In order to have developed in the country, it is necessary to fulfil the basic needs of the people like safety, security, protection from hunger, protection from violence, protection from discrimination. However, the truth is that there is no place left where there is no safety issue or no security issue, especially for women. Cities are expanding at a pace now these days due to various government initiatives like smart cities, affordable housing etc. It is anticipated that by 2030, 60% of the world's population would be living in urban areas (Credits, n.d.). A city can grow and progress with the essential condition of safety and security. Countries having more issues of safety and security are more prone to lag behind in the race of gaining the economic edge over other countries (Safety, n.d.). Comparatively women and girls are more prone to violence, sexual harassment, domestic violence, stalking, staring etc.

Urban Ambiance and Built Environment

The importance of the idea of 'environment' in research into architecture and urban planning can be explained in part by the absence of current urban studies, which had been intense in various specialities and which could not adequately deal with a broader view of the urban world. A tool overcomes the ties that make the world clearly explained objects by joining discoveries and defining concepts. This is therefore recognized that the processes of appropriating the urban place go by knowledge of the sensory experience in the city's surroundings. Becoming the body the responsive device that catches the experience of the environment, we might say that this realisation takes place not only when the body penetrates the town atmosphere but particularly when it penetrates our body (Dixit, Mandal, Thanikal, & Saurabh, 2019; Dixit, Sharma, & Singh, 2020; Dixit & Sharma, 2020). The distinctive sounds characterizing the urban areas feel their smells, light, colours, and temperature differences. Going deep, Gilling and Barton (2007) describe the satisfactory ambiance that consists of three pillars: Urban design and planning, social framework and law environment (Figure 1).

The built environment is defined as the human-made things to facilitate and enhance the quality of life by adopting them. For example, parks, roads, buildings, infrastructure, houses, reactors, ships, cars, and so on. Physical environment means the neighbourhood or the infrastructural environment within the city. It comprises of physical things whether it is natural or artificially created by human-like buildings, streets, parks, roads etc. as mentioned in Figure 2 below. These elements collectively may be termed as Built Environment. Some natural phenomenon like weather, physical terrain of the place, rain etc. also have an impact on the built environment (Brisky, 2014), (Table 1).

More specific research questions will be introduced and investigated in both the built environment and urban ambiance . This chapter addresses several questions:

- What is the pre-requisite of adopting a satisfactory ambiance model barriers in the implementation of satisfactory ambiance?
- What is the impact of satisfactory ambiance model on the efficiency, quality of life, working conditions, and the productivity of the cities?

Objective of the present study is to identify the attributes of satisfactory ambiance . A further object is to analysing the attributes using statistical tools. The overall goal of this chapter was to pursueProposing a framework model for the successful implementation of satisfactory ambiance in our built environment.

Figure 1. Satisfactory ambiance
Source: (Gilling & Barton, 2007)

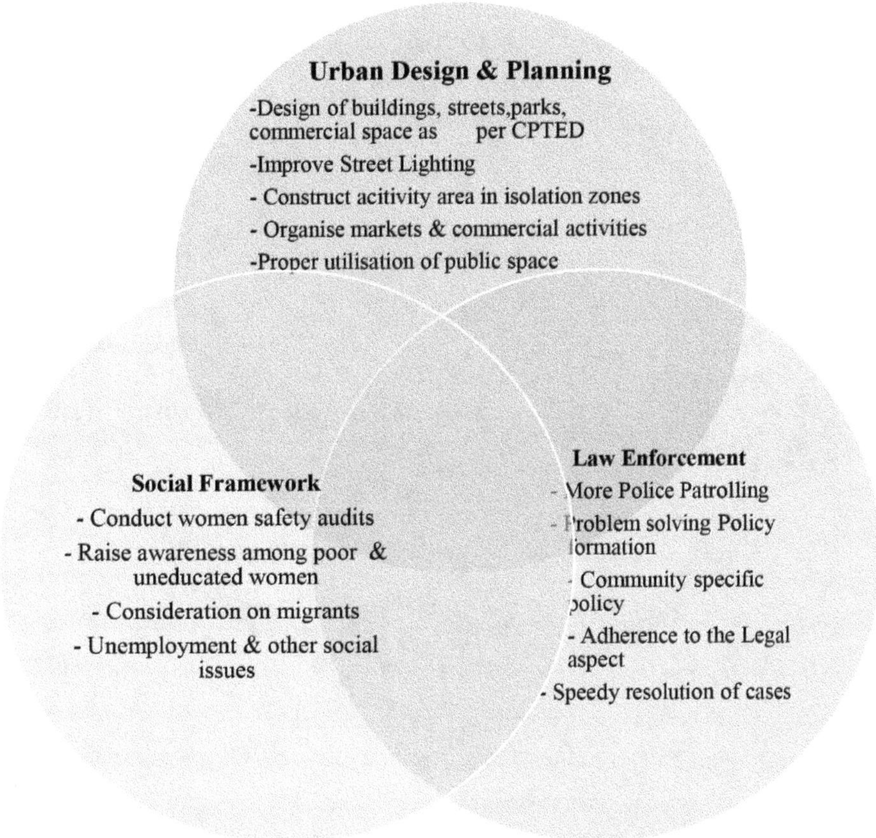

LITERATURE REVIEW

Physical environment means the neighbourhood or the infrastructural environment within the city, as mentioned in Figure 3. Comprises of physical things whether it is natural or artificially created by human-like buildings, streets, parks, roads etc. All these elements collectively may be termed as Built Environment. Some natural phenomenon like weather, physical terrain of the place, rain etc. may also have an impact on the behaviour of women. For example, in an audit, some women expressed weather as a hindrance, while other women do not find it as a barrier (Brisky, 2014).

The physical environment from the perspective of occupants can be perceived on the basis of the following (Brisky, 2014):

- Eyes on the street/road
- People's behaviour
- Weather condition
- Physical conditions of the sidewalk or path
- Commuter's behaviour
- The comfort of having a companion while engaging in physical activity

Figure 2. Satisfactory built environment

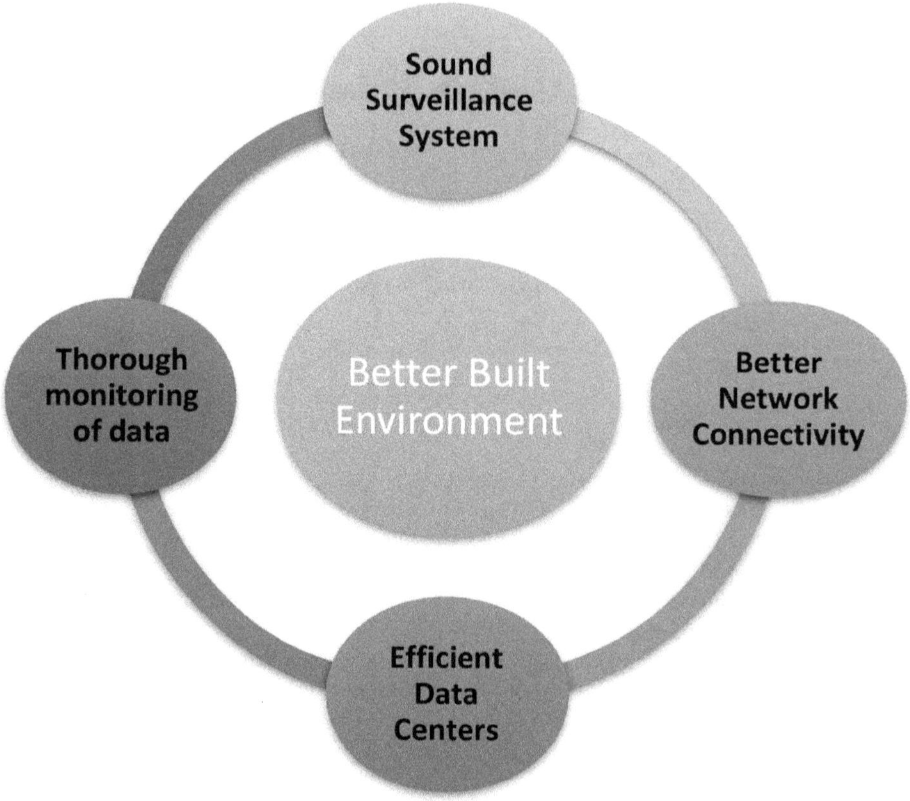

Rapoport (1969) recalls that, when researching cultural connections between people and space, it is less important to understand what people do than to learn how they conduct such activities. Ambiance refers more to the "how" of operation than to the "what" of the movement. This implies, that the atmosphere is not an object of perception: it defines the terms of perception that influence all kinds of behaviour. We may assume, then, that the atmosphere evokes our subjective sense of the collective experience, the consciousness of the part of an urban site whose sensations have their significance expressed by their users. Be sure that all structures have the potential to inspire their inhabitants' actions and interventions. For certain situations, we find the users crowded in a single atmosphere. The streets display few static attitudes, displacement is a significant phase in the correct negotiations to construct the role within the community (Abusaada & Elshater, 2020; Babina & Attaway, 2000). It is not only due to the sensitive characteristics of such environments (type of light, sounds, rhythms, climate and physical configuration) but also because of the presence of a motivating character that affects the occupant and encourages him to share the atmosphere of the region. Ambiance cannot be defined without the experience, and we are tired of the user and the researcher; here a different path opens up that environmental analysis seems to force or strengthen in the area and urban architecture: the urgency of reflection and of constructing research methodologies that are focused on participatory and responsive approaches by subjects and subjectivities of urban environments. Putting a common experience in the context ambiances enable us to shed new light on the issue of being together by facilitating a double dynamico-socialisation of sensory and sensitivity of social life.

Table 1. Parameters of the built environment

Built environment/ambiance factors	Attributes	References
Density	Built shape (height to the building perimeter ratio), Maximum construction area to site area, population density, outdoor thermal management	(Abdel-Galil, 2012; Anciaes, Nascimento, & Silva, 2017; Ercan, 2011; Greed, 2005; Ley & Murphy, 2001; Schaefer, 2012; Trumbull, 2013)
Road	The size and height of the lane, the breadth of the road (special zone of width) to the total area, the area to the site	(Abrego & Schmalzbauer, 2018; Cozens & Van Der Linde, 2015; Hekmat, Morgan, Soltani, & Gough, 2015; McGregor, 2016; G. Tiwari, 2002; P. Tiwari & Gulati, 2013; Week, Process, & Urban, 2013)
Greenspace	Open green area, the density of vegetation, walking area, others	(Abaya Gomez, 2008; Douglas, Lennon, & Scott, 2017; Foster, Wood, Christian, Knuiman, & Giles-Corti, 2013; Sundaresan, 2017; Temelová & Dvořáková, 2012; Wang, Chau, Ng, & Leung, 2016)
Clean environment	Sanitary conditions, number of skips (for garbage collection) to a number of households, average length of skipping from households, daily garbage recovery by the municipality, proximity to nallah or any other dismal services, quality of housing.	(Foster & Giles-Corti, 2008; Prevatt, 1998; Stummvoll, 2009)
Visual character	Glass façade to solid building surface ratio, building materials used in façade, Colour and texture	(Anciaes et al., 2017; Cozens & Van Der Linde, 2015; Erkip, 2005; P. Tiwari & Pandey, 2010)
Convenience	Range to an internet cafe, range from the playschool, bus stop/taxis stand, service store and day-to-day market.	(Duke, 2003; Ellis, 2002; P. Tiwari & Gulati, 2013; Zhan et al., 2018)
Noise and Safety	Feeling of walking safely at night, proximity to noise-generating activities in the vicinity, Average distance of households from the highway and the main road	(Ercan, 2011; Hedayati Marzbali, Abdullah, Ignatius, & Maghsoodi Tilaki, 2016; Timperio, Veitch, & Carver, 2015; Yuen, 2004)
Infrastructure	Basic services such as parks, buildings, recreational facilities, subways, highways, dedicated corridors, transportations.	(Anciaes et al., 2017; Dodman, 2007; Erkip, 2005; Greed, 2005)
Social environment and physical environment	Police, emergency services, disaster management, skill enhancement centres, employment centres	(Andersen, Gustat, & Becker, 2015; Ho & Cheung, 2011; Mason, 2007; McCarthy, 2002; R. Pain, 2000; R. H. Pain, 1997) (Koskela & Pain, 2000; McCarthy, 2002; Md Sakip & Abdullah, 2018; Paydar, Kamani-Fard, & Etminani-Ghasrodashti, 2017; Sevilla-Buitrago, 2013; Stummvoll, 2009; Wang et al., 2016)
Land use form	Greenfield, brownfield, marshy land, FSI, urban hubs.	(Greed, 2005; Niwa, 2007)
Aesthetics/sign of physical disorder	Lightning at night, abandoned area, littered, refused, vacant,	(Erkip, 2005; Foster & Giles-Corti, 2008; Foster et al., 2013; Freestone, 2014; Milan & Creutzig, 2017; Sundaresan, 2017)
Neighbourhood characteristics	Dead ends, blind corners, concealed areas, thick vegetation	
Faulty design	Accessibility, location, surrounding, size, and shape	(Ercan, 2011; Erkip, 2005; Greed, 2005; Mason, 2007)

Figure 3. Physical environment leads to urban ambiance
Source: (Chan et al., 2001)

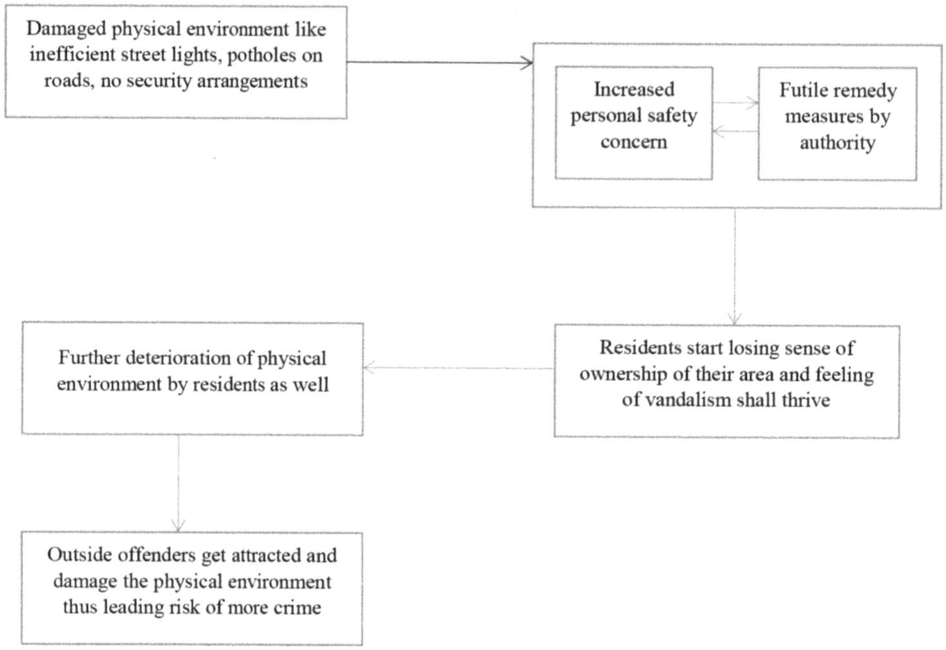

The word "productivity was first used in 1766 when it was first specified in an article by Quesnay" (Vaggi 1987). "Over a century later, in 1883, Littre characterised profitability as the personnel to deliver, that is, the want to create" (Jarkas, 2010). "In the mid-twentieth century, a more apt definition, the relationship amongst produced output and the ways utilized to deliver that output" was created". "In 1950, the Association for European Monetary Participation (OEEC) presented a formal meaning of the Productivity as a remainder acquired by separating yield by one of the creation factors" (Sumanth & Yavuz, 1984). Therefore, "it ended up conceivable to talk about the Productivity of investment, or crude materials as per whether the Output is being considered in connection to wealth, speculation, or basic materials" (Sumanth & Yavuz, 1984). Contingent upon the valuation destinations and the accessibility of information, a few profitability definitions remain experienced.

Rohini Pande (Padte, 2013), in a developing country like India, development is interlinked with the economic activities of men and women. In recent times women are also equally part of economic development as men are. This needs to be understood that the foundational idea in development economics is interlinked. For example, construction of a commercial firm is equally important as the construction of approach road towards it. Development is only possible if it is unbiased, cohesive and is done in a collaborative manner. Urban areas have always been looked like the places of freedom, new ideas, a better lifestyle, better facilities, and a better environment for self-development. In spite of all such natural expectations of people about cities, this shining perspective on the city does not apply to all in the same way.

Accordingly, the attributes of satisfactory ambiance are:

1. Crowd
2. Guardianship

3. Eyes on the street/road
4. Littered
5. Visibility
6. Weather condition
7. People's behaviour
8. Security
9. Thick vegetation
10. Surveillance
11. Openness
12. Availability of public transport
13. Neglected
14. Commuter's behaviour
15. The comfort of having a companion while engaging in physical activity
16. Lighting
17. Abandoned
18. Dead ends
19. Physical conditions of the sidewalk or path
20. Gender diversity

RESEARCH METHODOLOGY

The research methodology provides a way of consistently addressing the research issue (Kotari 2005, p.8) and aims to organize the research process through logical steps by appropriate steps. It helps to understand the information necessary to resolve the problem or accomplish the research goals. The key topics in this study are to analyse the impact of urban ambiance on the productivity of the built environment using cluster analysis. For the current research project questionnaire data collection approach is adopted. A structured questionnaire survey is used to collect the data for the study. And the received responses were analysed using quantitative research technique. The analysis for a complex topic is very difficult to develop and requires a lot of imagination. The goal was for research variables to be routinely categorized, theoretical structure identified, a theory formulated, and research methods developed for this reason.

A conceptual structure helps organize the research process into logical steps and appropriate phases. The creation of an action plan supports guiding and directing research to ensure that there is a clear connection between all phases, namely the formulation of research objectives and aims, the examination of literature, data collection and analysis during the study. The primary data shall be collected from the respondents covering the different parts of the world using a structured questionnaire survey, to know the impact of satisfactory ambiance on the productivity of the built environment. The population selected for the study is through-out USA, Canada, Brazil, and the UK, Europe, Asia, and the Middle East countries.

Cronbach's Alpha

The confidence of the research instrument was evaluated by the alpha (to) coefficient of Cronbach. The alpha of Cronbach is the consistency measure related to the variance of the true value of the' underlying structure.' The determined theoretical parameter is' A' Construction' (Hatcher in 1994 stated in Santos in 1999). The alpha coefficient varies from 0 to 1 and can be used to characterize accuracy in questionnaires or scales (i.e. ranking scale) based on variables derived from dichotomous factors. As the score increases, it is more reliable (Santos 1999). For the current study, the value of reliability analysis is 0.89, which is considered as highly reliable data (Table 2).

Table 2. Reliability analysis values of all the factors: Cronbach's alpha

Attributes	Cronbach's alpha
All attributes selected for the study	0.89
Social environment and physical environment	0.75
Productive and sustainable built environment	0.75
Convenience	0.74
Faulty design	0.88
Aesthetics	0.81

A formal survey of 800 respondents has been proposed and 315 correct and valid responses have been obtained at a response rate of 40 per cent. The respondents were asked to evaluate their responses to the impact of the urban environment on the productivity of the built environment on the Likert scale from 1 to 5. The rating of the Likert scale was described below: the interviewees had to measure the questions using five-point scale ' 5' which was the extremely high,' 4' medium,' 3' moderate,' 2' very low,' 1' quite low impact. The questionnaire header includes useful study information, and then some basic information about the study to the respondent. The last part of the questionnaire is having a section to capture any feedback of suggestion about the study as well as about the questionnaire. The questions are structured to be simple and easily understood by the respondents. The answers received by the interviewees have been summarised in an Excel datasheet and the data has been analysed using SPSS 23 software.

The answers received from the respondents were summarised in an Excel datasheet and the data analysed with SPSS software. The relative importance index (RII) is carried out to determine the priority of the major factors and the reliability analyses to check the consistency of the received data are followed.

$$Rii = \frac{\sum_{r=1}^{5} r * n_r}{5N} \quad (1)$$

R is the rating for a specific building cost-influencing element on the Likert scale (1-5), N is the number of respondents who have a specific Likert rating scale, N the total number of respondents to a specific issue

KEY FINDINGS

RII Values

Most significant attributes of urban ambiance are impacting the productivity of the built environment. The most significant attributes in urban ambiance have an impact on the productivity of the built environment where people's behaviour, eyes on the street/road, openness, security, and availability of public transport having an RII score of 81.4%, 81.1%, 80.2%, 79.7%, and 78.8% respectively (Table 3). The least significant attributes of urban ambiance impacting the productivity of the built environment. The least significant attributes of Urban Ambiance having an Impact on the Productivity of the built environment were neglected, surveillance, dead ends, thick vegetation, and crowd having an RII score of 75.6%, 75.2%, 75.0%, 75.0%, and 74.9% respectively (Table 3).

Table 3. Relative importance index ranking of attributes

Sr. No.	Attributes	No of responses	Score	RII value
1	People's behaviour	315	1282	81.4%
2	Eyes on the street/road	315	1277	81.1%
3	Openness	315	1263	80.2%
4	Security	315	1255	79.7%
5	Availability of public transport	315	1241	78.8%
6	Gender diversity	315	1238	78.6%
7	Commuter's behaviour	315	1235	78.4%
8	The comfort of having a companion while engaging in physical activity	315	1234	78.3%
9	Visibility	315	1234	78.3%
10	Lighting	315	1234	78.3%
11	Abandoned	315	1221	77.5%
12	Weather condition	315	1207	76.6%
13	Physical conditions of the sidewalk or path	315	1204	76.4%
14	Littered	315	1204	76.4%
15	Guardianship	315	1199	76.1%
16	Neglected	315	1191	75.6%
17	Surveillance	315	1185	75.2%
18	Dead ends	315	1182	75.0%
19	Thick vegetation	315	1182	75.0%
20	Crowd	315	1180	74.9%

Factor Analysis

Factor analysis enables us to reduce the number of data measurements and draw up a table based on the variances explained by the structures/factors and the load factor of the different factor attributes. The current study considered attributes with a variable load equal to and above 0.4[8]. The statistical factor narrowed 20 characteristics to 5 variables, describing a total variance of 56 per cent (Table 4) and (Figure 4).

Table 4. Factor analysis

Factor/Attribute	Factor loading	%age of variance explained
Social environment and physical environment		17%
People's behaviour	0.68	
Availability of public transport	0.88	
Gender diversity	0.83	
Commuter's behaviour	0.70	
The comfort of having a companion while engaging in physical activity	0.51	
Physical conditions of the sidewalk or path	0.60	
Eyes on the street/road	0.76	
Productive and sustainable built environment		12.4%
Openness	0.57	
Surveillance	0.49	
Visibility	0.65	
Guardianship	0.51	
Security	0.48	
Convenience		10.30%
Abandoned	0.76	
Weather condition	0.64	
Neglected	0.57	
Crowd	0.54	
Faulty design		9.6%
Dead ends	0.52	
Thick vegetation	0.65	
Aesthetics		6.5%
Littered	0.58	
Lighting	0.56	
Total variance explained		56%

Figure 4. Productivity and sustainable built environment

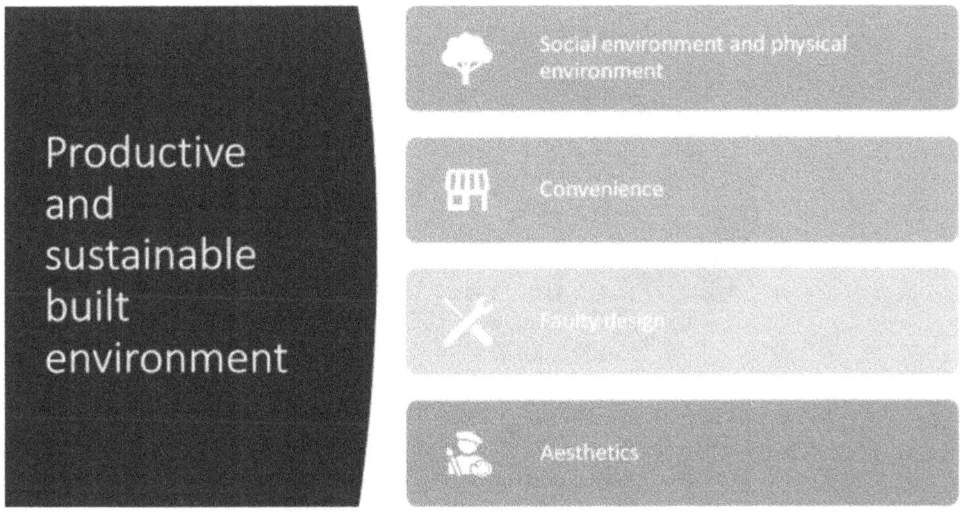

Cluster Analysis

Cluster analysis is an exploratory technique used for its partitioning capacity in various fields of management. The mechanisms in the software were defined by means of exploratory data analysis. It used to group the data according to inherent group homogeneity. For data analysis and testing using SPSS 20, two-stage cluster analysis was used. Two-stage cluster analysis allows both the categorical and ongoing variables to be captured. For the cluster analysis in groups, the attributes with significant RII value (i.e. above 0.78) have been considered. Table 2 lists the attributes selected for the study. Commuter's behaviour is the most significant predictor and shows an impact value of 100% on all the remaining nine attributes, followed by lightning, gender diversity, openness, and security. The first five attributes having an impact value above 0.4 on the predictor importance scale (see Figure 7). Two-step cluster analysis formed three major clusters sharing of 42%, 32.1%, and 26%, respectively (Figure 5, Figure 6, Figure 7, and Figure 8).

Figure 5. Cluster analysis model summary

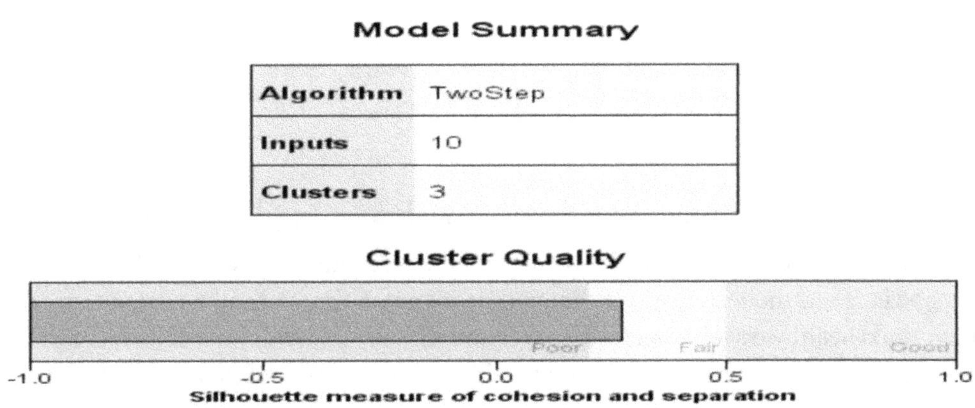

Figure 6. cluster analysis cluster sizes

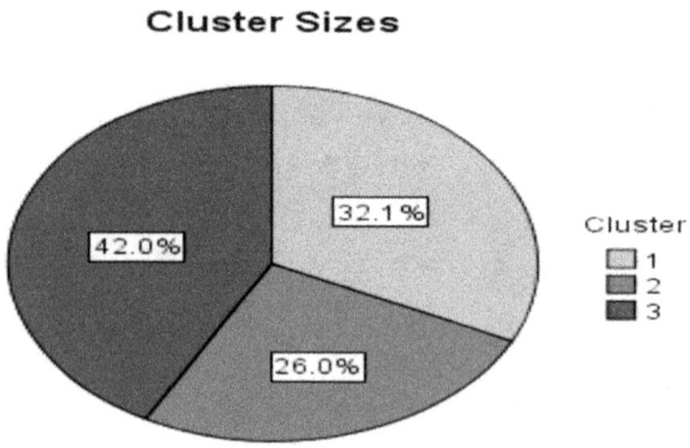

Size of Smallest Cluster	34 (26%)
Size of Largest Cluster	55 (42%)
Ratio of Sizes: Largest Cluster to Smallest Cluster	1.62

Figure 7. Cluster analysis predictor importance

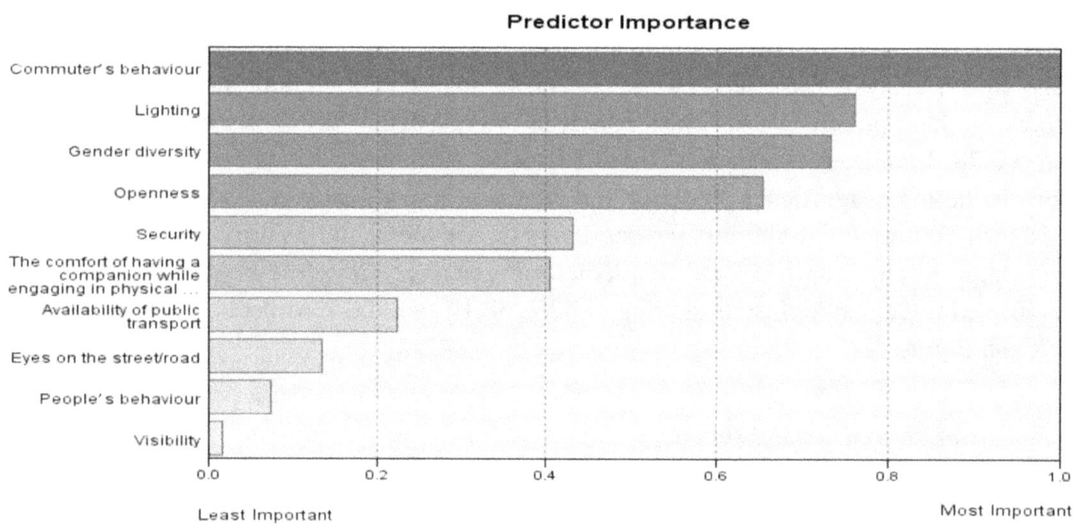

DISCUSSION AND CONCLUSION

As a species, we certainly value healthy experiences, whether they are manifested as a spiritual walk in the forest or a stimulating public market. However, not all perception frames are ideal experiences. The speed and business of our modern lives are dictated by many of our stressful environments. We go from A to B with minimal healthy travel experiences. We pour into dense towns on a daily basis through overly congested transport modes. The culprits of an even bigger economic polarity that determines "the ner-

Figure 8. Cluster wise predictor distribution

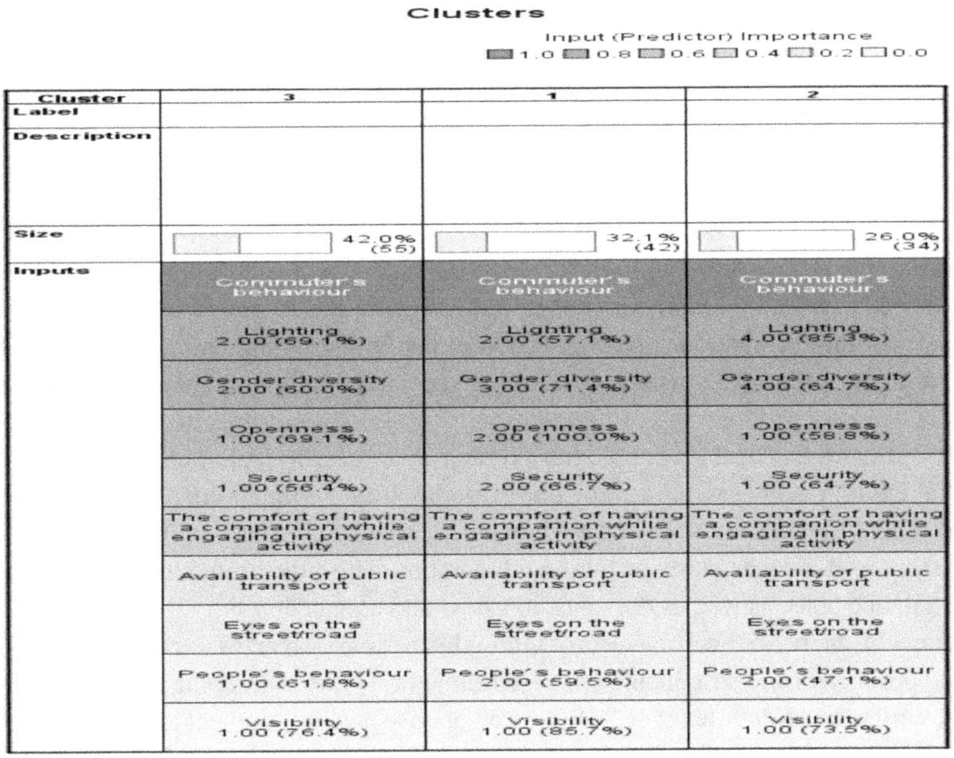

vous energy of a moving society"–are crowded trains, buses, and highways (Greenburg, 2011). Modern advertising strategies further assault the senses in any direction possible. Artificial lighting turns the "no end of the city" into a fluoro-century of over-stimulation. The obsession with online status, the emails and our need to be updated further entangles and alienates us from our technology. This description allows us to lose interest in our world both in the structure we live in and in the social connections we create.

There are physical and psychological elements that can help us relieve ourselves from our modern urban issues. These approaches are emerging from research into our modern hospitals and healing environments. These environments are extremely fragile and designed around the human inhabitant in space–environments that have extreme stress and anxiety that can ultimately lead to death. The field of environmental psychology shows the positive effect that healthy natural phenomena can have on the overall well-being of the family. This area, known as "evidence-based design," demonstrates that we can alleviate real, physical and psychological stressors by integrating nature into our environment. Through utilizing "good stimuli" as constructive sources of natural stimulus, physicians with therapeutic drugs have been able to cure complex operations quicker, minimize the need for treatment to change pain as well as alleviate elevated stress and anxiety rates.

Positive distractions can offer restorative healing in the design of our built environments. However, creating a multi-sensory experience in space is not so easy as designing from a catalogue of scientific components. A formulaic ingredient-based methodology approach architecture may not always strike an emotional chord with its audience. Subjective qualities of space, which are at the heart of the phenomenological architectural predictions, permit a masterful composition of the environment. A sense of spiritual and intimate connection to space must occur on a personal, subjective and objective level. Environmental studies focused, for example, on healthy attributes of natural light but managed to avoid the eternal relativity light with the shadow. The splendours of natural light cannot be appreciated without the situational filter that darkness encourages. Subjective architectural meanings have been used over time to foster the personal knowledge of an object in space.

The planners and architects for our sustainable contemporary environments are increasingly responsible for creating more human-driven spaces; spaces that eventually improve our rate of urban wellness experiences. However, the economic forces that shape our city counteract these thinking approaches. Development intensification overwhelms us into less than ideal living and working environments. Technology speeds our lives and through our connection with both the natural world and our relationships. To compensate for these conditions, the conception of the built environment must be seen through the lens of human health – especially with a view to a person's daily healthy space experiences. This can be achieved by incorporating a large range of "good stimuli" subjective and objective types. This helps reduce environmental stresses and enables people to find their own specific spatial balance. The development of new spatial environments can thus allow our everyday lives a vibrant encounter with the shape we have created. The different senses are engaged in and synthesised further into multisensory sensations, which arise from well-balanced architectural compositions. This ultimately increases our awareness of our existence in one place as well as our consciousness and connection.

The findings of the factor analysis clearly highlight the importance of socio and physical environment, and the productive and sustainable built environment. Do we use coping mechanisms to navigate our more stressful modern environments? Do we personalise our environment to suit our needs? What is an atmosphere, if anything? Until we begin to understand how for purposes of fostering better-designed environments, we metabolise our external settings, we need to demonstrate the resources and mechanisms we use to identify them.

REFERENCES

Abaya Gomez, J. E. (2008). Waterfront design without policy? The actual uses of Manila's Baywalk. *Cities (London, England)*, *25*(2), 86–106. doi:10.1016/j.cities.2008.03.001

Abdel-Galil, R. E. S. (2012). Desert reclamation, a management system for sustainable urban expansion. *Progress in Planning*, *78*(4), 151–206. doi:10.1016/j.progress.2012.04.003

Abrego, L. J., & Schmalzbauer, L. (2018). Illegality, motherhood, and place: Undocumented Latinas making meaning and negotiating daily life. *Women's Studies International Forum*, *67*(March), 10–17. doi:10.1016/j.wsif.2017.12.004

Abusaada, H., & Elshater, A. (2020). Effects of urban atmospheres on changing attitudes of crowded public places: An Action Plan. *International Journal of Community Well-Being*, *3*(2), 109–159. doi:10.100742413-019-00042-w

Anciaes, P. R., Nascimento, J., & Silva, S. (2017). The distribution of walkability in an African city: Praia, Cabo Verde. *Cities (London, England)*, *67*(April), 9–20. doi:10.1016/j.cities.2017.04.008

Andersen, L., Gustat, J., & Becker, A. B. (2015). The Relationship Between the Social Environment and Lifestyle-Related Physical Activity in a Low-Income African American Inner-City Southern Neighborhood. *Journal of Community Health*, *40*(5), 967–974. doi:10.100710900-015-0019-z PMID:25808676

Babina, B. J., & Attaway, J. S. (2000). Atmospheric affect as a tool for creating value and gaining share of customer. *Journal of Business Research*, *49*(2), 91–99.

Brisky, B. A. (2014). *Physical environment and women's physical activity behaviors*. ProQuest Dissertations and Theses.

Cozens, P., & Van Der Linde, T. (2015). Perceptions of Crime Prevention Through Environmental Design (CPTED) at Australian Railway Stations. *Journal of Public Transportation*, *18*(4), 73–92. doi:10.5038/2375-0901.18.4.5

Credits, P. (n.d.). *Safe Cities for Women and Girls*. Academic Press.

Dixit, S., Mandal, S. N., Thanikal, J. V, & Saurabh, K. (2019). *Study of Significant Factors Affecting Construction Productivity Using Relative Importance Index in Indian Construction Industry*. Academic Press.

Dixit, S., & Sharma, K. (2020). An Empirical Study of Major Factors Affecting Productivity of Construction Projects. In K. G. Babu, H. S. Rao, & Y. Amarnath (Eds.), *Emerging Trends in Civil Engineering* (pp. 121–129). Springer Singapore.

Dixit, S., Sharma, K., & Singh, S. (2020). Identifying and Analysing Key Factors Associated with Risks in Construction Projects. In K. G. Babu, H. S. Rao, & Y. Amarnath (Eds.), *Emerging Trends in Civil Engineering* (pp. 25–32). Springer Singapore.

Dodman, D. (2007). Post-independence optimism and the legacy of waterfront redevelopment in Kingston, Jamaica. *Cities (London, England)*, *24*(4), 273–284. doi:10.1016/j.cities.2006.09.001

Douglas, O., Lennon, M., & Scott, M. (2017). Green space benefits for health and well-being: A life-course approach for urban planning, design and management. *Cities (London, England)*, *66*, 53–62. doi:10.1016/j.cities.2017.03.011

Duke, D. (2003). Alzira Rufino's a casa de cultara de Mulher Negra as a form of female empowerment: A look at the dynamics of a black women's organizations in Brazil today. *Women's Studies International Forum*, *26*(4), 357–368. doi:10.1016/S0277-5395(03)00080-3

Ellis, C. (2002). The {New} {Urbanism}: {Critiques} and {Rebuttals}. *Journal of Urban Design*, *7*(3), 261–291. doi:10.1080/135748002200003933

Ercan, M. A. (2011). Developing sustainable communities in historic neighbourhoods of istanbul. *Tijdschrift voor Economische en Sociale Geografie*, *102*(2), 205–219. doi:10.1111/j.1467-9663.2010.00625.x

Erkip, F. (2005). The rise of the shopping mall in Turkey: The use and appeal of a mall in Ankara. *Cities (London, England)*, *22*(2), 89–108. doi:10.1016/j.cities.2004.10.001

Foster, S., & Giles-Corti, B. (2008). The built environment, neighborhood crime and constrained physical activity: An exploration of inconsistent findings. *Preventive Medicine*, *47*(3), 241–251. doi:10.1016/j.ypmed.2008.03.017 PMID:18499242

Foster, S., Wood, L., Christian, H., Knuiman, M., & Giles-Corti, B. (2013). Planning safer suburbs: Do changes in the built environment influence residents' perceptions of crime risk? *Social Science & Medicine*, *97*, 87–94. doi:10.1016/j.socscimed.2013.08.010 PMID:24161093

Freestone, R. (2014). Progress in Australian planning history: Traditions, themes and transformations. *Progress in Planning*, *91*, 1–29. doi:10.1016/j.progress.2013.03.005

Gilling, D., & Barton, A. (2007). Crime prevention and community safety. In Critical Social Policy (Vol. 17). doi:10.1177/026101839701705204

Greed, C. (2005). An investigation of the effectiveness of gender mainstreaming as a means of integrating the needs of women and men into spatial planning in the United Kingdom. *Progress in Planning*, *64*(4), 241–321. doi:10.1016/j.progress.2005.08.003

Hedayati Marzbali, M., Abdullah, A., Ignatius, J., & Maghsoodi Tilaki, M. J. (2016). Examining the effects of crime prevention through environmental design (CPTED) on Residential Burglary. *International Journal of Law, Crime and Justice*, *46*, 86–102. doi:10.1016/j.ijlcj.2016.04.001

Hekmat, S., Morgan, K., Soltani, M., & Gough, R. (2015). Sensory evaluation of locally-grown fruit purees and inulin fibre on probiotic yogurt in Mwanza, Tanzania and the microbial analysis of probiotic yogurt fortified with Moringa oleifera. *Journal of Health, Population and Nutrition*, *33*(1), 60–67. doi:10.1016/j.jenvp.2011.09.004 PMID:25995722

Ho, W. C., & Cheung, C. K. (2011). Social sustainability for mothers in Hong Kong's low-income communities. *Habitat International*, *35*(4), 529–536. doi:10.1016/j.habitatint.2011.03.001

Koskela, H., & Pain, R. (2000). Revisiting fear and place: Women's fear of attack and the built environment. *Geoforum*, *31*(2), 269–280. doi:10.1016/S0016-7185(99)00033-0

Ley, D., & Murphy, P. (2001). Immigration in gateway cities: Sydney and Vancouver in comparative perspective. *Progress in Planning*, *55*(3), 119–194. doi:10.1016/S0305-9006(00)00025-8

Mason, R. (2007). Building women's social citizenship: A five-point framework to conceptualise the work of women-specific services in rural Australia. *Women's Studies International Forum*, *30*(4), 299–312. doi:10.1016/j.wsif.2007.05.007

McCarthy, J. (2002). Entertainment-led regeneration: The case of Detroit. *Cities (London, England)*, *19*(2), 105–111. doi:10.1016/S0264-2751(02)00005-7

McGregor, K. (2016). Emotions and activism for former so-called 'comfort women' of the Japanese Occupation of the Netherlands East Indies. *Women's Studies International Forum*, *54*, 67–78. doi:10.1016/j.wsif.2015.11.002

Md Sakip, S. R., & Abdullah, A. (2018). An Evaluation of Crime Prevention through Environmental Design (CPTED) Measures in a Gated Residential Area: A Pilot Survey. *Asian Journal of Environment-Behaviour Studies*, *3*(6), 21. doi:10.21834/aje-bs.v3i6.232

Milan, B. F., & Creutzig, F. (2017). Lifting peripheral fortunes: Upgrading transit improves spatial, income and gender equity in Medellin. *Cities (London, England)*, *70*(February), 122–134. doi:10.1016/j.cities.2017.07.019

Niwa, H. (2007). Article. *Development*, *134*(4), 635–646. PMID:17215298

Padte, R. K. (2013, June 29). Keeping Women Safe? *Economic & Political Weekly, Vol*, *46*, 1–7.

Pain, R. (2000). Place, social relations and the fear of crime: A review. *Progress in Human Geography*, *24*(3), 365–387. doi:10.1191/030913200701540474

Pain, R. H. (1997). Social geographies of women's fear of crime. *Trans Inst Br Geogr NS*, *22*(2), 231–244. doi:10.1109/PRIME.2014.6872711

Paydar, M., Kamani-Fard, A., & Etminani-Ghasrodashti, R. (2017). Perceived security of women in relation to their path choice toward sustainable neighborhood in Santiago, Chile. *Cities (London, England)*, *60*, 289–300. doi:10.1016/j.cities.2016.10.002

Prevatt, J. (1998). *Crime Prevention Through Environmental Design (CPTED) and the role of facilities planning in force protection*. Academic Press.

Safety, P. (n.d.). Article. *Public Safety*, 162–189.

Schaefer, D. R. (2012). Youth co-offending networks: An investigation of social and spatial effects. *Social Networks*, *34*(1), 141–149. doi:10.1016/j.socnet.2011.02.001

Sevilla-Buitrago, A. (2013). Debating contemporary urban conflicts: A survey of selected scholars. *Cities (London, England)*, *31*, 454–468. doi:10.1016/j.cities.2012.08.006

Stummvoll, G. (2009). Environmental criminology and crime analysis. *Crime Prevention and Community Safety*, *11*(2), 144–146. doi:10.1057/cpcs.2008.22

Sundaresan, J. (2017). Urban planning in vernacular governance: Land use planning and violations in Bangalore, India. *Progress in Planning*, *0–1*(October). Advance online publication. doi:10.1016/j.progress.2017.10.001

Temelová, J., & Dvořáková, N. (2012). Residential satisfaction of elderly in the city centre: The case of revitalizing neighbourhoods in Prague. *Cities (London, England)*, *29*(5), 310–317. doi:10.1016/j.cities.2011.11.015

Timperio, A., Veitch, J., & Carver, A. (2015). Safety in numbers: Does perceived safety mediate associations between the neighborhood social environment and physical activity among women living in disadvantaged neighborhoods? *Preventive Medicine*, *74*, 49–54. doi:10.1016/j.ypmed.2015.02.012 PMID:25735603

Tiwari, G. (2002). Urban Transport Priorities. *Cities (London, England)*, *19*(2), 95–103. doi:10.1016/S0264-2751(02)00004-5

Tiwari, P., & Gulati, M. (2013). An analysis of trends in passenger and freight transport energy consumption in India. *Research in Transportation Economics*, *38*(1), 84–90. doi:10.1016/j.retrec.2012.05.003

Trumbull, N. S. (2013). St. Petersburg, Russian Federation. *Cities (London, England)*, *31*, 469–490. doi:10.1016/j.cities.2011.05.003

Wang, Y., Chau, C. K., Ng, W. Y., & Leung, T. M. (2016). A review on the effects of physical built environment attributes on enhancing walking and cycling activity levels within residential neighborhoods. *Cities (London, England)*, *50*, 1–15. doi:10.1016/j.cities.2015.08.004

Yuen, B. (2004). Safety and dwelling in Singapore. *Cities (London, England)*, *21*(1), 19–28. doi:10.1016/j.cities.2003.10.005

Zhan, D., Kwan, M. P., Zhang, W., Fan, J., Yu, J., & Dang, Y. (2018). Assessment and determinants of satisfaction with urban livability in China. *Cities (London, England)*, *79*(March), 92–101. doi:10.1016/j.cities.2018.02.025

Chapter 8
Theorizing the Critical Placemaking:
Impact of Creative Ambiance on Social Activities

Aya Elgobashi
Arab Academy for Science, Technology, and Maritime Transport, Egypt

Yasmeen Mohamed Elsemary
Arab Academy for Science, Technology, and Maritime Transport, Egypt

ABSTRACT

All over the world, lots of challenges are affecting the urban development of spaces in the cities, especially spaces attached to the heritage site. It provides various ambiances that touch the senses of the users. Although the attached public space to heritage sites is full of rich, deep knowledge and plays an active role in the community, it was neglected. Those spaces contribute to the entrenchment of culture and social heritage of the community to let future generations get benefit from it. In this chapter, the researchers theorize the "critical placemaking" with creative ambiance as a tool for successful attached to public spaces. This has been drawn from key concepts in environmental psychology of the users such as physical environment, atmosphere, and services environment. The proposed framework seeks to theorize the processes of placemaking and its key concepts and how it helps in the preservation process. The argument that the placemaking process needs a creative ambiance, which contributes to transforming community narratives and that space, becomes more inclusive.

INTRODUCTION

Recently, many heritage and attraction sites have undergone a new type of reformation as a step for the conservation and protection of heritage. In the past decade, a lot of heritage attraction was transformed into a place to be an educational center, more beneficial, and attractive for the community (Schwartz & Cook, 2002). Heritage sites can be a motivation for visitors if transformed into educational elements. It

DOI: 10.4018/978-1-7998-3856-2.ch008

was applied worldwide and successfully to a lot of the heritage sites (Poria et al., 2003). In this study, the researcher states a literature review study which is mainly based on theorizing ideal and creative experiences for the user by creating the right atmosphere in which to view the display, exhibit, or activities for attraction in the open space with the same identity of the historic building. Users have the option to enter the historical buildings or know about the historical monuments from the activities, and exhibit area surrounded the historical building or heritage site to give them more varieties with a different atmosphere.

This chapter will first go through the importance of historic preservation and its types. Second, it will examine the role of physical environment/atmospherics within heritage sites and how it will have a significant impact on the social and perception of the users. The main aim here is to explore how users may feel about historical preservation among the historical buildings and open spaces with the same ambiance and atmosphere. Third, the study will seek for framework for the placemaking with a creative ambiance to make interrelation between the existing atmosphere and by adding a creative ambiance that will make those spaces more attractive to the users and know the overall attitude of the user's perception toward the new attraction. Finally, the fourth contribution to the theorizing of placemaking with creative ambiance seeks to assess the relationship between the new attraction with unique creative ambiance and its impact of social dimensions on the users.

Pieces of research defines positive and negative behavior in two basic ways: approach and avoidance (Bakker et al. 2014). In marketing, the researcher found that there is a linkage between the surrounding atmosphere and the user's attitude to buying a particular brand for an example. These marketing practices should be achieved by successful historical preservation, even though not only the atmosphere can achieve this. These practices need a creative ambiance to be more attractive and satisfactory for the users. (Mary J Bitner & Obermiller, 1984).Ultimately, the exterior and interior designs of heritage sites should have the same weight of importance and become fundamental in shaping the attraction's identity as well as its patronage. The above discussion highlights the benefits of having more enticing space. The following section will highlight the historic preservation and its importance, and then the chapter will go through discussion, which explores the specifics of a pleasing physical environment and atmosphere within open spaces in heritage sites toward placemaking with creative ambiance and its impact on the social life of the user's customers.

BACKGROUND

History is the backbone of every societal development, and historic buildings and spaces attached are an integral part of history (Heintzelman & Altieri, 2013). Historic preservation is to sustain the process to get the place to be the same as it was in the past to let the future generations get a lot of benefit from it in the form of historical and cultural resources to include elements from the built environment such as buildings, structures, objects, and districts; landscape features, including significant trees and plantings, hardscape, fountains, lighting, sculptures, signs and other natural or designed features (Heintzelman & Altieri, 2013).

Sense of place, shaping its meaning and collective identity are behind a significant dynamics power called preservation (Adams et al., 2001) One of the vital preservation experts' decisions concerning notes that if the historic preservation is done in an uncritical way, it will freeze the sense of place creating a warped nostalgia (Farrar, 2011). An argument such as Farrar's also concluded that the places did not reflect history or period only although places are flowed by a lot of multiple temporalities; for example, understanding the value of the place, which is the historical background to our present life (Farrar, 2011). Those arguments of Farrar were close enough to the argument of Lynch in livable and memorable cities which were evidence of time, helping us to introduce our history to our present life, which will affect our culture and life (Lynch, 2005).

The importance of preservation is not only to see the place as things, it should also be as a process to make relationships and practice with the historical building in heritage sites and that is delineated on walking tour maps (Farrar, 2011). If historic preservation is applied as a process to make relationships and practice, it would help a city become smarter because of its history, a unique building with it all activities that would help the user's perception to be unique (Angelidou et al., 2017).

Five senses of quality communities are reinforced from the preservation of the historic built environment which are, sense of place, sense of identity, sense of evolution, sense of ownership, and sense of community (Iliyasu, 2014). Angelidou took the argument in a different direction which is a comparison of those sensitive buildings to the homogeneous skyscrapers in many large cities. We find that those senses have been lost, and this affects the perception of the users (Angelidou et al., 2017). Moreover, in Feliden's (2005) argument, he observed that preservation where values are attached to the cultural property, and which are emotional values (wonder, identity, continuity, spiritual and symbolic), Cultural values (documentary, historical, archaeological age, and scarcity, aesthetic and symbolic, architectural, townscape, landscape and ecological, technological and scientific), and Use values (functional, economic, social, political and ethnic) (Feilden, 2005). Preservation involves five necessary steps to reach success: Identify, Investigate, Develop, Execute, and Educate.

Frequent consultation with a variety of organizations and close collaboration among technical specialists, architects, owner/occupants, and preservation professionals together can be a successful method for Preservation (The WBDG Historic Preservation Subcommittee 2019). Before starting in any preservation process of any historical element, the preservationist should understand its type well to know how to work with it. The following discussion includes an explanation of the types of historic preservation, its approaches and criteria for the chosen specific type in this chapter.

Historic Preservation and Sense of Place

Beyond history, historic preservation has moved in a very real sense. Historic preservation is no longer preserving the past only. It is widely understood as rapid change and how to manage it (Hyden & Sheckels, 2016) as well as deterioration of the old building or sites and how to make people aware of what they are allowed to do and what not to do (Meeks & Murphy, 2016). A lot of awareness was done for the American people to know and realize that when they cut the craft of the historic buildup environment, they will lose a lot of meaning as those places form their identity and community, so the historic preservation process will be as the past, which is the foundation of our future (Meeks & Murphy, 2016). Therefore, before starting preservation, types of historical elements should be clear to know how to preserve each type whether if preservation is element of structure, its human-made structure or standalone structure, for example, houses, monuments, commercials that have some historical background (Community toolbox 2019), or if what preservation has done for neighborhoods with historical buildings dating from centuries and that still function as residences and shops.

The most important example in the Egyptian context is Islamic Cairo. Islamic Cairo is the name commonly given to the historical neighborhoods in Cairo, which is a part of the city remarkably different from the downtown of Cairo (Yeomans, 2006). If preservation is done for Landscapes features, the historical value is not only in buildings or monuments; it can be in the landscape in historical sites. It is scenes of a turning point in history with a lot of painting colors, patterns, and valuable items. Historic preservation is not commonly referring to the building structure and landscape; it mostly refers to the physical places which have a lot of cultural heritage aspects that focus on the physical and environmental impact of the users of historical places. According to The WBDG Historic Preservation Subcommittee argument, the main cumulative effect changes over time, so while attempting preservation, one of the main concerns of the designers and preservationists is the integrity of a historic building with the surrounding area. Then they start to set up a major guideline and design goals while preserving.

Nowadays, while updating the historic building structures, it requires integrity between the historical background of the building and adding new technologies. The historical atmosphere remains clear and the same as it was. In updating the building, it requires sensitive creativity to redesign the spaces by using materials that match with the general historical atmosphere and codes of the city (The WBDG Historic Preservation Subcommitte, 2019). While adding new functions, activities, and events to the historic building, which are indoor or outdoor, a new level of protection should be added according to those functions. While attempting preservation of the building, the preservationists should address that user's safety and security as the main issue in the success of preservation of those historic sites or buildings (The WBDG Historic Preservation Subcommittee, 2019).

Recently, according to the rapidly increasing urban population, which demands a lot of urban settlement so accessibility around the historic site can change, it affects sites negatively. The WBDG Historic Preservation Subcommittee recommends achieving preservation goals needs, clear accessibility should be provided to users and users with disabilities too.

There is a lot of impact and expanding in the cites, so we lost a lot of historical sites such as streetscape, heritage buildings, and even micro-culture of local public space. A lot of urban challenges are going on to overcome this, and it is essential to re-conceptualize the relationship between urban development and historic preservation. They are instead dividing every element in the heritage site. Preservationist should seek to explore how all of this complicated relationship operates and interacts to identify opportunities for design interventions (Zhong, 2012).

In preservation, most of the historical buildings surround landscape or open spaces that are neglected while it is an integral part of a historic property (The WBDG Historic Preservation Subcommittee 2019). Those surround spaces have great cultural resources to the community because they are places for social interaction of the user (The WBDG Historic Preservation Subcommittee 2019).Preservation of heritage faces a lot of unique problems which can be in terms of the cultural, economic, and political values of the society in which it exists, although financing problem being the main issue in preservation (Faris Hmood, 2019).

A lot of researchers emphasize that all these problems can be solved by overlapping those several aspects to achieve successful preservation. But this overlapping in fields cannot be done only by governance, architect, urbanists, and researcher; a considerable part goes to the environment, and community participation that activates the sense of responsibility, awareness of the importance of the city, and its cultural and historical value (Faris Hmood, 2019). The most important place where the community can participate in historical sites is the public spaces, in that the open spaces are attached to historic buildings. Those open spaces have a significant value, which will add cultural heritage to the community (Carr, 1992). It is regarded that cultural heritage is a combination of various elements: human processes, practices, and urban elements (Abdel Tawab, 2014). Public space gives a definite meaning to the community, such as local culture, identity, and shape. Therefore, the built environment and culture are tightly related to each other (Nas, 2011). "When public spaces are successful, they will increase opportunities to participate in communal activity" (Carr, 1992).

In this research, the focus will be on the Public spaces attached to the historic building, which is named 'Historical settings.' They include natural or designed spaces, urban, rural, and cultural landscapes (Parsons, 2010). 'Historic settings 'is one of the most critical approaches in the historic urban landscape. The Historic Urban Landscape is a new preservation philosophy for public spaces, which was adopted on the 10th of November 2011 by the 36th session of UNESCO's General Conference in Paris. Figure 1 shows transformation over the years between constructions and built environment leading to the concept of historic urban landscapes and their subsequent evolution in 2011. After the Paris Recommendations, the meetings to define and settle the idea of historic urban landscapes will be discussed in the further section to come.

NEW PRESERVATION PHILOSOPHY 'POWER OF PLACE'

A new preservation idea introduced at the beginning of the 21st century, which was changes in public policies for the historic environment and added a democratization atmosphere to heritage under the banner of Heritage Protection Review (HPR). This idea introduced 'Power of Place' which underpinned by the public policy resources of the Heritage Lottery Fund (Flora Gathorne-Hardy 2004). Despite that the Department for Digital, Culture, Media, and Sport is a department of the United Kingdom government and mentions that the Historic Environment is: A Force for Our Future (DCMS, 2010). It stated in understanding the value of heritage Power of Place that it should be in consideration as emphasized, the importance of this heritage. This heritage is more than bricks and mortar. Heritage is an integration between the community who made this heritage building and public space where the community participates (DCMS, 2010).

In 2013, the World Heritage Center announced in a conference that the important approaches of the historic urban landscapes move beyond the preservation of the physical environment with focuses on the surrounding which interrelated with the human environment (Atmosphere). This atmosphere can be of tangible or intangible qualities (World Heritage Center, 2013). The following section focuses on the significance of Historic Urban Landscape (HUL) by introducing the Historic Urban Landscape (HUL), it's physical environment/atmospherics role within heritage sites and how it will have a significant impact on the social and perception of the users and its major levels to reach its approach.

Significance of Preserving Historic Urban Landscape

The historic urban landscape is the historical layering of cultural and natural values and attributes which extend to the urban context of the city, not only the historic center. The natural benefits of the site are infrastructures above and below ground, topography, geomorphology, hydrology, and natural features, although the built environment can be open spaces and gardens and its visual relationships with all other elements of the urban structure (UNESCO, 2011). Cultural values are the social practices of the users of the site and economic process, which reflect diversity and identity (UNESCO, 2011). Historic Urban Landscapes (HUL) are a basis for a comprehensive and integrated approach to identification and assessment, conservation, and management of historical sites. This basis goes through sustainable frameworks with aims while preserving the quality of the human environment (UNESCO, 2011). If the Historic Urban

Figure 1. The diagram presents the fundamental referential milestones that have marked the convergence between constructions and their environment to land on the concept of historic urban landscapes and their subsequent evolution (Source: A. Azkarate, A. Azpeitia, 2016)

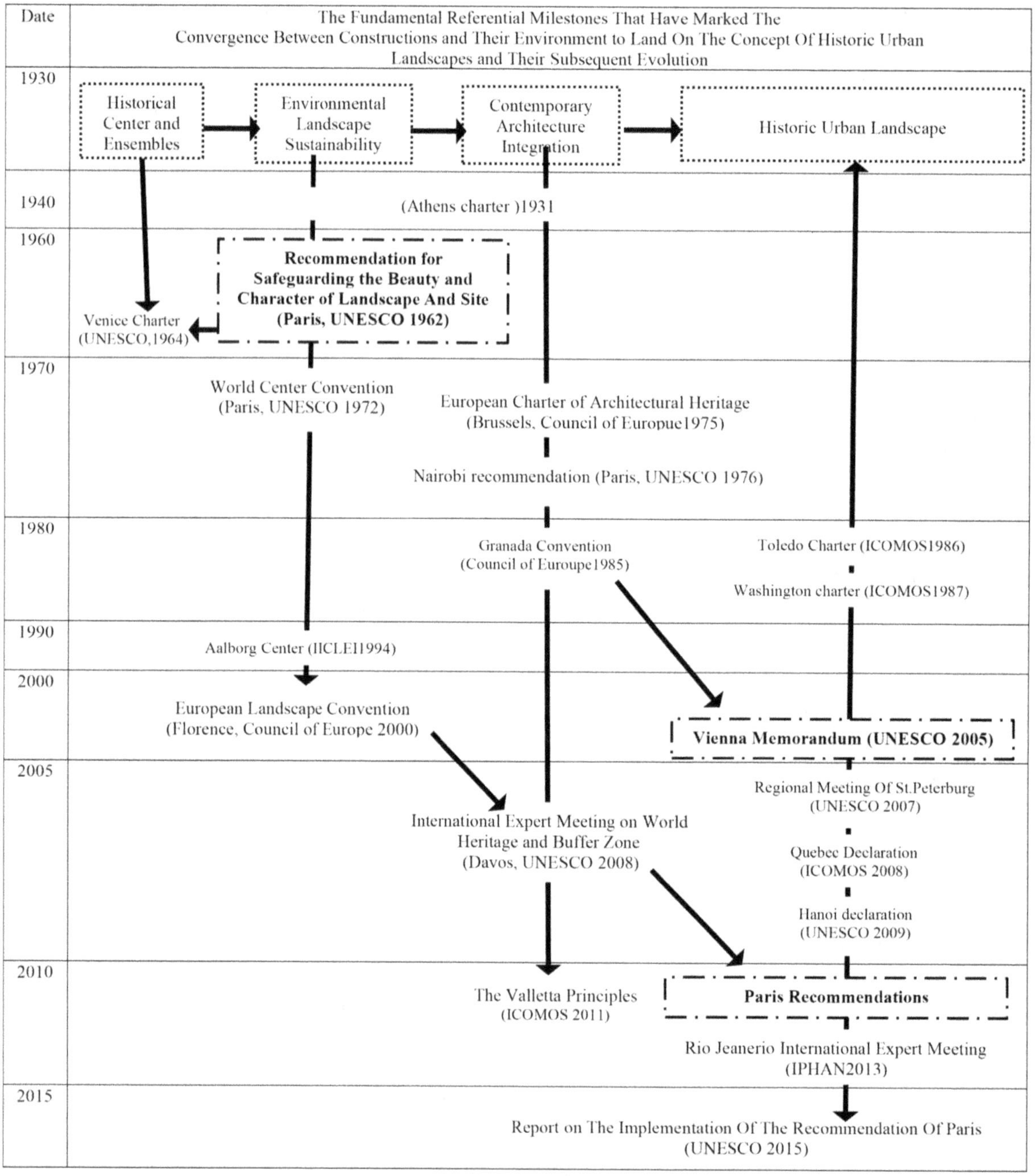

Landscape (HUL) enhances the quality of the human environment, it reflects on the productivity and sustainability of urban spaces by promoting social and functional diversity (UNESCO, 2011). Historic Urban Landscape (HUL) is a turning point between history, present, and future generations between the urban and natural environment relationship (UNESCO, 2011). Finally, the main approach of the Historic Urban Landscape(HUL) is an interpretation and conclusion of tradition and perception between different communities with respecting the values of the national and international communities (UNESCO, 2011).

In the Vienna Memorandum International Conference on the subject of "World Heritage and Contemporary Architecture," by the World Heritage Committee at its 27th session 2005 in Vienna, Austria the expert starts to point out a proposal for the categories of Historic Urban Landscape (HUL) (UNESCO, 2005). It consists of two main categories: tangible and intangible. The classification model was tested in the Vienna Memorandum International Conference by two experts. A tangible asset is a photo that describes the historic city with one of these sub-categories: urban scenery, landscape scenery (context/setting), form, buildings/monuments, details, and natural features (Ginzarly et al., 2019). The historical setting is taking from ICOMOS Xi'an Declaration 2005 is defined as immediate and extended of the environment surrounding the heritage structure (ICOMOS, 2005). An intangible asset is a photo of practices, expressions, activities, or knowledge of users, which in 2013 ICOMOS Australia Burra Charter was known by Cultural Significance. It means aesthetic, historical, scientific, social, or spiritual value for past, present, or future generations. Although it embodied in the place itself, its fabric, setting, use, associations, meanings, records, related places, and related objects, values for different ranges can from individuals or groups in the places (Australia ICOMOS & International Council on Monuments and Sites, 2013).

In this research, the focus will be on the 'Historic settings' as it played a role in enhancing the public policy objectives, including education, sustainable economic development, health and well-being, community cohesiveness, and placemaking (Parsons, 2010). Finally, this research will work on the spatial and visual components and how proposed new development can tap into the historical setting and make this proposal achieve. While studying a wide range of attraction for the visitor or users, it concluded that an estimate of 37% of the visits include a visit to a cultural/heritage. Examples of this historical attraction are museums, conservatories, zoos, aquariums, galleries, performing arts centers, reservation sites, historical monuments, buildings, and cemeteries (Boyd, 2002). But by preserving the historical aspect, is what makes this place different from others and attractive to the users. Undoubtedly, each user is searching for a specific feature and environment on the heritage site, so each attraction must have an emotional and creative atmosphere or ambiance to meet the satisfaction of the users. This emotional uniqueness can serve as a form of competitive differentiation for that specific attraction (Bonn et al., 2007).

The most important challenges while creating an ideal experience in each historic site are to create the right atmosphere and physical environment, which reflect the historical background of the place by creative ambiances such as exhibition area and events. This attraction can be an indoor and outdoor space to define what space wants to say to users and what benefits they will get. Spaces that will be on focus in this research are the outdoor spaces attached to the historic building (Boyd, 2002). The following discussion explores the specifics of a pleasing physical environment and atmosphere within those outdoor spaces attached to heritage sites. This research will study the impact of "critical placemaking" with creative ambiance framework on the social quality of users; but before studying impacts, we have to study the role of the physical environment and atmosphere as it contributes to placemaking and creative ambiance of attached spaces in Heritage Sites.

Placemaking and Environmentalism

It is important to examine the role of atmospherics and physical environment within heritage sites before proposing a framework for placemaking as "A major part of creating this 'ideal' experience lies in creating the right atmosphere or physical environment in which to view the display, exhibit, or attraction" (Bonn et al., 2005). An argument for Bonn et al. (2005) concluded in his research (International versus Domestic Visitors: An Examination of Destination Image Perceptions) which studies 'ideal' experience and its effect on visitors to heritage sites and how these visitors could make a relevant contribution to helping in enhancing the position of the heritage site (Bonn et al., 2005). Consequentially, the researcher will highlight the main factors that affect atmospherics and physical Environment Within Heritage Sites based on Theoretical foundation.

The Atmosphere as an Attractive Tool

It was the first time to describe the term atmospherics into a design of retail environments by Kolter (1974); consequently, it was defined as the "conscious designing of space to create certain effects in buyers" (Kotler, 1974). Kolter gives the term of the atmosphere a very specific description that is a "silent language," which goes to the body language of the users. While studying this commutation, he found that there are such intangible forms called ambiance. The main influence in the atmosphere term takes place through the sense and emotional mechanism of the users (Kotler, 1974). He predicted that the significance of the atmosphere will be a base of the pleasure consumption rather than purchases of necessity, and in-market segments in which there is little to differentiate based on price or quality alone (Kotler, 1974).

A new singularity argument had been drawn by Kotler (1974) for the atmosphere term, which intended atmosphere (the sensory qualities conferred to space by design choices) and the perceived atmosphere, which may vary for individual customers depending on their selective attention and specific goals (Kotler, 1974). The physical environment is a salient feature which informs the customer's perceived environment, which in turn to the behavior of the users through the attention (standing out from the crowd); message (regarding the kind of establishment and its intended market); and affect (cues such as colors, sounds, and textures) to subtly influence the desire to purchase (Kotler, 1974).

The atmosphere has several variables such as layout, crowding, product location, music, color, lighting, and aroma that have been tested for their impacts on retail consumers' affective responses, perceptions of value and quality, and behavioral responses such as time spent, traffic flow and purchasing patterns(Turley & Milliman, 2000). According to those variables, Kotler (1974) drew four dimensions of the atmosphere: Visual (sight) dimensions – color, brightness, size, and shape; Aural (sound) dimensions – volume, pitch, and tempo; Olfactory (smell) dimensions – scent and freshness; Tactile (touch) dimensions – softness, smoothness, and temperature (Kotler, 1974). Finally, Kolter drew the connection and purchase probability, which is illustrated in the following Figure 2, which is applied to any spaces which target the users' satisfaction.

In various studies, spaces, and environments, the visitor experience is influenced by the atmosphere. As an example, if there is a festival, the atmosphere of the festival gives the place the uniqueness and makes it attractive to the users (Van Vliet, 2018). The atmosphere dimensions are different from the point of view of one researcher to another. Berman and Evans (2013)identified four dimensions for atmosphere term: Exterior – storefront which includes marquee, entrances, windows, lighting, and con-

Figure 2. The causal chain connecting the atmosphere and purchase probability
source ((Kotler, 1974).

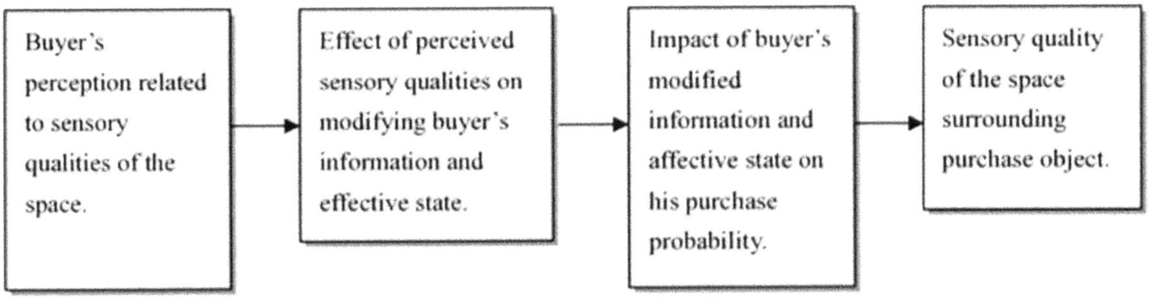

struction materials; General interior – flooring, bright, color, scents, sounds, store fixtures, wall textures, and dressing facilities; layout – allocation of floor space (selling space, merchandise space, personnel space, customer space), classification of origins, determination of a traffic-flow pattern, mapping out locations, the arrangement of individual products; Interior (Point-Of-Purchase) Displays – the principal type of displays are assortment display, theme-setting display and rack display (Berman & Evans, 2013).

Kottasz's conceptual model of the effect of atmosphere stimuli on emotions and behavior applied in 140 museum visitors collected over ten museums, showed that environmental stimuli also influence the PAD dimensions in museums. Some variables during this study were selected. Also, it was mentioned in the above studies (light, temperature, and decorative elements) (Van Vliet 2018). It was observed from this model how museum visitors react to the physical museum space, despite the recognized importance of this and the (financial) interests involved, such as in the redesign of museum spaces, provide insight into how visitors perceive different exhibition spaces and its relationship with different aspects of the visitor experience (Van Vliet 2018). Figure 3 shows Kottasz's conceptual model with all variables which affect the users of the museums.

These studies made by many researchers revealed the importance of atmospherics in-store and applied in any space interact with the user and influence with purchase behavior of users. The indicators which constitute atmospherics in this study were selected based on research studies quoted above, and their applicability to any space selected for this study. Nowadays, the atmosphere is widely acknowledged as an important component of experience quality in a wide range of leisure settings (Chang & Horng, 2010).

The marketer used the atmosphere term widely as an umbrella's term overall design and ambiance of public space, historical sites, and commercial spaces to create a great service environment for the users (Bakker et al. 2014). Finally, at the end of the atmospherics literature, Stimulus-Organism-Response (S-O-R) model by Mehrabian & Russell (1974) is the most influential theoretical framework which measures the user response to the environment (Mehrabian & Russell, 1980). Although this model has a lot of criticism, it refers to a unidirectional environment-behavior relation and for inadequately taking consumers' goals and motivations into consideration. Yet it is the most commonly used paradigm in atmospheric research (Massara et al., 2010). While in Baker's model, they found evidence that atmosphere depends mainly on the influence of ambiance (lighting, color), design (layout, routing) and social factors (friendly personnel) on visitor attitude towards the heritage attraction, as well as on the intentions to revisit the site and recommend it to friends and others (Bakker et al. 2014).

Figure 3. Kottasz's conceptual model of the effect of atmosphere stimuli on emotions and behavior
Source: (Van Vliet, 2018).

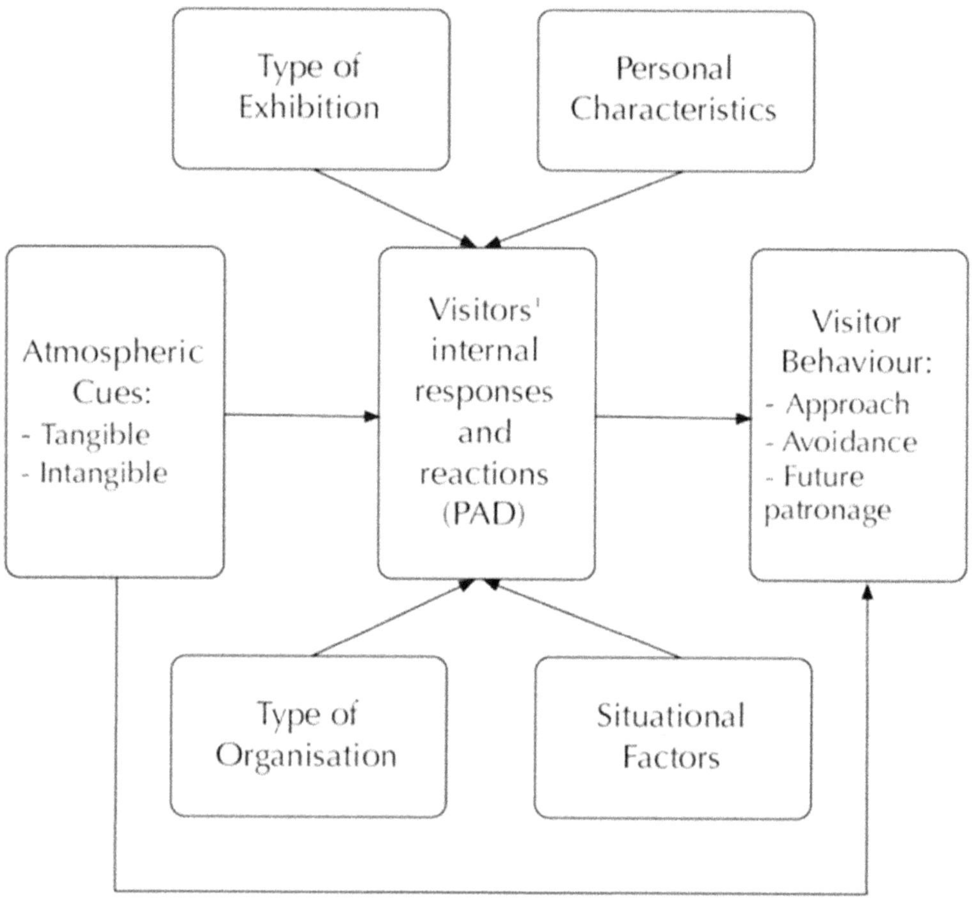

The following section will highlight the physical environment and how they affect the users based on the theoretical foundation starting with the Stimulus-Organism-Response (S-O-R) and ending with Baker's model which will be the base of this research to help the researcher put guidelines to create a creative ambiance in the open spaces attached to the historic building.

The Physical Environment

Mehrabian and Russell (1974) Environmental psychologists suggest that there are two basic ways when people react to their environment: approach and avoidance (Mehrabian & Russell, 1980). Stimulus-Organism-Response (S-O-R) model, derived from environmental psychology, emphasizes the sensory input for the surrounding environment and combines with the factor which affects the users personally; all this factor subsequently results in behavioral outcomes (Mehrabian & Russell, 1980).

The S-O-R model, derived from environmental psychology, asserts that sensory inputs from the environment (the Stimulus) combine with personality factors to trigger an internal, primarily emotional reaction (the Organism); this subsequently results in behavioral outcomes (the Response) (Mehrabian & Russell, 1980). Users' responses result in either approach or avoidance behaviors. Approach behaviors

Figure 4. Original ideas of Mehrabian and Russell
Source: (Mehrabian & Russell, 1980).

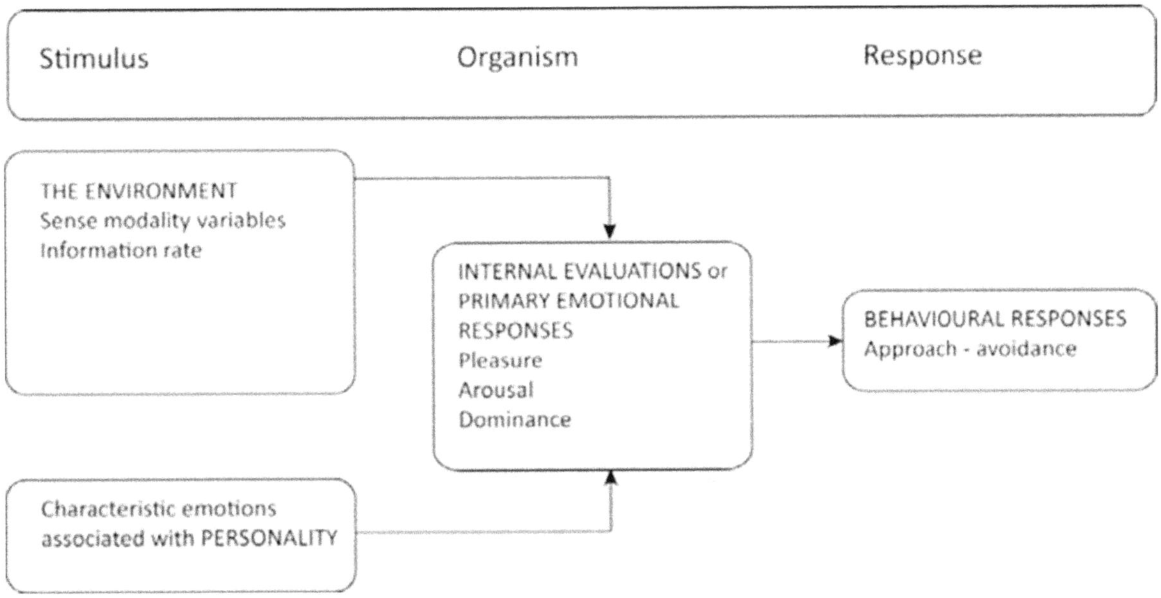

are affected by the surrounding environment or the ambiance such as explore, work, and cooperate with others, while avoidance behaviors include retreating, aversion to others, and desire to leave the situation (Mehrabian & Russell, 1980).

In 1980, the S-O-R model widely used by a lot of researchers, but it did not incorporate a comprehensive framework for characterizing the service environment (Bakker et al. 2014). The S-O-R model was developed by the Bakker and Bitner to overcome this gap (Bakker et al. 2014). While the service environment model was proposed by Baker to overcome the problems of the S-O-R model, the service environment comprises three main dimensions: ambient, design, and social. The ambient dimension includes factors such as noise, cleanliness, and air quality – environmental features that we seldom consciously recognize unless they are inappropriate to the setting. Conversely, design factors are those features that are at "the forefront of our awareness," comprising both aesthetic and functional elements (Bakker et al. 2014). In the social dimension, a dramaturgical framework applied to classify customers who share the service environment as the "audience." For example, the experience enhanced by sharing it with a crowd. Service personnel form the other part of the social dimension (Bakker et al. 2014).While in the Bitner model, the service environment was different. This model was given a different name by the servicescape model (Bitner 1992). In the servicescape model, Bitner (1992) added some changes in the dimensions, which are environmental dimensions that were characterized as "ambient conditions" (nonvisual sensory stimuli), "space/function" (layout and furnishings) and "signs, symbols, and artifacts." These dimensions come together to create the "perceived servicescape," to which both staff and customers respond "cognitively, emotionally and physiologically" (Bitner 1992). In turn, it influences their behavior and social interactions, which is illustrated in Figure 5.

Figure 5. The original ideas of Mehrabian and Russell (1974) about pleasure, arousal, and dominance connected to ABC psychology and the distinction between feeling, thinking, and acting that is used for ages and are still useful to describe environmental experiences
Source: (Bakker et al. 2014).

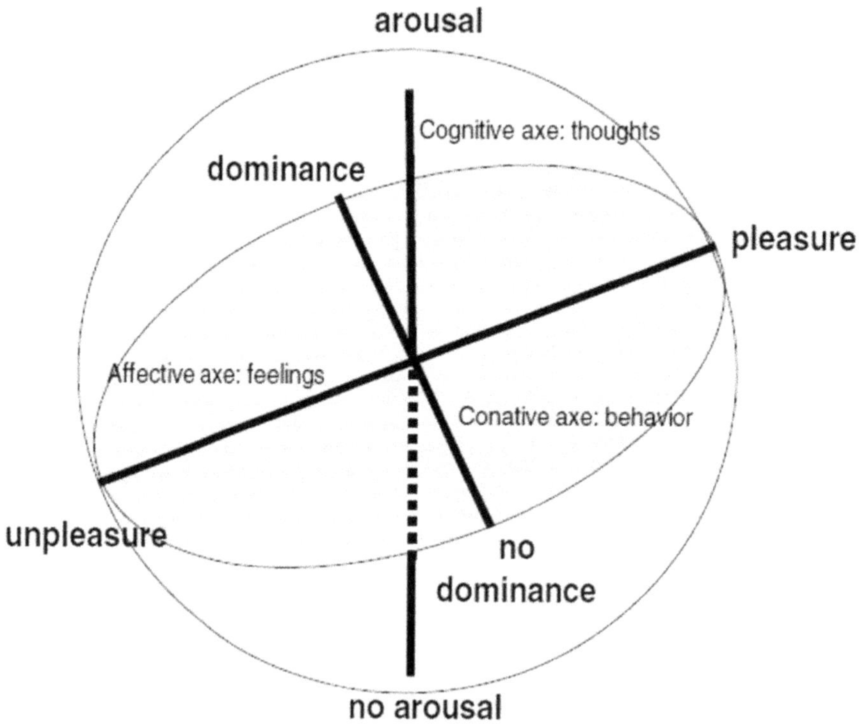

As a complement to Kotler's model for atmospherics, Baker developed a typology of the service. Atmospheric elements have been categorized along with two main schools of thought, Baker (1986) and Bitner (1992).In this study, a new dimension is proposed by Baker (1986) that affects the atmosphere or environment of the heritage sites, which is called "service environment" and categorized the "elements" into three groups:

- **Ambient Elements:** It directly affects the nonvisual senses and, in some cases, may have a relatively subconscious effect such as lighting, temperature, and music.
- **Design Elements:** It directly affects the visual and more tangible in the surrounding environment, such as color, furnish, and circulation of the space.
- **Social Elements:** It directly connects to the users in the public space and interaction with each other, and it impacts their perception (Bakker et al. 2014).

Another school of studying the atmosphere term, Bitner (1992) proposed a framework of understanding the environment and its relationship with users. This framework is a combination of environmental dimension and user perception, which are Ambient conditions - temperature, air quality, noise, music, and odor. Space/Function - layout, equipment, furnishings Signs, Symbols, and Artifacts – signage, personal artifacts, style of décor (Bitner 1992).

Table 1. Characterizing the service environment. Comparison between the parallels between the environmental variables described in Bitner's (1992) servicescape model and the dimensions of the service environment characterized by Baker (1987)

Environmental Dimensions of the 'Servicescape' (Bitner, 1992)	Components of the Service Environment (Baker et al., 1992; Baker, 1987)
Ambient	
• Temperature • Air Quality • Noise • Music	• Air Quality • Temperature • Humidity • Circulation/Ventilation • Noise (level, pitch) • Scent • Cleanliness
Space/Function	
• Layout • Equipment • Furnishings • Signs, Symbols, and Artefacts • Signage • Personal Artefacts • Style of décor	• Design • Functional • Layout • Comfort • Signage • Aesthetic • Architecture • Colour • Scale • Materials • Texture, pattern • Shape • Style • Accessories
Social	
• The servicescape places staff and • consumers as 'moderators' rather • than integral to the environment or "perceived servicescape"	• Audience (other customers) • Number • Appearance • Behavior • Service Personnel • Number • Appearance

Source: Components of the Service Environment (Baker et al., 1992; Baker, 1987).

Finally, it was concluded that physical environment and atmospheric dimensions affect the users and their perception in the public space, so it is important to apply any of the proposed frameworks by research to attached public space to fulfil the service environment. The focus on place and placemaking also provided a focal point of reference for achieving this service environment based on the capacity of organizations and individuals in their community to deliver services (Hes, Dominique et al., 2019). Local governments meet the expectation to open up processes and decision-making to community participation by placemaking, but to be successful, it should be done in creative ways. The following section will discuss the benefits of placemaking in preserving the attached space in heritage sites and the proposed framework.

Place-Making Meets Preservation

Since the mid-1970s, placemaking has been an interdisciplinary, multifaceted concept that has been widely researched (Andrews, 1974). In 1989 William Whyte started to trace the concept of placemaking in spatial planning with key authors such as Jane Jacobs and George Andrews. He wrote in his book "The Last Landscape" about the negative impact of urban sprawl on people's behavior in an urban setting (Whyte, 1968). Then he started to explore the liveliness of the urban setting through some observations which were studied in his book City: Rediscovering the Center, which mainly focuses on the behavior of the users (Whyte, 1989).

Whyte regarded that concept which Jane Jacobs mentioned in her contributions entitled The Death and Life of Great American Cities, which suggested that small-scale changes in an environment may change a neighborhood. It was totally right as the manner in the urban development changed the impact on the setting of the social environment of three users. But the placemaking is not restricted to a single discipline; it extends to Spatial and Design Disciplines, Social Science, Art, Education, Music, Tourism, etc. (Alvarez et al., 2017).

Fred Kent, Project for Public Spaces Founder, mentioned that "Placemaking is an act of doing something. It's not planning; it's doing. That's what is so powerful about it." (Silberberg et al., 2013). The most powerful major part is creating this 'ideal' experience, which lies in creating the right atmosphere or physical environment in which to view the display, exhibit, or attraction" (Bonn et al., 2005). Bonn concluded this in his research (International versus Domestic Visitors: An Examination of Destination Image Perceptions) which studied 'ideal' experience and its effect on visitors to heritage sites and how these visitors could make a relevant contribution to helping in the enhancement of the position of the heritage site (Bonn et al., 2005).

There was an emphasis on applying this ideal experience, the Project for Public Spaces, the National Trust for Historic Preservation, and its National Main Street Center collaborated to integrate the value of the historic preservation movement and placemaking as a core principle. It was founded by Anne T. and Robert M. with long term foundation support for both organizations. They proposed in the form of the question: What if we built our communities around places? (PPS, 2010). This collaboration emphasized that placemaking gave public space to reinvention and reimagination to be the heart of the community. It also strengthens the connection between place and users interacting with it. As placemaking is a collaborative process that helps to promote better urban design, it also creates creative patterns for the users, giving attention the physical and cultural and social identities that define the place (PPS 2010).

Each city faces a lot of urban challenges type of placemaking program is different from one urban context to another (PPS 2010). The main focuses of this research are the lack of participation and poor design of the heritage sites, and for this reason, the transformation will be applied based on four forms of placemaking of funding, problems, and outcomes (PPS, 2010). These forms of place-making meet together in a focal point targeting how to make ''Successful Places" (PPS, 2010).

The first type aims for the improvement of the spaces to develop the community. This community suffers from a lot of liveliness public spaces, and it is known by ''Standard place-making. The process of this type starts by promoting long term small projects of activities for the communities (Strydom et al., 2018).

The second type aims to a different goal, which is more specific to the ''an economic or infrastructural development," and it is known as ''Strategic Place-making," which targets to attract more business, which helps in the improvement of the spaces. For example, for this business project, green pathways, waterfronts, entertainment facilities, or social gatherings place and construct rapid transit lines (Wyckoff, 2017).

The third type specifies which has been chosen for this research as it depends on promoting and restoring the community identity by creating creative environment for the users through cultural activities in the heritage site museums, art exhibitions, and outdoor public art displays (Strydom et al., 2018). Finally, the last type is ''Tactical place-making''; it targets the communities which suffer from a lack of funding the spaces and adopting a new policy of low-cost without any political and financial commitments (Bauer et al., 2010). Figure 6 illustrates the relationship of three specialized types of placemaking to create Quality in Places (Wyckoff, 2017).

Initially, all types of placemaking must pass through four main steps with a comprehensive vision to help the community (PPS, 2010). The first step is to Define Place & Identify Stakeholders to detect the main concerns, obstacles, and different stakeholders, such as residents, organizations, and civil bodies. The second step is to Evaluate Space & Identify Issues by holding an observation workshop to define the current use of space, its dimensions, and how to develop it. The third step is to outline the place vision on the bases of the ''selected form of placemaking and missing activities'' while respecting the value of space simultaneously. This also activates an experimental step based on the triangulation principle. The last step is an on-going re-evaluation and long term improvements, which is accomplished through running continuous evaluation to reassure stakeholders and develop a long–term management and maintenance plan (PPS, 2010).

Figure 6. The relationship of three specialized types of placemaking to create quality in places
Source: (wyckoff 2017)

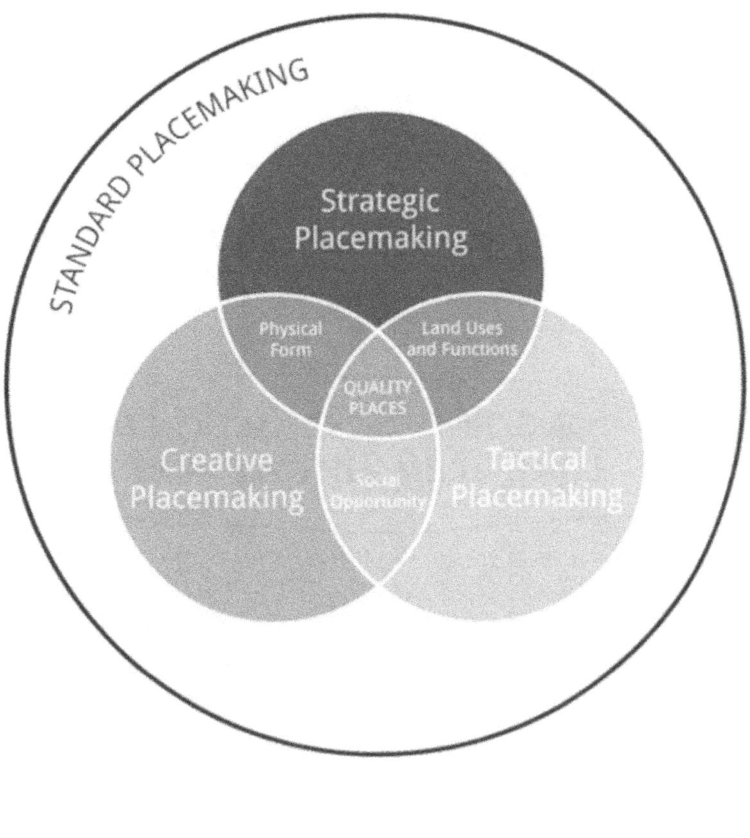

Figure 7 illustrates the steps of placemaking base on Wyckoff (2017). Placemaking is not only a process for transforming public space, it also has a philosophy that concentrated on observing to meet the demand of the users by asking a question to the users interacting with these spaces and listening to their inspiration for the place (PPS, 2010). This philosophy can be applied in existing space or a new one but with differences in steps and orders to get more people involved in observing, planning, and shaping a place (PPS, 2010).

Figure 7. The Steps of placemaking based on Wyckoff (2017)

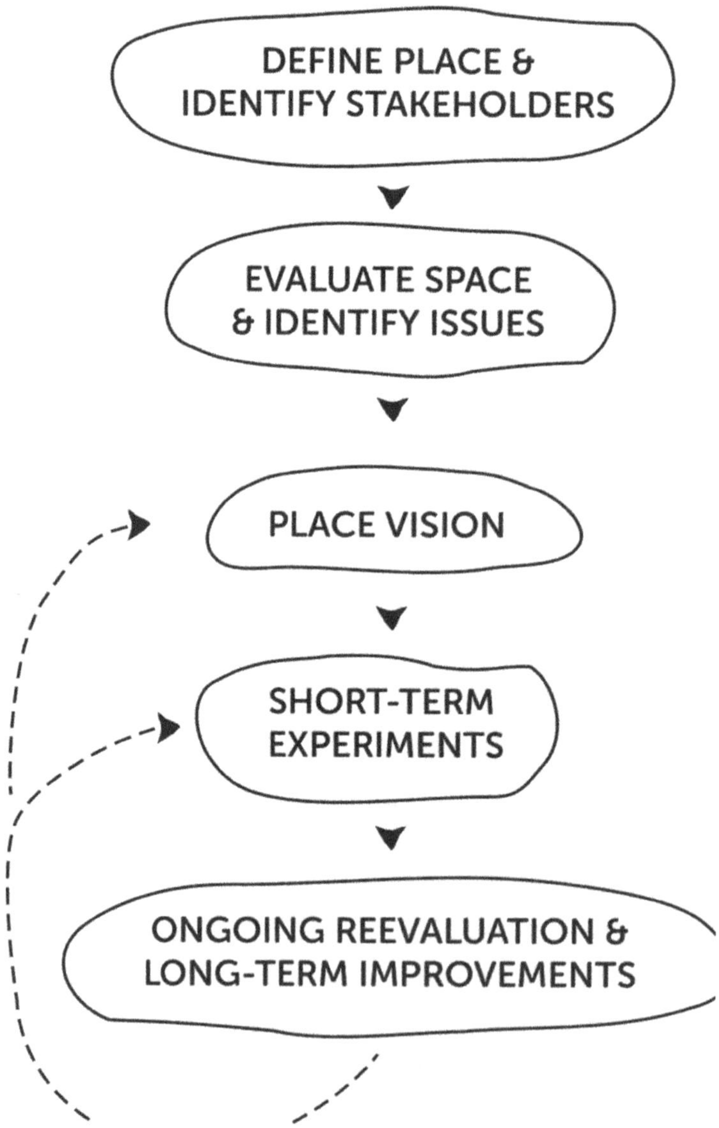

PLACEMAKING AS A NEW APPROACH TO CREATE SUCCESSFUL PLACES

In 1993, relying on William H. Whyte's study, Carr published "Public Space," where he analyzes various public spaces throughout the U.S. He ends up in explaining human behavior within public spaces structured on three pillars: needs (comfort, relaxation, passive /active engagement, and discovery), rights (access, freedom of act, claim and change) and meanings. The focus on meanings, identity and therefore the way people understand public spaces is related to Kevin Lynch's A Theory of Good City Form; "A good place is one which, in some way appropriate to the person and her culture, makes her aware of her community (Lynch, 2005).

The spirit of the place also has high relevance in the methodology compared to Whyte and Car, PPS added another feature in the portrait of successful public spaces: accessibility and connectivity. Figure 8 shows the basic criteria for a successful public space. Project for Public Spaces (PPS) is the central hub of the global Placemaking movement that started to evaluated thousands of public spaces. It found that to be a successful place they should be following four qualities: they are accessible; people are engaged in activities there; the space is comfortable and has a good image; and finally, it is a sociable place: one where people meet each other and when they come to visit. Figure 8 shows Placemaking Criteria Required To be Successful and also it helps people in judging any place (PPS, 2010).

There are many different ways of considering, enabling, and implementing placemaking, and the anticipated goals may be different for each of the stakeholder groups (Hes, Dominique et al., 2019). A Leadership in Placemaking Reflective Tool illustrated in Figure 9 has been developed to assist peacemakers in overcoming this challenge of reading the many situations that arise during placemaking projects band to strengthen their skills in responding accordingly. This leadership has types of actions that may take place within placemaking projects and the different ways in which engagement can occur (Hes, Dominique et al., 2019). This type of action referred to building relationships with the users to help in making decisions. These domains are placed equally around the circular diagram, as each Domain has equal value within the tool (Hes, Dominique et al., 2019). The outermost, relationship-oriented level of engagement requires those in leadership roles to step back from active participation in achieving the given tasks, focusing instead on facilitating others to build new relationships and strengthen existing connections between and within stakeholder groups (Hes, Dominique et al., 2019). Consequentially, the researcher will focus on the criteria of place-making and then the main factors that affect atmospherics and physical environment, which will help in the evaluation of public space in heritage sites.

The Criteria of Placemaking

Fred Kent, founder of the placemaking idea, mentions that everyone has the right to live in a place, and this place should be a great place. That makes the place where they already live great (PPS, 2010). PPS developed the place diagram on criteria divided into four sections which are ''Access & Linkage, Comfort & Image Uses & Activities, and Sociability. "This section helps in applying placemaking to any public space to be a successful and great place, as shown in Figure 10 (PPS, 2010).

The following discussion will explain each section and how can the researcher or the designer apply it in any public space. First, the access and linkage section; this section is concerned with the connection of the space by the surrounding visually and physically (PPS, 2010). This section investigates the diversity of transportation and the available option to help this public space. The main approach of this section is to make public space seen from far and remarkable for the users, although to reach this ap-

Figure 8. Basic criteria for a successful public space
Source: (Whyte, 1989)

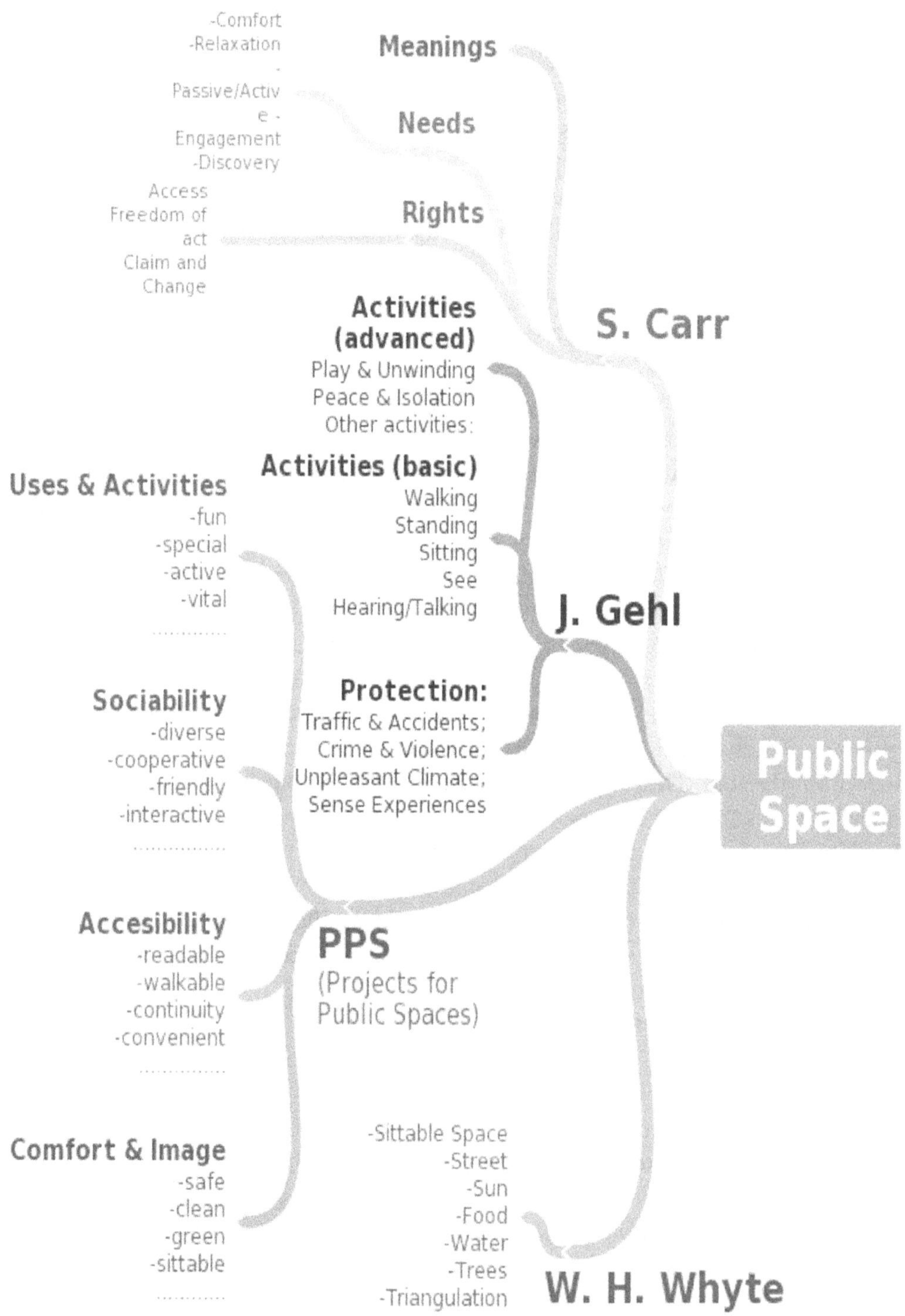

proach it should be connected by the surrounding building by sidewalks, streets and bicycle lanes which will help to change the perception of the users (PPS, 2010). Second, this section focuses on Comfort & Image, which help in the success of the public space. If the public space is comfortable and reflects a good image on the perception of the users, it will be attractive, and it will be successful.

This section involves the facilities given to the users such as the seating area, its location near the sun or in the shade, safety and cleanness. All these factors affect the comfort and image of the users (PPS, 2010). Third, this section focuses on Uses & Activities as they are considered the main reasons why people often visit this place and revisit it. The main approach here is to make a balance between Uses & Activities (PPS, 2010). Finally, the last section focuses on the sociability of the public space; it involves the feeling of readiness to interact comfortably with people, particularly strangers. In this section, a lot of studies should be done on the impact on social classes of users on their sense of place and attachment. The sociability will help the researcher to know how the people use this space if they use it in groups or individually. It also focuses on the participation of user in the space as it keeps the space in good conditions (PPS, 2010). Although Empirical research on placemaking has been a relatively small body, there is a significant volume of theoretical research on placemaking, such as that of Whyte's work. The theoretical research helps to better understand the place phenomenon, placelessness, and placemaking in a board of sense, and it often claims the traditional environment to provide the users better sense of place (PPS, 2010). While the book of Place and Placelessness of Edward Relph drives the concept of

Figure 9. Leadership in placemaking reflective tool
Source: (Image by Lara Mackintosh,2018)

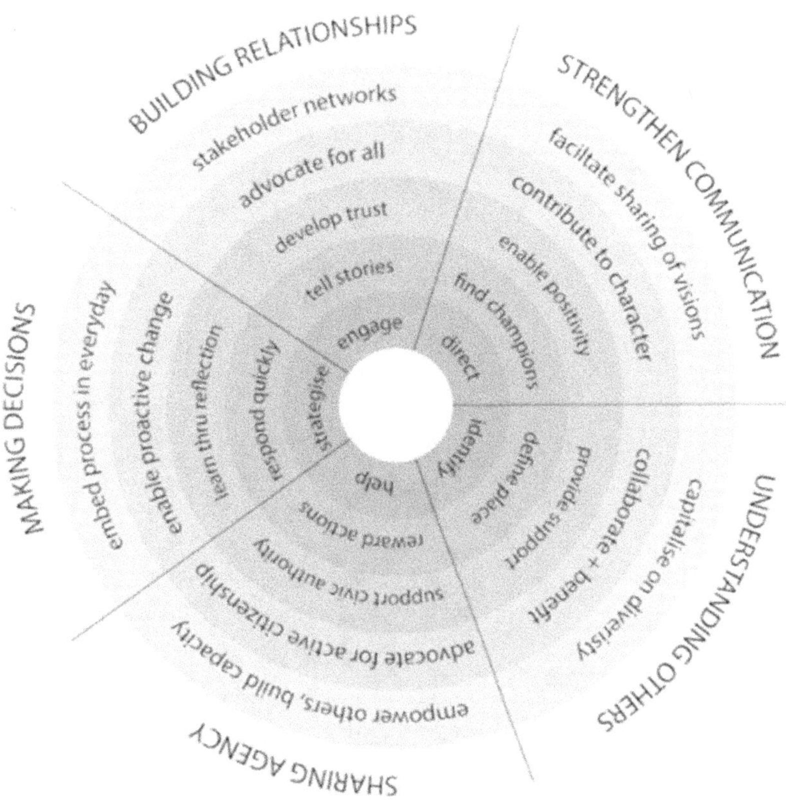

the place in a new way called "directly experienced phenomena of the lived-world and hence is full with meanings, with real objects, and with ongoing activities" (Relph, 2008). This meaning was translated to several dimensions of placemaking including the physical and non-physical (psychological and social), and it is known by the ambiance . It is important to note that despite categorization, the physical and non-physical aspects of a place are highly interdependent and rare (Redi et al., 2018).

Ambiance and Social Quality of Life

Ambiance has a great impact on the social quality of life of the users. The concept (QOL) is a multidisciplinary one with various aspects. However, it is full of meaning and layers that are divided into two aspects: objective and subjective. Its objective aspects such as living standards with cost regulations, the economic productions, and infrastructure of the facilities. Health, entertainment, leisure, culture, and art added to these aspects with many concerns. The subjective aspect of QOL is the users, the experiences, and the satisfaction in real life (Dajian & Peter, 2006). It is applied to the historic public park, and its impacts were significant in the social life of users (Elgobashi & Elsemary, 2019). In this research, the focus will be on the social dimensions as they act as a major factor in the QOL and urban design. Hypothetically, social content affects the public spaces to the extent that without social interaction, there will not be a spatial component (Carmona et al. 2003). Social Quality is an advanced aspect of quality of life, but it is grounded in a theory of 'The Social.' The Social Quality approach focuses on the individual; it focuses on the social dimensions and the acting capacities of social beings. Consequently, it reflects on the self-realization and emerge of oriented identities to the community (Wallace & Abbott, 2009).

Figure 10. The placemaking criteria required to create responsive places, Source: (PPS 2010)

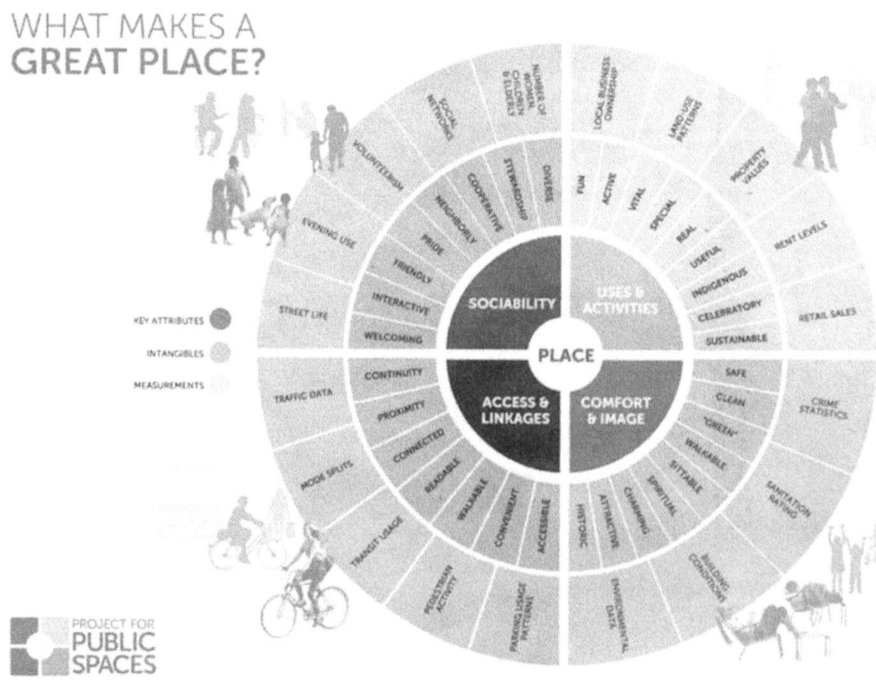

Table 2. The ambiance is a multi-faceted concept and characterized by physical, physiological, socio-logical, cultural, and psychological aspects; each of them has a negative and positive impact

Dimension		Terms	
		Negative	**Positive**
Physical	Land use (Greenness)	Constructed, Built, Cemented.	Green.
	Location (Urbanization)	Isolated, Rural	Suburban, Urban, Central
	Size	Tiny, Small Medium-sized.	Panoramic
	Status	Under Construction, Developing, Re-Developing.	Well-Established
Physiological	Touch→Temperature	Cold	Warm
	Sight→Light	Dark	Shiny, Bright
	Sight→Colors Colorless, Gray Colorful	Colorless	Gray Colorful
	Hearing→Sound	Sound Silent, Quiet	Noisy, Loud
sociological	Cleanliness	Dirty.	Clean, Immaculate.
	Income	Poor, Cheap, Sustainable, Accessible, Gentrified	Wealthy, Pricey, Fancy, Rich, Luxurious, Expensive, Exclusive
	Ethnicity	Monocultural, Homogeneous	Mixed, Diverse, Eclective, Ethnic, Exotic, Multicultural, international, Wordly
	Population	Density Deserted, Empty	Busy, Crowdy, Crowded, Dense, Overcrowded
	Sense of the place	No feeling belonging	Feeling belonging
	Type of Activity (Night→Day)	No nightlife and not safe.	Leisure, Touristy, Sightseeing.
	Social activities	Not attractive to people.	Popular and different types of socio-economic target it
cultural	Education	Vulgar, Simple, Gritty	Smart, Intellectual, Sophisticated.
	Informality	Formal, Decorous, Conforming, Official	Frivolous, Casual, Spontaneous, Easygoing, Familiar, Free, Unofficial, Informal, Folksy

Source: (Redi et al., 2018).

In addition to environmental atmospheres, there are the cultural, social, workplace, family, and interpersonal atmospheres. A social situation can be supportive or discouraging, liberating or stifling, inspiring, or dull, depending on the atmosphere and ambiance of the space. Spirit of the place is a similarly ephemeral, unfocused, and non-material experiential character that is directly affected by the atmosphere. Indeed, this atmosphere cannot be successfully reflecting on the user's perception without creative ambiance. As ambiance taxonomy focuses mainly on the physiological satisfaction and attraction of the users which gives Space a unique perceptual character and identity (Zalta, 2006).

Project for Public Space (PPS) and The National Trust for Historic Preservation team up to promote a new vision for the future (and past) of American communities. National Trust starts a mission to share a deep resonance to give the users a unique experience while visiting the historical site by its different categories. This principle will bring people together to protect, enhance, and enjoy the places that matter to them (PPS, 2010). It helps restore the Historical and social functions of a building, space attached, or historic district (PPS, 2010). A consistent trend follows only that Public Space in the heritage site had a positive impact on their communities. PPS supposed the importance of protecting historic buildings with attached open spaces as those places stand as vital centers of activity in the community (PPS, 2010).

Architecture plays an important role in people's belonging to spaces to reflect an understanding that great buildings are the backbone of any great community. This is only one particular lesson involving building with the community, although there is much to learn from the success of the preservation movement (PPS, 2010). We defined a 3-level taxonomy of ambiance: the first and more general level consists of categories of factors impacting ambiance (e.g., physiological, psychological); the second and intermediate level consists of dimensions (e.g., senses – "sound," perceptual characteristics – "joyful"); and the third and more specific level consists of terms characterizing each dimension (e.g., "loud," "cheerful"). Let us detail each level next in Figure 11 as each color reflects the effect of ambiance dimension.

Figure 11. Wheel representation of psychological ambiance taxonomy. The color of each ambiance dimension reflects the satisfaction and sense of the users
Source:(Redi et al., 2018)

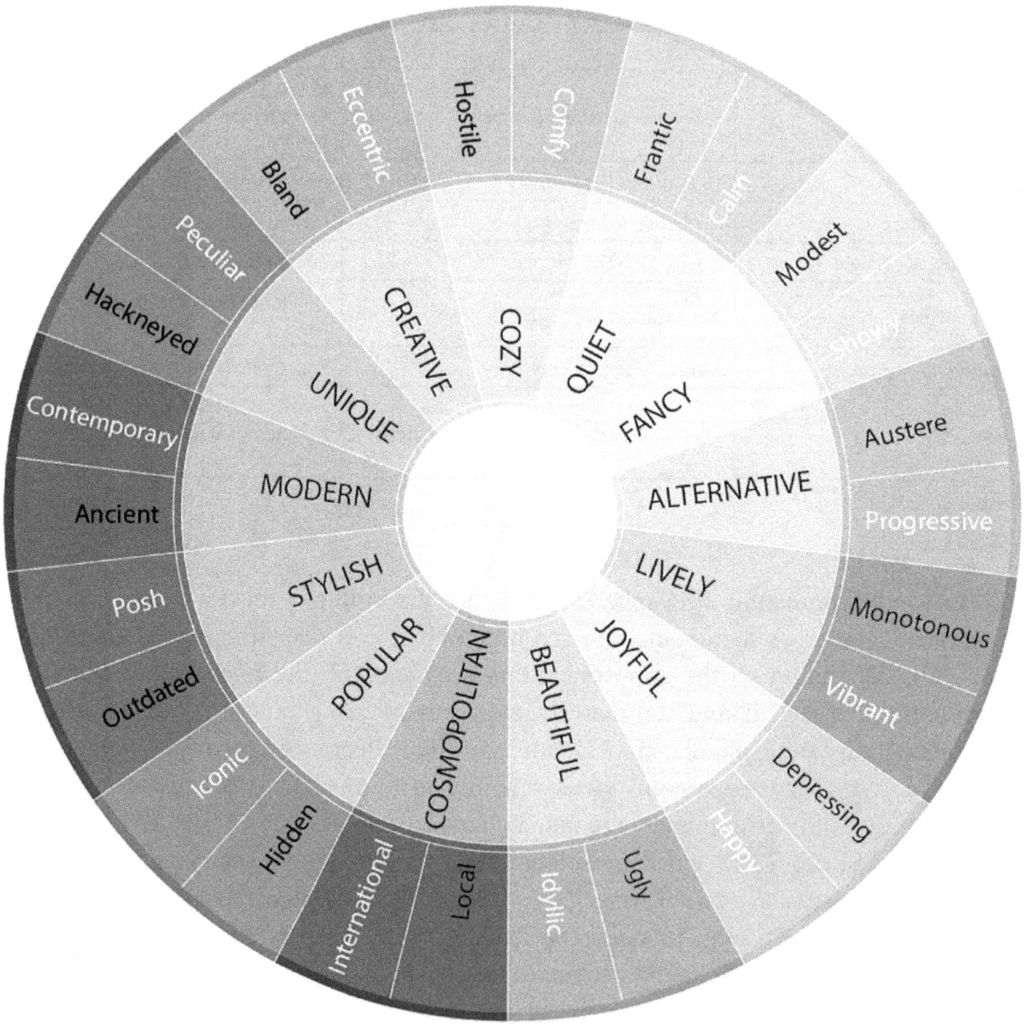

From the atmosphere or ambiance, the researcher can read automatically behavioral and social aspects as it is an epic experiential dimension or prediction. In the atmosphere image, a temporal layering of narrative into the setting appeared, while in ambiance emotionally, the layering of temporal traces, as well as images of past life in our settings, appeared (Pallasmaa, 2014). It is connected to signs of life instead of being isolated in hermetic and artificial conditions. As ambient element seeks for historically dense settings because they connect us experientially and imaginatively (Pallasmaa, 2014). The following examples will explore the significance of adding creative ambiance to the attached spaces of the existing atmosphere of the historical building to users and searching for a specific goal of enhancing visitor experience and attraction. This idea was applied in a lot of historic sites by adding activities, events, or facilities to be more attractive—an example of these implications here to the historic building.

Finally, it concluded from these successful examples that the facility's atmospherics becomes critical in shaping individual user attitude evaluations for the space attached to the historic building, whatever the function of this building. While the ambient elements are based on the premise that comfort, satisfaction, and delight results (Osmond 2007).

DISCUSSION OF FINGINGS AND FUTURE RESERCH DIRECTIONs

From the comprehensive review, Carmona (2019) identified over 20 ways through which place can lead to social, ecological, and economic benefits based studies(Hes, Dominique et al., 2019). Each study applied very different evaluation strategies, often presenting a partial narrative of place benefits.

Figure 12. The newly renovated Shed 3 at Detroit's historic eastern market
Source: ("Eastern Market Corporation," 2018).

Figure 13. Louvre museum, Paris
Source: ("Louvre museum, Paris.," 2019)

Figure 14. Palace of Versailles, Paris
Source: ("Palace of Versailles, Paris," 2019).

Figure 15. A mixture of activities, car-free spaces, and a strong cultural identity add to the success of Nyhavn in Copenhage
Source: (Smith, 2016).

Placemaking practices were done as initial steps to measure the success of each space. All over the world, a lot of hosts seek how to assess the better quality of those spaces. For example, Place Standard, Public Space Toolkit, Happy Maps). In the market, there are over 75 indices, frameworks, guidelines, and engagement tools that can inform place evaluation strategies(Hes, Dominique et al., 2019). However, the current approaches to evaluating place have various limitations:

The assessment goes through as it is an end product, and it involves participation so it has a lot of problems in future development. Standard indicators should be used, such as crime rate and population statistics, which give a unique narrative of a particular UN-HABITAT place or the emotional connection developed between the visitor and that place. The evaluation processes tend to focus on the outputs and outcomes of the project, failing to incorporate measures that link to the overarching objectives of place processes. Current approaches are not holistic and often only focus on human outcomes, yet we know that ecological or non-human elements are essential for human well-being (Hes, Dominique et al., 2019).

Throughout this analysis, we have drawn from key concepts in the Physical environment and atmosphere to theorize the process of what I have termed "critical placemaking." This provides new ground for empirical research. Empirically, assessing placemaking processes is one line of future research that examine how aspects of the placemaking process. This direction is linked to the physical environment and atmosphere, which is called service environment processes and targeted outcomes, such as social quality of life.

These assessments would have important practical implications in evaluating and strengthening the positive effects of placemaking for individuals and communities, as well as identifying potential barriers. This assessment needs to be across with some input, which includes people, time, resources, strategies applied as part of the process. It essentially documents "how" the placemaking initiative was developed. One practical outcome may be the creation of evaluation rubrics to examine what specific environmental

Figure 16. Four dimensions of place framework for relationship-based evaluation
Source: (Image by Cristina Hernandez-Santin)

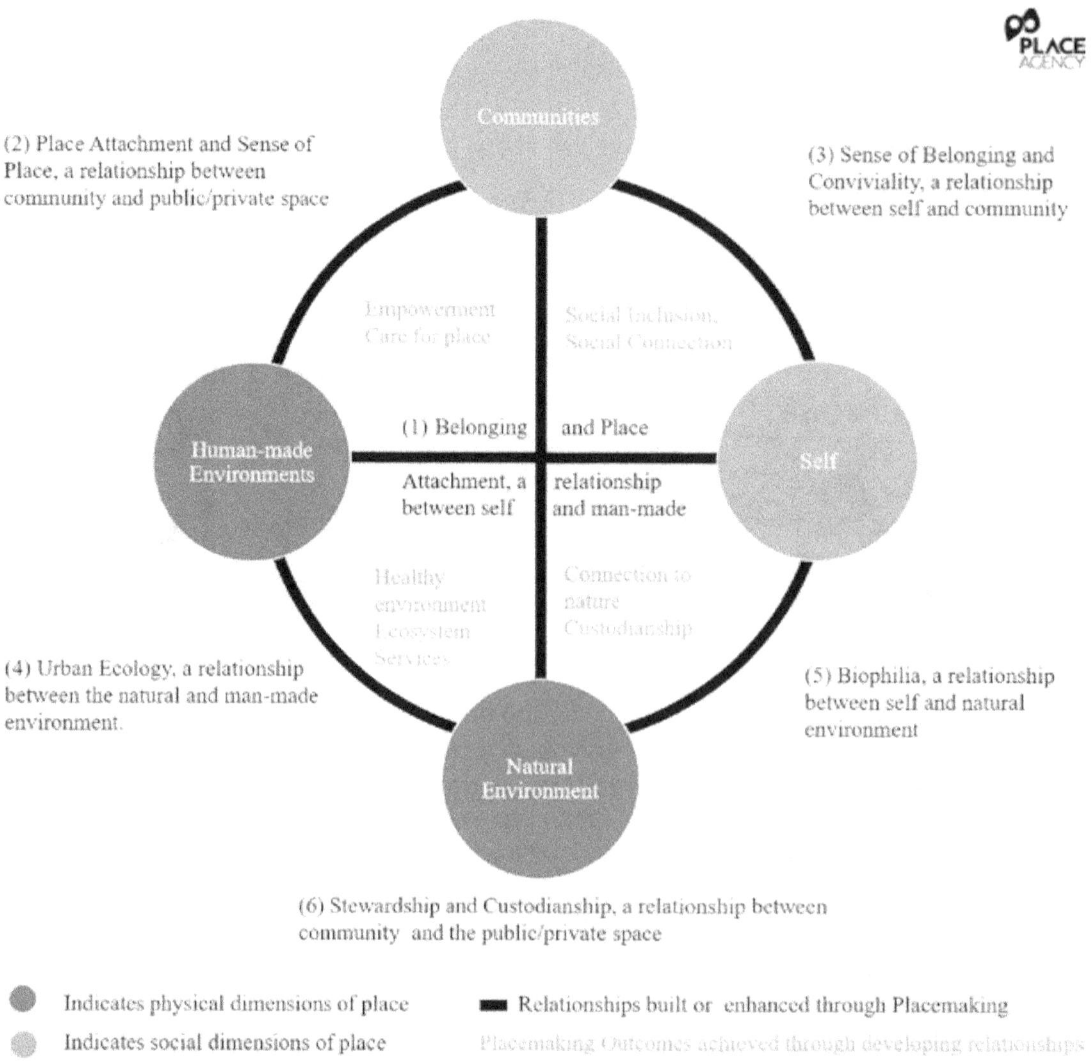

and social features of placemaking support targeted outcomes such as social bridging and democratic participation. The ability to identify and measure tangible indicators of placemaking would help community organizers and public institutions seeking to increase social cohesion and access to resources, such as community centers and museums, evaluate their progress.

CONCLUSION

Placemaking with creative ambiance has a vital role to play in exploring and promoting processes that can affect social activities. This chapter identifies the attached public space from a social problem that contributes to social, economic, and racial inequality, and advances a theoretical framework for "critical placemaking" as a way to make attracted public space for public use and be more attractive, synthesizing key concepts from physical environment and atmosphere. Throughout this chapter, we have focused on three questions: Who can access and participate in attached public places? Whose stories and histories are told in public places? And finally, what kinds of interactions and behaviors are afforded by public places?

First, this chapter considers who can access and participate in attached public places. Researchers suggest that this attached public space cannot be accessed by all the participants as it's not targeting low-income people's perception. It is imperative to consider the role of placemaking in catalyzing change. Sutton and Kemp (2011) affirm this point, stating, "Bringing place into consciousness and realizing that everyone has the potential to be makers of places changes not only the place but the people as well" (Hes, Dominique et al., 2019).

Second, this chapter addresses whose stories and histories are told in public places. Using a narrative lens in the proposed framework illustrates how inclusion and participation in attracted public places are communicated through material and symbolic features of the physical environment and atmosphere through dialogue in everyday social interactions critical placemaking is theorized as a way to make space for a plurality of voices and histories and create a more inclusive community narrative. Finally, the present analysis attends to what kinds of interactions and behaviors are afforded by attached public places. The research tried to validate that if preservationist and Urban designers put into consideration some guidelines for those attached public space, it will affect the interactions and behaviors of the users.

Drawing from community psychology, we argue that critical placemaking can facilitate a lot of benefits for the attached public spaces and empowerment. The reimagining and transforming our relationship with place, and relationships community members is possible with applying placemaking with creative ambiance. Critical placemaking can work to transform public places into more inclusive and democratic settings of everyday learning and critical education, planting the seeds for social change. Placemaking is the continuous process of shaping contributing to 'place' (Hes, Dominique et al., 2019). Moreover, one feature or a key for right placemaking with a creative ambiance to have different strategies that can be applied for the process of finding long-term solutions which will have great benefits on participants that make them involved in the place (Hes, Dominique et al., 2019). It is worth saying that this argument was done as an initial step seeking further studies to be done as a trial for upgrading attached public spaces, which will contribute to the social quality of life. Future research should consider case studies in different communities. These case studies should follow the assessments or framework, which was an argument in the literature review to validate if those attached public spaces are attractive or not.

REFERENCES

Adams, P., Hoelscher, S., & Till, K. (Eds.). (2001). Texture of place. Univ of Minnesota Press.

Alvarez, L., Borsi, K., & Rodrigues, L. (2017). The role of social network analysis on participation and placemaking. *Sustainable Cities and Society*, *28*, 118–126. doi:10.1016/j.scs.2016.06.017

Andrews, G. F. (1974). *Maya Cities: Placemaking and Urbanization* (1st ed.). Univ of Oklahoma Pr.

Angelidou, M., Karachaliou, E., Angelidou, T., & Stylianidis, E. (2017). Cultural heritage in smart city environments. *ISPRS - The International Archives of the Photogrammetry, Remote Sensing and Spatial Information Sciences*, *42*(W5), 27–32. doi:10.5194/isprs-archives-XLII-2-W5-27-2017

Australia ICOMOS & International Council on Monuments and Sites. (2013). *The Burra Charter: The Australia ICOMOS charter for places of cultural significance 2013*. http://australia.icomos.org/wp-content/uploads/The-Burra-Charter-2013-Adopted-31.10.2013.pdf

Bakker, I., van der Voordt, T., Vink, P., & de Boon, J. (2014a). Pleasure, Arousal, Dominance: Mehrabian and Russell revisited. *Current Psychology (New Brunswick, N.J.)*, *33*(3), 405–421. doi:10.100712144-014-9219-4

Bauer, C., Recreation, S. C., Fletcher, R., Travieso, C., & Woodward, D. (2010). High performance landscape guidelines: 21st century parks for NYC. In *Design Trust for Public Space : City of New York. Parks & Recreation*.

Berman, B., & Evans, J. R. (2013). *Retail management: A strategic approach* (12th ed.). Pearson.

Bitner, M. J., & Obermiller, C. (1984). *The elaboration likelihood model: Limitations and extensions in marketing*. Academic Press.

Bitner, M. J. (1992). The Impact of Physical Surroundings on Customers and Employees. *Journal of Marketing*, *56*(2), 57–71. Advance online publication. doi:10.1177/002224299205600205

Bonn, M. A., Joseph-Mathews, S. M., & Dai, M. (2005). *International versus Domestic Visitors: An Examination of Destination Image Perceptions*. doi:10.1177/0047287504272033

Bonn, M. A., Joseph-Mathews, S. M., Dai, M., Hayes, S., & Cave, J. (2007). Heritage/Cultural Attraction Atmospherics: Creating the Right Environment for the Heritage/Cultural Visitor. *Journal of Travel Research*, *45*(3), 345–354. doi:10.1177/0047287506295947

Boyd, S. (2002). Cultural and Heritage Tourism in Canada: Opportunities, Principles and Challenges. *Tourism and Hospitality Research*, *3*(3), 211–233. doi:10.1177/146735840200300303

Carmona, M., Heath, T., Oc, T., & Tiesde, S. (2003). *Public places, urban spaces: The dimensions of urban design*. Architectural Press.

Carr, S. (1992). Public spaces. Cambridge, UK: Cambridge University Press.

Chang, T.-Y., & Horng, S.-C. (2010). *Conceptualizing and measuring experience quality: The customer's perspective*. doi:10.1080/02642060802629919

Community toolbox. (2019). *Section 7. Changing the Physical and Social Environment*. Implementing Promising Community Interventions. https://ctb.ku.edu/en/table-of-contents/implement/physical-social-environment

Dajian, Z., & Peter, R. (2006). 2010 World Expo and Urban Life Quality in Shanghai in Terms of Sustainable Development. *Zhongguo Renkou Ziyuan Yu Huanjing*, *4*(1), 15–22. doi:10.1080/10042857.2006.10677445

DCMS. (2010). *Principles of selection for listed buildings*. The Department for Digital, Culture, Media and Sport is a Department of the United Kingdom Government.

Elgobashi, A., & Elsemary, Y. (2019). *Towards Social Quality of life Through Cairene Public Parks*. Architecture and Urban Planning Departments.

Faris Hmood, K. (2019). Introductory Chapter: Heritage Conservation - Rehabilitation of Architectural and Urban Heritage. In K. Hmood (Ed.), *Urban and Architectural Heritage Conservation within Sustainability*. IntechOpen. doi:10.5772/intechopen.86670

Farrar, M. E. (2011). Amnesia, Nostalgia, and the Politics of Place Memory. *Political Research Quarterly, 64*(4), 723–735. doi:10.1177/1065912910373553

Feilden, B. M. (2005). Conservation of historic buildings (3rd ed.). Elsevier.

Ginzarly, M., Pereira Roders, A., & Teller, J. (2019). Mapping historic urban landscape values through social media. *Journal of Cultural Heritage, 36*, 1–11. doi:10.1016/j.culher.2018.10.002

Heintzelman, M. D., & Altieri, J. A. (2013). Historic Preservation: Preserving Value? *The Journal of Real Estate Finance and Economics, 46*(3), 543–563. doi:10.100711146-011-9338-8

Hes, D., & Hernandez-Santin, C. (Eds.). (2019). Placemaking Fundamentals for the Built Environment. Academic Press.

Hyden, C. T., & Sheckels, T. F. (2016). *Public Places: Sites of Political Communication*. London: Lexington Books.

ICOMOS. (2005). *Xi'an declaration on the conservation of the setting of heritage structures, sites and areas*. International Council on Monuments and Sites(ICOMOS).

Iliyasu, I. I. (2014). *Challenges of preservation of cultural landscapes in traditional cities: Case study of Kano Ancient City*. Academic Press.

Kotler, P. (1974). *Atmospherics as a Marketing Tool*. Academic Press.

Lynch, K. (2005). The image of the city. MIT Press.

Massara, F., Liu, S. S., & Melara, R. D. (2010). *Adapting to a retail environment: Modeling consumer–environment interactions*. Academic Press.

Meeks, S., & Murphy, K. (2016). *The Past and Future City: How Historic Preservation Is Reviving America's Communities* (2nd ed.). Island Press. doi:10.5822/978-1-61091-709-4

Mehrabian, A., & Russell, J. A. (1980). *An Approach to Environmental Psychology*. Academic Press.

Pallasmaa, J. (2014). *Space, place and atmosphere. Emotion and peripheral perception in architectural experience*. Academic Press.

Parsons, R. (2010). *New Design in historic setting*. Academic Press.

Poria, Y., Butler, R., & Airey, D. (2003). The core of heritage tourism. *Annals of Tourism Research, 30*(1), 238–254. doi:10.1016/S0160-7383(02)00064-6

PPS. (2010). *Placemaking Meets Preservation*. https://www.pps.org/article/placemaking-meets-preservation

Redi, M., Aiello, L. M., Schifanella, R., & Quercia, D. (2018). The Spirit of the City: Using Social Media to Capture Neighborhood Ambiance. *Proceedings of the ACM on Human-Computer Interaction, 2*, 1–18. 10.1145/3274413

Relph, E. (2008). *Place and placelessness*. Pion Ltd.

Schwartz, J. M., & Cook, T. (2002). Archives, records, and power: The making of modern memory. *Archival Science*, *2*(1–2), 1–19. doi:10.1007/BF02435628

Silberberg, S., Lorah, K., Disbrow, R., Muessig, A., & Naparstek, A. (2013). How placemaking builds places and communities. Massachusetts Institute of Technology.

Strydom, W., Puren, K., & Drewes, E. (2018). Exploring theoretical trends in placemaking: Towards new perspectives in spatial planning. *Journal of Place Management and Development, 11*(2), 165–180. doi:10.1108/JPMD-11-2017-0113

The WBDG Historic Preservation Subcommitte. (2019). *Historic Preservation*. Author.

Turley, L. W., & Milliman, R. E. (2000). *Atmospheric Effects on Shopping Behavior: A Review of the Experimental Evidence*. Academic Press.

UNESCO. (2005). *Vienna Memorandum on "World Heritage and Contemporary Architecture – Managing the Historic Urban Landscape."* Author.

UNESCO. (2011). Proposals concerning the desirability of a standard-setting instrument on historic urban landscapes 36 C/23. *Records of the General Conference, 36th Session, Paris, 25 October - 10 November 2011, v. 1: Resolutions*. UNESCO.

Van Vliet, H. (2018). *The measurement of atmospherics*. Plan B.

Wallace, C., & Abbott, P. (2009). *Social Quailty in Europe*. Academic Press.

Whyte, W. (1968). *The Last Landscape*. University of Pennsylvania Press.

Whyte, W. H. (1989). *City: Rediscovering the Center*. University of Pennsylvania Press.

World Heritage Center. (2013). *New life for historic cities: The historic urban landscape approach explained*. https://unesdoc.unesco.org/ark:/48223/pf0000220957_chi

Wyckoff, M. A. (2017). *Placemaking as an Economic Development Tool: A Placemaking Guidebook*. Academic Press.

Yeomans, R. (2006). *The Art And Architecture of Islamic Cairo*. Garnet Publishing Limited.

Zalta, E. N. (2006). Dewey's Aesthetics. In *The Stanford Encyclopedia of Philosophy*. Metaphysics Research Lab, Stanford University. https://plato.stanford.edu/entries/dewey-aesthetics/

Zhong, J. (2012). Reconnecting lost place: Comprehensive green space design based on historic preservation principles in compact Changsha City. *China*, 53.

Chapter 9
The Gastronomy of the Eye:
A Tale of Two Cities in India

Vrushali Dhage
ArtForum SF, USA

ABSTRACT

Works of art can be read at various levels: from being objects of simple retinal pleasure to the other extreme of being significant critical statements of their time. This chapter aims to strike a cerebral dialogue through the works of art. The current study shall consider the latter function of art and analyze the methods in which contemporary Indian artists have made attempts to provide a critique of the early initiatives towards developing Delhi and Mumbai as 'smart cities'. The review of works from India concludes the essential role of infrastructural projects and envisioned spaces built in the era of economic liberalization. The study aims at drawing a methodological approach, with an art historical perspective, with the artists analysing and translating the urban experiential phenomenon, into artworks.

INTRODUCTION

The concept of the "gastronomy of the eye" was discussed by Honoré de Balzac, in *The Physiology of Marriage* as, "to saunter is a science; the gastronomy of the eye. To take a walk is to vegetate; to saunter is to live" (p. 31); based on the art of knowledgeable observation, over mere retinality. Attesting the belief and necessity of a knowledgeable, tasteful (Bouvier, 2005, pp. 95-111) and critical observer of one's surroundings, a questioning individual, as a positive contributor to the society; adding a social and a scientific dimension to it. A historic parallel in defining the urban-experiencing-questioning-human, as neutral/detached, keen/unbiased, yet a knowledgeable thinker, is of the 19th century *flâneur,* through the urbanscape of Paris; turning sauntering into a discipline of perception. Baudelaire further insisted on a creative aspect to *flânerie* by associating it to modern painters and writers, as individuals with discerning capacities who could "see" better (Benjamin, 2006) whereas Aimeé Boutin highlighted the expansion of the sensory perceptions while discussing the "sensual turn" which complemented "the rise of visual and cultural studies which positioned the *flâneur's* modernist gaze as a painter of modern life

DOI: 10.4018/978-1-7998-3856-2.ch009

at the centre of many of its inquiries" (Rethinking the Flâneur: Flânerie and the Senses, 2012, pp. 124-132). Though one couldn't overlook the capacity of some capable commoners to the same, Baudelaire granted the painters and writers a higher pedestal. He regarded them individuals capable of translating their observations into creative expressions of the same, thereby extending the very experience, and granting the works of art/writing a status of documents supporting the observations. The validity of this approach can be tested with respect to contemporary times too.

The premise of this study aims at analysing the transforming urban character of two metropolises of India namely, Delhi and Mumbai, through various smart city initiatives, corresponding to the adoption of a neo-liberal economic approach by the country; palpable to the common people, primarily through the growing number of constructed forms like – the Special Economic Zones (SEZs), business parks and various infrastructural activities. Customarily, objective, rational and empirical proofs help analyse the success or failure of such transformations. But the current study aims at devising a *qualitative method* with an *art historical approach* based on "human observation and experience", with the consequent "works of art" as expressions of the experience (Leddy, 2019). Therefore, the "observant experiencing humans" under consideration, consist of some select contemporary Indian artists, as recontextualized *flâneurs,* navigating through the new manifest forms of the new economic approach. These artists are perceiving, conceiving and translating the "new urban ambiance" into works of art; thereby, modelling the artworks as critiques of the given premise.

The study provides a brief overview of the neo-liberal economic scenario in India, as a step towards globalization, beginning from the late 1980s. Followed by mentioning the transformations and attempts towards upgradations in Delhi and Mumbai. This is done by primarily focusing on the built forms, which drastically changed the characters of the respective cities, and provided a cue for the artists to base their works on. Simultaneously, aiming at re-contextualizing the notion of the artist as a *flâneur* with respect to the current approach. Finally, the study focuses on an in-depth analysis of artworks, which address various aspects of these transformations. The review of this transformation goes through multi-faceted ways – from a personal-physical sensory experience of walking through spaces to the macro-vision of the restructuring of the cityscape. The review also investigates the new notions of updating and up-gradating of the influence of advertisements and the media in the mental-shaping of one's identity within the larger globalizing urban framework. This investigation of the spaces has been marginalised or completely excluded from the grand projects of projected progress.

TOWARDS ECONOMIC GLOBALIZATION

The influence of the neo-liberal economic stand of the West started having a significant impact on India by the 1980s. Consequently, policies on liberalization were framed. Their implementation took place in phases, beginning with industrial liberalization, and later of the financial markets. This gradual approach ensured minimum trauma and a smooth economic transition in the respective sectors (Nayak, 2008, pp. 122-127). The newly liberal Indian market was then looked upon by various stronger capitalist economies as an emerging market with a high potential for returns. This led to direct investments from foreign players in various sectors. The rising competitive capital investments made it evident that India would have to equip itself with the necessary infrastructure, to hold the interest of these entities.

Then along with the local economic factors, the globally experienced technological revolution of Information and Communication Technology (ICT), catalysed the process of global integration. The new satellite systems and fiber-optic cables that provided a mechanism for the internet-based technologies to run (Steger, 2013, p. 43), accelerated and eased the integration of financial markets across the globe, best explained as *technocapitalism* – representing its inseparable aspects (Kellner, 2002, pp. 285-305). This phenomenon affected India only by the 1990s. Through economic liberalization, India's intellectual capital found venues towards democratic participation at an international level. These individuals, with their great skillsets attracted many Multinational and Transnational Companies (MNCs and TNCs) to India. It is during this time that India emerged as an outsourcing hub and China as a manufacturing hub (Dickens, 2001, pp. 2-4). On the one hand, the web-technology evidently reduced the notion of locational-rootedness. Individuals could be employed in any part of the world. Businesses no longer needed one fixed space and could be outsourced or spread across the globe, giving an illusion that cities would no longer matter. On the other hand, this very spatial dispersion gave rise to multiple nodes across the globe. These nodes would be an integral part of the global networks and systems, characterized by urban agglomeration and hyper-concentration, (Sassen, Cities in a World Economy, 2006) of the intellectual capital – as individuals, and the material facilities – as the "updated" physical spaces; each reflecting a global ethos.

In the 90s, very few Indian cities fulfilled these two criteria. Then the two metropolises namely, Delhi the political capital, and Mumbai the financial capital of the country, with an already established global reputation, with a relatively updated infrastructure, economic base and talent pool, were a preferred choice for the international investors to set their physical presence. Considering the immense potential from global participants, the Indian government started developing these two cities on the lines of their likely global counterparts. Thereby, entering a new phase of urban development which would reflect the technocapitalist character of globalization.

Smart City Vision in Naissance

Driven by the objective of becoming a strong and integral participant in the global economic order, the government made efforts at strengthening the network of global integration at various levels and through various fields, especially in transforming the two cities (Jayasuriya, 2009). Consequently, colloquial terminologies like developing of "world-class" cities, "smart cities" or cities of "global cadre", gained circulation to describe the efforts, highlighting the development of the new urban ecosystem, through technology and economy/entrepreneurial tendencies, as the foundational elements. It is a known fact that these terminologies have no rigid definition (McClellan, Jimenez, & Koutitas, 2018). In the first world countries, it could spell the adoption of the latest technologies to upgrade the pre-existing ones or inventing newer ones (Pelton & Singh, 2019). Whereas in some other spaces it could spell, that of providing better infrastructure. Ranging from, facilitating mobility through the urban spaces through road or rail connectivity, having hospitals with better equipment, schools with improved facilities or providing basic internet access to small-scale businesses or cutting-edge technology to larger set-ups. The study shall consider the following developmental efforts – first, the creation of the Special Economic Zones (SEZ) and Business Parks, purely from an economic perspective; and second, of providing better infrastructural facilities, especially for commutation, in the cities. A cascading effect of these changes was traceable in the soaring of the realty and housing sector too, the growth of which gave a different dimension of performativity, cosmopolitanism, and diversity to the growing cities. Evidently, the mentioned developmental efforts taken by the government form a small part of the totality, but these were the only efforts that were understandable and strongly palpable to the common man and the artist-*flâneur,* within the city as its lived experience. The other efforts, for example, the international trade ties, monetary policies, or labour policies, etcetera, were relatively complex and distant to the citizens.

The SEZs and well-spaced Business Parks were characterized by the state-of-the-art structures housing global business. The uses involved with updated technology assuring smooth functioning, saturated with a high-skilled and high-income talent pool, new connectivity through metro-rails, flyovers and wide roads. The dotted peripherally by luxury homes, gated communities, multi-speciality hospitals, international schools, and malls were also included. These uses would house the global brands, which had entered India in its post-liberalization phase. Most of these large "template-like" plans, materialized on the outskirts of the characteristic "new phases" of the metropolises, like New Delhi and New Mumbai, over the expensive and cluttered older locations. A move could be traced from the national industrial modernity towards the post-industrial, postmodern globality (Lyon, 1999, pp. 46-47) (Bell, 1973). All of which gave a general perception of sustainable and well-distributed progress.

Yet some early efforts in the 1990s, towards attaining these traits in Delhi and Mumbai, exhibited not just economic aspirations but political agendas and event-based hasty targets of exhibitionism, aiming for "instant development". For example, the completion of Phase II of the Delhi Metro Railway, during the 2010 Commonwealth Games hosted in the capital, to provide easy access to the sports venues, marked the third category. Series of project violations, mishaps and deaths took place while attempting to complete the project just before the Games. Similarly, the desire of transforming Mumbai into Shanghai (Tripathi, 2007, pp. 45-49), led to an array of project implementations, beginning with the construction of "fifty flyovers" in 1997, followed by elevated metro and monorails, elevated footbridges or skywalks and the ongoing disputed coastal road project. During their conceptualization, these structures were to be considered "proud emblems" of the "tiers of smartness". Even as these parameters seemed realistic and practical, the lived experience of the same wasn't uniform. To some critical experiencers, the "utopian dream" itself was drip-irrigating an "anti-utopian reality".

On the one hand, the infrastructural up-grading of the cities did facilitate public transport yet overlooked the threats to the natural resources and environmental impacts, with their constant encroachment of the greenfield. Various smart technologies purporting low emission methods were causing detrimental environmental effects in the very process of their implementation. For example, the construction of flyovers and metro rail resulted in indiscriminate felling of greenery. These "template-like" structures radically transformed the earlier character of the cities, bringing a sense of unnecessary uniformity across cities. For example, Bangalore transformed from a beautiful "Garden City", to an "IT City", irreversibly destroying the ecological and environmental levels, of the city known for its botanical charm. Large spaces cleared on the pretext of development seldom benefitted the locals, many of which destroyed the greenfield areas surrounding the cities.

Most importantly, the profits of these efforts were shared in an extremely skewed manner, predominantly leading to "the growth of an enlightened type of a workforce" (Sassen, 1991, pp. 195-196). Stiglitz stated that even as the quality of jobs and life raised for a sizeable Indian population, a palpable contrast exists within and around the cities, with the non-participants or those excluded from the global economic network (Globalization and its Discontents, 2003). Therefore, the consequent alterations of the economic, topographic and demographic characters in the cities were met with views of extreme polarities; bringing the very notion of the "vision" of development under question. These contradictions, fissures, and polarities seemed justifiable and dangerously magnified considering the scope and the unstoppably rapid rate of the transitions as a lived reality.

Contextualizing *Flâneur*

The usual ways of validating the success of the developmental visions are based on the empirical data, as provable parameters, like employment figures, pollution levels, number of project completions, commutation facilities, etcetera. Seldom is a qualitative "human experience" considered as a logical indicator, which itself holds the potential to be an alternative process of analysis, questioning and validation. Prioritizing human experience as a primary parameter necessitates efficiently defining the "analysing-questioning human". Fundamental to which would be – the politics, the position, and the process, as factors conditioning the emergence of the questions. The individual then has to be read through his/her – ideological and political affiliations, the position of being aspirants or participants in the globalized scenario, of falling under the category of beneficiaries or of being marginalized, worse excluded, expelled or exploited by the same, proponents or opponents of the same, an insider/outsider, etcetera, as factors influencing the manner and direction of reading the continually transitioning urbanscapes.

Bringing back the artist- *flâneur,* as the detached, knowledgeable, experiencing human in focus, capable of reading the co-existing contradictions of one's time unbiasedly. The *flâneur* could then be closely compared to Dickens' omnipresent and analytical narrator in *A Tale of Two Cities*; capable of analysing opposing – *situations* and *people*. *Situations* of the coexistence of hope and despair, light and darkness, an age of wisdom and foolishness (Dickens & Shruckburg, 1949), not as mental constructs but live social situations (Sanders, 1988); and the *people*, some as beneficiaries and remaining as the marginalized, overlooked others.

With the urban smart-spaces at the centre of the inquiry, the position of the contemporary artist-*flâneur* needs re-contextualization. Further, understanding the positions and perspectives of the artists, at the same time considering their identity as individuals in the growing network of the art world. Most of the artists in India benefitted from globalization, as it led to the expansion of its art market. Delhi and Mumbai in the 1990s were already thriving art-cities, having numerous patrons, private/public galleries, buyers, collectors, etcetera, and the required art-infrastructure. The establishment of international auction houses namely, Christies and Sotheby's, resulted in the further inclusion of Indian artists, in the global art networks, further gaining visibility and opportunities and consequently making them beneficiaries of the same. The global peripatetic character of their life, countered/balanced by a rooted experience of the local cityscapes through their works and interactions, granted the artists an insider/outsider perspective towards the same, and thereby granting multiple perspectives to the process of reading and analysing the spaces under consideration, before materializing them into artworks.

With the *artist-flâneur* as the experiencing human, navigating through *urban spaces* of interest/inquiry, employing the analytical mind towards the creation of relevant *artworks*, the methodological approach considers these three as prime entities of analysis. The approach borrows from Archigram and Lefebvre. Archigram's thematic sections of the exhibition Living City, 1963 – Man, Survival, Crowd, Movements, Communication, Place and Situations (Sadler & Archigram (Group), 2005) (Shepard, 2011, pp. 16-37), as ways/parameters of reading, locating and perceiving the relation between the space and the lived experience. And the dialectical relation between Lefebvre's perceived, conceived and lived experiences to assert the importance of its third moment of the social space, that is, "representational space" (Plan of the Present Work, 2000, pp. 260-265) (The Production of Space, 1991), as the core of the study. According to Lefebvre, this third moment embraces the "loci of passion", moving from an exact replication of situations/spaces towards unbound imaginations. The following sections explicate how various artists have employed different ways of expression, ricocheting between duplication of spaces to their conjured illusions, of stating locational specificities and generalizations, attempting to trigger a dialogue with the "viewer", rather the "experiencer of the work".

Technocapitalism and Manifest Forms

From the 1990s new architectural styles for the commercial-corporate structures emerged, aiming to exhibit an "updated-global" character. The numbers of which increased at a phenomenal rate. One of the prominent traits was the creation of glass façade buildings. Some early exemplars of these types of structures were erected in business districts located in the heart of the Mumbai and Delhi in the 90s. Over the years these started spreading in the not-so-developed peripheries too. Sudhir Patwardhan's work titled *The Emergent*, 2012, Figure 1, takes a predictive stance on this situation. The image can be read as two parts. One, of the numerous angular roofed and differently coloured older shanties, evidently depicting high-density habitation, with no trace of a planned arrangement or even of a commutable road. The other, of a new, posh, glass façade building stands out towering beyond the picture frame. Both the elements dominating the visual equally could be interpreted as co-existing entities in a first glance; but the title favours the latter, implying its dominance over the prior. The "emergent" in the title, signals the emergence of a temporally new incongruent entity, which could be read as an insignia of development, as a beginning to a process which shall eventually annihilate the prior. At the same time, raising a question, whether the earlier residents could be absorbed in the new system. The absence of people in the visual consequently grants focus to the built forms, which are emblematic of the process of social erasures and inscriptions (Siva Kumar, 2012, p. 11). As a Mumbai-based artist, Patwardhan witnessed the changes in the cityscape from the decade of the 1980s. Requiring a regular commute between his suburban home to the elite, plush Art District of South Mumbai, helped him to not just "see" the contrasts through the physical emblems, but trace the process spreading the same contrast over various regions, using the "built-forms" as parameters to measure the same.

Pooja Iranna, a Delhi-based, artist employs a similar methodological approach, to read the "templatization" of urban spaces through standardized structures, overwriting the earlier discrete character of various cities. In *The Twist*, 2010, Figure 2, three glamorous buildings with a golden sheen and varying heights stand together in a cluster, suggesting a space of grandeur. Unlike, Patwardhan's work, these buildings have no surrounding or a backdrop as a referential frame, they are alone-standing; insinuating at the increasing number of their clones in innumerable urban locations, usually in greenfield regions, such that their count counts as a parameter of the prosperity of the location. A closer look at the work reveals distortions in the structures, justifying the title, which is camouflaged from a distance. The work is constructed from carefully balanced and attached staple pins, giving a sense of unease, such that even as the structure looks attractive and admirable, there is an inherent fragility and uncertainty of stability. Iranna metaphorically raises multiple questions, from the erasure of the established local architectural styles to the very feasibility, practicality, and life of such structures. For example, the Indo-Saracenic style evolved under the British Raj drawing from local and regional traits, styles and materials. Similar, was the adaptation of Art Deco structures to befit different Indian cities, each exhibiting a sense of a positive eclecticism in architecture. Contrastingly, the physical emblems of the globalized technocapitalism exhibit no such attempts of absorption of local traits but spell an assertion of a set-standard. Iranna draws a parallel between the notion of stylistic assertion, to the inclusion/exclusion of the earlier "people/occupants" of the land and the circumstantial assertions as a part of the process.

Figure 1. Sudhir Patwardhan, The Emergent, 2012. Acrylic on canvas, 30x30 in
(Source: Photo Courtesy - © 2012, Sudhir Patwardhan, used with permission)

Figure 2. Pooja Iranna, The Twist, 2010. Staple pins
(Source: © 2010, Pooja Iranna, used with permission)

One case study to explicate such acts of gentrification and painful displacements of the earlier residents is through the pockets of Lalbaug and Lower Parel in Mumbai. In the 19th century, with the establishment of textile mills in these areas, they characterized "colonial Bombay's industrial modernity" (Nallathiga, 2001), which brought about a major transition in the skyline of then-Bombay, with massive mills and chimneys dotting it. These mills were surrounded by low-cost, high-density residential buildings called the *chawls,* for the millworkers. The area was consequently named as *girangaon* in Marathi, literally translated as a mill-land/village/precinct. By the 1980s, with the rise of the power-looms and many several management issues, the mills became defunct, followed by a long strike by the mill workers, and consequently were declared "sick" in the 1990s (Harris, 2008, p. 2408). Within a decade, they were gradually opened to redevelopment, on the pretext of providing employment to the unemployed mill workers. In no time the mill-land transformed into a *global business district*, turning into a "new expensive area", benefiting from its proximity to the uber-expensive business district of South Mumbai (Dhage, 2013, pp. 110-113). Malls, shopping complexes and various ancillary businesses too were rapidly setup. The glass façade skyscrapers started replacing the mills, and the surrounding *chawls* too were being torn down to be replaced by luxury, low-density housing; thereby largely displacing the earlier residents and also erasing the historic identity of the space. The towering chimneys found competition from skyscrapers, again re-inscribing the skyline of the locality.

Figure 3. Sudhir Patwardhan, Lower Parel, 2001. Acrylic on canvas, 48x96 in
(Source: © 2001, Sudhir Patwardhan, used with permission)

In *Lower Parel,* 2001, Figure 3, Sudhir Patwardhan explores the mentioned piece of history through its remnants and the palpable layers of time. The foreground is filled with people, from a non-elite class, possibly residents of the *chawls*, engaged in various activities, in a space which has simple shops. Many of which are housed in a grey-stone structure, which is a local railway station built during the colonial times. The other layer is that of a mill, with a chimney towering out of the visual format, depicting the mentioned industrial modernity of the space. And finally, the background is a multi-stories structure under construction, this represents the contemporaneity of the space. Even as the image shows the presence of the earlier residents, like *The Emergent*, *Lower Parel* too, takes a near predictive stance, insinuating the possibility of these people becoming one of the layers of history of this space. The nature of these changes evidently implied a rise in the need for skilled and educated professionals over unskilled labourers in the mentioned locality. Even the simple profession of janitors and cleaners, for the new structures, meant outsourcing the task to specific companies, with a uniformed force, over employing local individuals. On the one hand, there was the justification given, that such outsourcing aided the erasure of the age-old discriminatory caste system, which forced the lower classes to undertake unhygienic tasks. Even while being a certainly positive step, it evidently addressed only a minute fraction of the larger issue of economic displacement and distress. Over the years, it has been proven that the opening of mill land for redevelopment did not benefit the mill-workers but simply led to gentrification (Adarkar & Menon, 2004, p. 3). These artworks can, therefore, be categorized as archival material, documenting the phases, people and causal factors, by the artist who is oxymoronically a "distant-participant"; an individual who personally witnessed the historic fall of the mills and efforts towards their redevelopment, and has physically walked through the older congested localities, to enter the sanitized environment of the spacious, glamorous new structures, to re-read these layers of history and contemporaneity.

RECONFIGURING/REIMAGINING URBAN SPACES

The infrastructural projects mentioned earlier, dealing with the upgradation of the then existing, or the introduction of the new technologies, both as a process and a product, affected the citizens closely right from its inception. For example, the projects of "commutation and connectivity", through the construction of flyovers, metro and monorail projects in both the cities were considered crucial towards aiding a well-dispersed growth. Resulting in the restructuring of the cities through new nodal points and networks and setting renewed notions of spatial hierarchies. One is reminded of Archigram's categories of "the crowd" as possible navigators of the space and "movement as the generator of form" (Sadler & Archigram (Group), 2005, pp. 52-89), for this case. In *Face to Face*, 2010, Figure 4, the Gigi Scaria, a Delhi-based artist, as a regular commuter, stands in the middle of a flyover, possibly a new addition to the city, aiming to capture the completion of a metro-line – a newer "retrofitted" layer of smartness added above the existing one. The two converging ends are loaded, each with buildings from Shanghai and India respectively, an obvious artistic manipulation. The symmetry of the visual seems intentionally staged. On the one hand, he addresses the political intent of equating Indian cities to Shanghai, through projected technological adaptations and parity; such that the convergence of the ends could hypothetically imply a status of parity. And on the other hand, he explores the enormous desire for vertical growth of the realty developers, as a tactic to maximize the return on investment for a given piece of land.

Another work titled, *Amusement Park*, 2009, Figure 5, is an outcome of the multiplicity, simultaneity and the sheer lifelessness of such construction projects. Drawing inspiration from the mechanical amusement rides of a park, Scaria substitutes the rides through what could be considered as the modern-day urban replicants; letting skyscrapers, glass façade buildings, the metro rails, large digital hoardings, flyovers, structures under construction, some demolish, and makeshift structures replace the space. The synoptic visual is saturated with "aspirational structures", aiding the motorized economy, elite lifestyle, exhibitionism and hypermobility. Further, replacing the sounds in a children's park with those of unrhythmic and deafening noises of construction machines, traffic and machinery. On the one hand, with the sparse depiction of greenery, he addresses the issue of the overwriting of the green spaces by concrete structures, which were met with massive protests by the citizens and environmentalists, and of the decreasing number of natural spaces for children's recreation. And on the other, he addresses the naturalization of the unnatural under the pretext of development. Such that felling of green spaces, or the filling of underground water-tables, to construct emblems of progress gets legitimized with impunity; a myopic approach causing irreversible ecological damages.

Anant Joshi, creates a *faux terrain* (Adajania, 2008), to emulate the lived experience and innate character of Mumbai, by recreating its skyline. The metaphorical scape is constructed from wooden objects of varying sizes and shapes resembling firecrackers, each towering to represent the vertical growth of the city and the undulating skyline, in the *Untitled,* 2012, Figure 6, installation. These are painted in

Figure 4. Gigi Scaria, Face to Face, 2010. Digital print on Archival Paper
(Source: © 2010, Gigi Scaria, used with permission)

Figure 5. Gigi Scaria, Amusement Park, 2009. Video stills (total video duration - 5 mins)
(Source: © 2009, Gigi Scaria, used with permission)

various bright, attractive colours, representing the quotient of glamour and attraction towards the city, and its multi-cultural character. The overcrowding and close placement of the objects stands emblematic of the dense demographics of the city and of the government's efforts at constantly adding tiers of smartness to the city through redevelopment and retrofitting, nearly testing its brink of saturation. From a distance, the installation looks like an attractive visual, which to the artist is a projected image – an illusion of the city's attractiveness. A closer look reveals a wick on every firecracker, placed on a table-cover with sources of ignition. Once set into motion through the underlying mechanical circuitry, each object of the work gyrates ferociously, creating a deafening sound, suddenly turning from an attractive and vibrant assemblage to a threatening, overpowering and repulsive object; nearly forcing the viewers to distance themselves from the work. Joshi builds the notions of precarity by placing the firecrackers on the sources of ignition. Joshi also proposes an evident of the possibility of unmanageable emergencies as an outcome of the rapid and uncontrolled/unplanned growth. Both Scaria and Joshi, by emulating the urban experiential phenomenon in a gallery space, presented a magnified or a fast-approaching version of the reality; by merging the visual excess of the urban spaces with its overpowering "auditory cacophony", adding a new dimension to the earlier discussed "sensual turn".

Figure 6. Anant Joshi, Untitled, 2012. Sculptural installation, 100 objects fixed and rotating on a 16 x 4 feet table
(Source: © 2012, Anant Joshi, used with permission)

MEDIATIC INFLUENCE

The discussion about the palpable transitions of the public spaces needs to be turned towards the private/personal aspect, that is, towards the citizens/participants. As individuals migrated to urban areas for employment the demand for housing grew in tandem. To be at parity with the world-class working spaces, the homes too were projected to possess such characters. The notion of housing was altered from that of mere homes to "smart homes", as spaces reflecting the updated character of its residents. It could be said that within a short span of nearly two decades, everything was taking a "smart" form or turn (Wessner, 2011). The marketing tools for the same, too, had expanded from the general and conventional, newspaper, magazines, television advertisements, and hoardings, to personal connections through mobile messaging, telemarketing, emails, suggestive advertisements based on individual data generation, through one's internet surfing interests. These additional methods, though sometimes intrusive, had successfully created a class of potential investors, based on their desires towards the adaptation of an updated global identity. Anant Joshi addresses the same with respect to the lavishness and rapid rise of such structures, whereas Kausik Mukhopadhyay, looks at the changing notions of smartness projected through the decades.

Figure 7. Anant Joshi, May Look Closer Than They Appear (detail), 2007. Golden paper, newspaper and gold leaf on canvas
(Source: © 2007, Anant Joshi, used with permission)

May Look Closer Than They Appear, 2007, Figure 7, consists of images of newspaper advertisements of uber-rich structures, to be constructed soon, instigating buyers to become a part of such state-of-the-art buildings. Each structure described strictly through superlatives, to assert its uniqueness. The artist uses gold paper and gold leaves to punctuate the images of the structures to assert its affluence. Further, giving a feel that these structures are perched within heavenly golden clouds, distant from the mediocrity and harshness of the real world, nearly creating a special pedestal for a particular economic class. The title of the work is inspired by the standard note on the rear-view mirrors of cars, indicating that the objects that appear in the mirror could be closer. The artist hints that even as one would think about the impossibility of the numerous projects reaching fruition in one city. Due to various impeding factors like scarcity of water, space that is extremely high prices infers a delay or total abandoning of the same.

However, actuality countless projects were taking shape, at an incredible pace, despite the scarcity of basic amenities, attesting the title of the work, that these structures might be materializing in a physical form faster than one might wonder. Along with the rate of growth, by considering elite residential spaces, Joshi simultaneously raises a question about the economic stratum of the participants, as it excludes the masses.

Kausik Mukhopadhyay analyses the newspaper advertisements for ideal housing spaces from two perspectives, one, of the sheer number of the proposed projects and the other of the changing nature of the same. For example, in the 1990s the advertisements were directed to creating ideal spaces which never existed before for ordinary people. Skyscrapers presented a panoramic view of the city. This virtual image also includes the gated-communities that were built with materials and technology of international cadre, with facilities like 24 hours water supply. The superior security systems for entertainment facilities were also provided. All traits which would catapult an average earner from his/her mediocre home into a luxury one.

The lure plays with the notions of the Lacanian lack. Just over a decade, with rapid development and lowering green spaces, the newspapers were carrying articles on the ecological and environmental hazards of the growing concrete jungles, further proven through the evident depletion of water reserves. This brought in the new phase of advertisements, focusing on environmentally friendly spaces/homes, providing a lush green view, placing the resident away from the pollution of the city into an environmental bliss. In reality, as these projects were moving toward the outskirts, and merely getting closer to the greener parts of the periphery, an aspect which would get erased with further development. The work *City of Spectacle*, 2006, Figure 8, borrows from Guy Debord's book *The Society of the Spectacle*. A city is recreated as a miniature model. With the assemblage of many independent "advertised homes", Mukhopadhyay, created a near-ideal city.

Furthermore, placing it on a higher pedestal, making it difficult to get a clear view, hinting at its exclusivity. The view from the ground level was possible only through a central glass basin, magnifying the light. For a closer look, one would have to climb up, implying the need to rise in the one financial status, to experience the same.

Figure 8. Kausik Mukhopadhyay, City of Spectacle, 2006
(Source: © 2006, Kausik Mukhopahyay, used with permission)

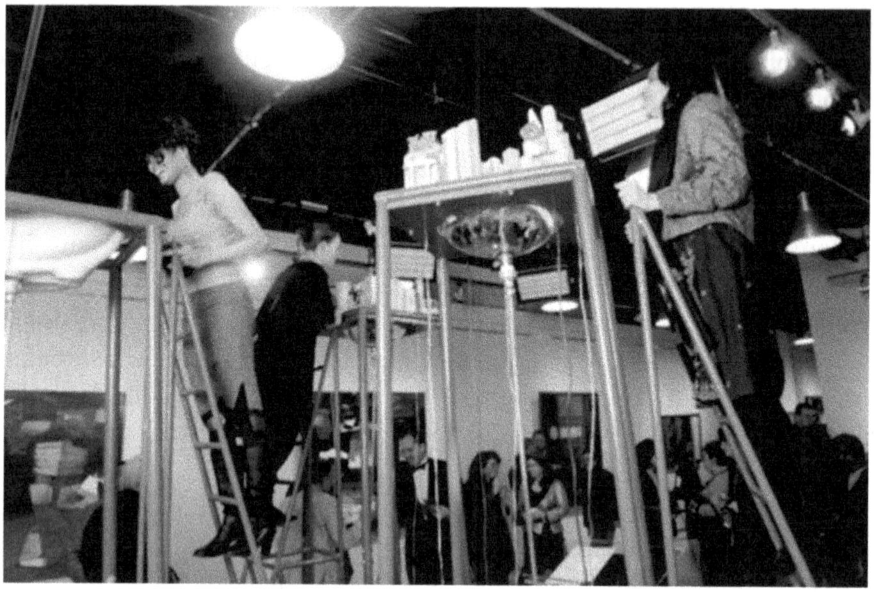

ZONES OF DETERIORATION / BLEMISHES

Ernest Burgess, while considering a city as a "living laboratory" explains the process of disorganization and reorganization, contributing to the growth of the city through different zones (The Growth of the City: An Introduction to a Research Project, 2008). His "zone of deterioration" or "bad lands", take a near mandatory presence in the configuration of the city. Different interpretations of such zones have been re-read by a few artists, not just as visual blemishes in the "world-class cities" under discussion, but as detached micro-scapes, with a near independent functional identity. This scene is like the slum of Dharavi, or the garbage dumping grounds, which ironically form a crucial part of the support system of cities. Prajakta Palav, through a series of works titled *Kachra,* 2005, Figure 9, translated as garbage from Hindi, paints an otherwise overlooked garbage-scape, the municipal dumping grounds. Her works are based on the grounds in Mumbai, a generalization of the same is possible for every growing city. These spaces have been of concern over the past decade. Once marked as outer spaces, these grounds have now been incorporated within the growing city limits.

The corresponding rise in the population has directly spelt rising volumes of waste. A critical management issue recognises the local municipal bodies, towards which investors have been largely unequipped, causing not just olfactory inconvenience. However, the focus was for a serious of health hazards for the (now) surrounding residents. Palav paints these dumping areas with near photo-realistic precision. With an odd choice of a landscape, she addresses various issues. By making a realistic work of art, she forces the viewers to pay visual attention to these detestable and overlooked areas, which while forming an essential aspect of the city's infrastructure, have been failing to cater to its escalating levels of waste. Secondly, transforming them from blemishes of the city into artworks appreciated for what it represents. Here Palav brings out a play of the – *aim* and the *process*.

The aim was to draw the attention of the individuals, a gallery visiting class, who would be mostly distant/unaware of the same and thereby initiating a dialogue over the overlooked. Besides, the process of validating a visual/view of the dumping grounds was within a context of its circulation. Consequently, that is the art world, by inverting its visuality and not reality, by transforming it into an object of aesthetic interest. A more profound irony comes forth as notions of purchase, procure-ability and a considerable financial value get assigned to *kachra* or garbage; underlining the commoditization of the "otherwise repulsive", within the global art market circuitry.

Figure 9. Prajakta Palav, Kachra I, 2005. Mix media on paper
(Source: © 2005, Prajakta Palav, used with permission)

The other aspect of the "bad land" can be traced through difficult spaces of inhabitations, the slums. In *Dream a Wish, Wish a Dream,* 2006, Figure 10, Hema Upadhyay, recreates the largest slum of Asia, Dharavi, seated in the prime locality of Mumbai. Dharavi needs to be understood as an independent microcosm, diametrically opposite to the high-end airport and plush districts surrounding it. As flights land in Mumbai, the first view while descending is of a vast horizontal area, cluttered with shanties covered in a blue tarp. Towering skyscrapers contrast the flatness of the space at its periphery, which could be read as a border, segregating the spaces of development from the impenetrability of the slum. The impenetrability best understood, as the earlier mentioned massive infrastructural projects in Mumbai, seem to bypass this patch. Upadhyay's work addresses two issues. One, of the intense impenetrability of the slum, not as much for individuals, as for the larger bodies like the government, developers and realty giants and similar bodies hoping to reap financial benefits from the prime location of the slum. The other, of the increasing size and population, and the capacity of the slum to absorb people, the majority of whom migrate to Mumbai in search of a decent livelihood. The slum is known to possess a cosmopolitan character, providing space for migrants flooding Mumbai, a city known for providing basic means of livelihood to many. A macro view helps to draw two parallel situations, that is, just like the metropolis through economic globalisation. The case also provided employment opportunities for the intellectual capital from the various part of the country, Dharavi is indicative of the desires, dreams, and aspirations of a particular lower/deprived economic stratum hoping to attain a better life or at least mere subsistence, as encapsulated in the title of the work. Here, employment could be found through a wide range of occupations, ranging from small domestic means of employment started by housewives, small-scale businesses, to manufacturing units, to illegal and nefarious activities.

Figure 10. Hema Upadhyay, Dream a Wish, Wish a Dream, 2006. Aluminum sheets, car scrap, enamel paint, plastic sheets, found objects, m-seal, resin and hardware material
(Source: © 2006, The Chemould Prescott Road, used with permission)

The installation seems like a scaled-down version of a portion of the slum, holding strong similitude to its aerial view. The shanties cover the terrain carefully. Unlike Patwardhan's *Emergent*, there isn't any trace towards a possibility for gentrification. Material like aluminium sheets, car scrap, enamel paint, plastic sheets, found objects, m-seal, resin and hardware material are employed in its creation, as article found or produced in the very slum. The absence of humans withdraws the connotation of a mere replication or a visual narration, further adding focus to the sheer physicality of the space.

Accordingly, locating the notion of experience of the space through a strict "position of externality", by making the viewers walk around the work. Such that, even the closest viewing point from the periphery of the work, wouldn't grant clarity of its intestinal depth. This could be related to the local tourist-packages providing guided tours of Dharavi, turning the blemish into a spectacle, where parts of the slum are treated as a tourist spot, providing an opportunity to witness the lives of its dwellers; and yet keeping these "aliens visitors" at a safe distance from its interiors. Since scaled-down versions of such pockets are present in most of the cities, it brings forward, questions regarding the scope, span and the penetrability of the smart city project. Evidently, each such slum holds a different function and character, demographic composition and density, and a highly skewed Human Resource Index with its neighbouring areas. Whereas the spaces like the dumps, even while having a few governmental bodies involved with respect to its management, hold a substantial environmental impact, especially on the people residing in its proximity. Therefore, the challenge remains of devising policies towards management and maintenance of these spaces, assuring equitable benefits, within the broader vision of the smart cities.

NATURAL RESOURCES AND SUSTENANCE

The earlier section focused on the bad-lands within the cities as man-made blemishes. Another aspect of the same, is of tracing the transformation of the earlier pure natural resources, especially the water bodies, into now degrading or disappearing spaces/bodies, as an outcome of uncontrolled and unplanned urban growth. To understand the relation of these water bodies with the one has to take a historic look. For Mumbai, the usually heavy monsoons would fill its large lakes, water tables, and small rivers, for yearly usage; as a set annual monsoon cycle. For Delhi, the river Yamuna has been the lifeline. The annual swelling or flooding of the river would help saturate aquifers and water tables in the capital. Delhi's relation to its water bodies has not been restricted to its functional usage, but also to notions of community service, political power and aesthetic structures. Various rulers constructed *serais* or caravansaries, beautiful *baolis* or step wells, *piau* or spaces to distribute water for free. The construction of the Taj Mahal, though not in Delhi, on the banks of Yamuna, gives an understanding of the notions of merging the aesthetic beauty of the structure with the magnificence of the water body. The mentioned overview is certainly not exhaustive, not accounting for the droughts or flooding or similar climatic calamities but provides a mere snapshot, to help draw a comparison with the current scenario where the notions of sustenance seem extremely jeopardized. Under the pretext of urban growth, many lakes, aquifers, water tables were filled to make space for construction activities, whereas some larger lakes and rivers became outlets for industrial effluents. The volume of these activities has converted these precious resources into unhygienic and polluted spaces, thereby drastically decreasing the level of the available potable water for the citizens, turning a necessity into a luxury.

Prajakta Palav in manuscript titled *Swimming Pool,* 2004, Figure 11, presents a sardonic visual play. A photo-realistically painted pristine blue swimming pool, to her is a reminder of a necessity as marketed in the advertisements of the increasing number of luxury homes. A closer look at the work reveals an irony, as the pool is lined with sewage pipes and decrepit shanties. On the one hand, she questions the very feasibility of providing clean water for recreation to the luxury homes, with rising scarcity for essential consumption. And one the other, she addresses a more significant issue of environmental sustenance. The reference to this work is the irreversible degradation of once a sweet-water Mithi River in Mumbai into a river saturated with toxins and industrial waste. The unregulated management and unplanned growth of the small and mid-scale industries on its banks, pouring effluents, led to the irreversible destruction of the river. By painting a pool instead of the actual polluted water, surrounded by sewage pipes and waste outlets, Palav reminds us of the loss and most importantly of the negligence of the local government bodies to understand the gravity and urgency of the problem. Another aspect of the issue involved the people of effect. Since lower-income groups and poor occupied the spaces around the river, there seemed a negligible need for accountability. Haughton (2000, pp. 461-475) are bringing into question the notion of integration of various social and economic categories/strata within the smart city vision.

Atul Bhalla a Delhi based artist's works analyse the current state of the natural resources, especially of the water sources in the growing urban spaces. A series of works titled *Looking for Dvaipayana,* 2014, Figure 12, he aims at posing a question like "what will be my defeat?" not just for himself but to all. As a part of the photo-performance, the artist becomes a near-abstract part of the visual, as he stands with his head lowered, intending to converse with the river. Just like the Mithi River, Yamuna too has faced the rising levels of toxins, foaming and accumulation of industrial waste. With no responsibility taken for the damaging actions, and with insufficient efforts taken to minimise further decay, the artist refuses to be a mute witness to the situation and opens a multitude of questions; ranging from – who

Figure 11. Prajakta Palav, Swimming Pool, 2004. Mix media on paper
(Source: © 2004, Prajakta Palav, used with permission)

Figure 12. Atul Bhalla, Looking for Dvaipayana, 2014. Archival pigment print
(Source: © 2014, Atul Bhalla, used with permission)

could the questioning individual be and whose defeat is it? Is it the people, failing to preserve the river, or the river failing to cater to the people? Who could be implicated in this crime? Thereby, it essential by now to move the artistic works from being a visual narrative to a dialectic.

Bhalla's effort to trace the lost sources of water take another form in the series *Looking for Dvaipayana,* 2014, Figure 13, translated as "looking for – that which is surrounded by water"; in this case with reference to Delhi, where such bodies according to the artist are disappearing with time, with increasing human greed. A body that seems like an abstract black entity stands near a *Janata Piau* – a public space for free water distribution. Evidently, an old structure, to the artist it represents a part of the history, when the rulers made provision for free and clean drinking water for the public, as mentioned earlier. Some of these spaces are still functional and hold no discriminatory approach. All present in the old part of Delhi, but absent in the newer, upcoming and smartly developed spaces, where only commercially bottled water is available. Though some metro stations, hospitals, and recreation spaces have these provisions, the entry to them is restrictive, thereby erasing a possibility of a provision for the general masses. To the artist, the loss of such public spaces is not just a loss of sources of water but is indicative of the loss of a humanitarian approach in planning cities.

Clean air and water were enlisted as provisions in the Smart City Mission (Smart City Mission Statement & Guidelines, 2015). Through this vision, it might seem obvious the insistence of the sameness that can be understood on experiencing its current degraded state to creating a permanent hazardous condition for people. In *Breath by Breath,* 2016-17, Figure 14, Vibha Galhotra, takes into account the rising level of pollution in Delhi. In the staged photographs, the artist tries to capture air into a glass case, possibly for some scientific research. The backdrop is of a polluted water body, forming with toxins, further degrading the local air quality. She strikes different notes in time.

Figure 13. Atul Bhalla, Looking for Dvaipayana, 2014. 16x24 inches Archival pigment print
(Source: © 2014, Atul Bhalla, used with permission)

Figure 14. Vibha Galhotra, Breath by Breath, 2016-17. Behind the camera - Rajesh Kumar Singh. Staged photograph, digital print on archival paper
(Source: © 2016-17, Vibha Galhotra, used with permission)

On the one hand, the work of Galhotra addresses "fresh air as a memory", which once existed. On the other hand, she looks at the changing or improved gadgets used by people to protect themselves from pollution on the suggestion/persuasion of the government and the commercial forces. Such that every year a new device gets naturalized in one's daily life. Beginning from the inexpensive masks, like the one worn by the artist, to the air and water purifiers, to the latest ones shown in *Negotiated Necessities, 2008,* Figure 15, Figure 16, choices purely based on one's financial capacities. The space for negotiation thereby moves from dealing with the causal factors of the environmental hazard to the ease of adapting consumer products for self-protection, as necessities. A similar approach from the government can be seen with the development of mobile applications, calculating the Air Quality Index (AQI) to help citizens plan their outdoor activities. Labelled as a "progressive" approach, the realistic remedial approach still seems distant. Delhi's air quality has been a long-standing issue, but in recent years with the rising infrastructural projects aiding private vehicle mobility have aggravated the situation. This situation includes the felling of green spaces, smog and the annual burning of the crop, drive the AQI to Very Unhealthy Levels very often. Evidently, over the years, increased cases have been reported with heart disease, stroke, chronic obstructive pulmonary disease and lung cancer, and acute respiratory problems. Even with the city successfully achieving high standards in terms of economic opportunities and the projected developments, the lived experience of the same differs radically.

Figure 15. Vibha Galhotra, Negotiated Necessities, 2008. Digital print on archival paper
(Source: © 2008, Vibha Galhotra, used with permission)

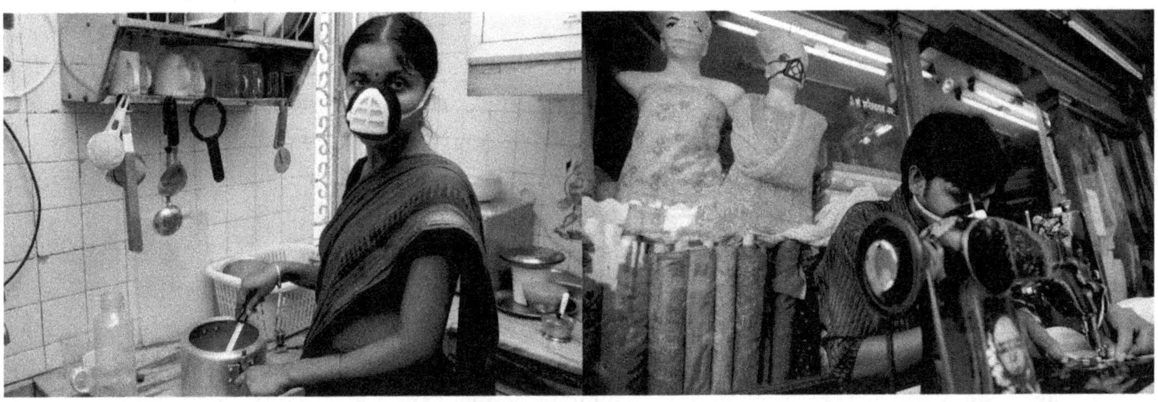

Figure 16. Vibha Galhotra, Negotiated Necessities, 2008. Digital print on archival paper
(Source: © 2016-17, Vibha Galhotra, used with permission)

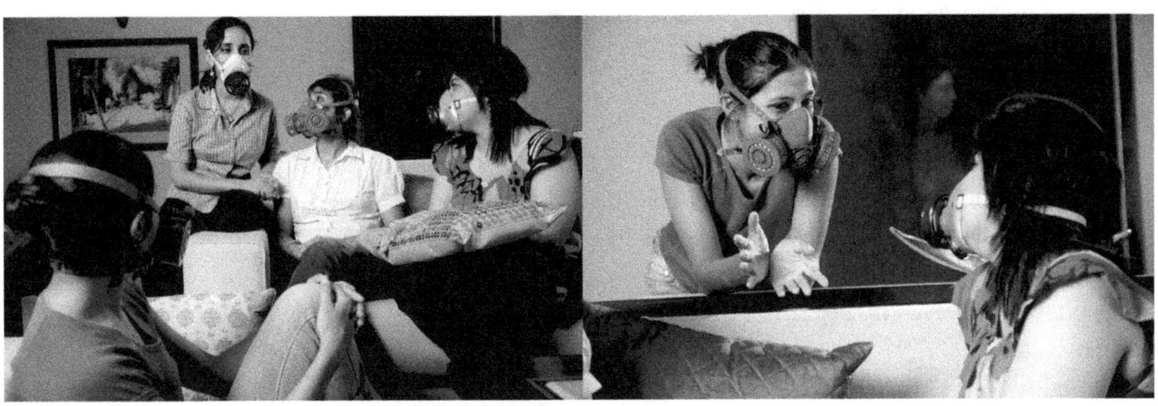

CONCLUSION

As a general overview, the study has to be understood as an attempt at presenting valid and varied viewpoints and perspectives of creative individuals, based on their respective perceptions, as a method of expanding the understanding of the contemporary urban lived experience. Many artists have played the role of the detached and knowledgeable contemporary urban *flâneur*, exercising the gastronomy of the senses, not just to gather a repository of experiences, but of employing them thoughtfully to conjure creative statements – the representational spaces, based on rational actualities. The artists and the artworks discussed, though not exhaustive, address a wide range of issues. The restriction could be understood, as most of them have based their works on the actual personal-physical experiences through the mentioned/ respective spaces and situations; therefore, bringing a first-hand understanding of the same forwards. The detailed discussion of the works does bring out a tone of pessimism, but the works have to be read as critiques of the actuality, over mere criticism.

It could be said that the resistance of a laudatory or even an optimistic tone exists, due to the discerning capabilities of these individuals, in reading the coexistence of the polarities, and their skewed ratios of benefits. It would be a gross error in overlooking or demeaning the contributions of the globalized economy of the country, as it has evidently provided a strong impetus to the national economy, aiding its stabilization, while providing a better quality of life of many. At the same time, one cannot ignore the masses which have not been benefited from the same, and worse ignored or exploited. Therefore, some works which voiced the concerns of economic groups/stratum which are usually ignored, have to be understood not just as attempts of representing them, but as ones exhibiting an "inclusive methodological approach", which the actual developmental efforts have been criticized for having failed to attain; aiming to address the recurring question, "for whom are the smart cities going to be?" The same is traceable with the "bad lands" and the lost or depleting natural resources too. Given which, the study disclaims any attempt towards forming conclusive judgements of the success or failure of the mentioned attempts towards development, but proposes an approach, based on the sentient aspects of not just individuals but the very urban spaces under consideration, with the artist employing their capacities and conceptualization towards creating a visual dialogism. Thereby, moving the works from the sphere of mere retinal pleasure towards stirring cerebral dialogues.

ACKNOWLEDGMENT

The author would like to thank every artist discussed in the study for sparing time for detailed discussions and for sharing the images of the works too. The art galleries – The Guild Art Gallery and the Chemould Prescott Road for sharing images and catalogs.

This chapter received no specific grant from any funding agency in the public, commercial, or not-for-profit sectors.

REFERENCES

Adajania, N. (2008). *Anant Joshi*. Chemould Prescott Road.

Adarkar, N., & Menon, M. (2004). *One Hundred Years, One Hundred Voices: The Millworkers of Girangaon: An Oral History*. Seagull Books.

Balzac, H., & Saintsbury, G. (1901). *The Works of Honoré de Balzac.* Retrieved November 26, 2019, from Google Books: shorturl.at/joCNT

Bell, D. (1973). *The Coming of Postindustrial Society*. Basic Books.

Benjamin, W. (2006). *The Writer of Modern Life: Essays on Charles Baudelaire* (M. Jennings, Ed.). Harvard University Press.

Boutin, A. (2012). Rethinking the Flâneur: Flânerie and the Senses. *Dix-Neuf, 16*(2).

Bouvier, L. (2005, September). A Taste for Words: Gastronomy and the Writing of Loss in Brillat-Savarin's Physiologie du Goût. *Mosaic, 38*(3).

Burgess, E. W. (2008). The Growth of the City: An Introduction to a Research Project. In J. M. Marzluff, E. Shulenberger, W. Endlicher, M. Alberti, G. Bradley, C. Ryan, & C. Zumbrunnen (Eds.), *Urban Ecology: An International Perspective on the Interaction Between Humans and Nature* (pp. 71–78). Springer US. doi:10.1007/978-0-387-73412-5_5

Dhage, V. (2013). Transnationalism and the Transformation of Architecture of Textile Mill-lands in Mumbai. *History Today: Journal of History and Historical Archaeology,* (14), 110-113.

Dickens, C., & Shruckburg, J. (1949). *A Tale of Two Cities (Oxford illustrated Dickens).* Oxford University Press.

Dickens, P. (2001). *Global Shifts: Mapping the Changing Contours of the World Economy* (6th ed.). Sage Publications Ltd.

Harris, A. (2008). From London to Mumbai and Back Again: Gentrification and Public Policy in Comparative Perspective. *Urban Studies (Edinburgh, Scotland), 45*(12), 2408. doi:10.1177/0042098008097100

Haughton, G. (2000). Environmental Justice and the Sustainable City. In M. Miles, T. Hall, & I. Borden (Eds.), *The City Cultures Reader*. Routledge.

Jayasuriya, K. (2009, January). Asian Regional Governance: Crisis and Change. *ASEAN Economic Bulletin, 26*(2), 227. doi:10.1355/ae26-2g

Kellner, D. (2002, November). Theorizing Globalization. *Sociological Theory, 20*(3), 285–305. doi:10.1111/0735-2751.00165

Leddy, T. (2019). *Dewey's Aesthetics* (N. E. Zalta, Ed.). Retrieved January 2020, from The Stanford Encyclopedia of Philosophy (Winter Edition 2019): https://plato.stanford.edu/archives/win2019/entries/dewey-aesthetics/

Lefebvre, H., Nicholson-Smith, & Harvey, D. (1991). The Production of Space. Oxford, UK: Blackwell Publishing.

Lefebvre, H. (2000). Plan of the Present Work. In M. Miles, T. Hall, & I. Borden (Eds.), *The City Cultures Reader*. Routledge.

Lyon, D. (1999). *Postmodernity*. University of Minnesota Press.

McClellan, S., Jimenez, J. A., & Koutitas, G. (Eds.). (2018). Smart Cities: Applications, Technologies, Standards, and Driving Factors (1st ed.). Cham: Springer International Publishing: Imprint: Springer.

Nallathiga, R. (2001, January-March). Redevelopment of Industrial Land in Urban Areas: A Case Study of Textile Mill Land Redevelopment in Mumbai. *Institute of Town Planners, India Journal*, 95-106. Retrieved November 2012, from http://itpi.org.in/pdfs/jan9_11.pdf

Nayak, S. S. (2008). *Globalization and the Indian Economy: Roadmap to Convertible Rupee*. Routledge.

Pelton, J. N., & Singh, I. B. (2019). *Smart Cities of Today and Tomorrow: Better Technology, Infrastructure and Security*. Springer International Publishing. doi:10.1007/978-3-319-95822-4

Sadler, S., & Archigram (Group). (2005). *Archigram: Architecture without architecture*. Cambridge, MA: MIT Press. Retrieved from Monoskop: https://monoskop.org/log/?p=1055

Sanders, A. (1988). *The Companion to A Tale of Two Cities (Dickens companions; 4)*. Unwin Hyman.

Sassen, S. (1991). *The Global City: New York, London, Tokyo*. Princeton University Press.

Sassen, S. (2006). *Cities in a World Economy* (3rd ed.). Pine Forge Press.

Shepard, M. (2011). Toward The Sentient City. In M. Shepard (Ed.), Sentient City: Ubiquitous Computing, Architecture, and the Future of Urban Space. Architectural League of New York, The MIT Press.

Siva Kumar, R. (2012). *An Extraordinary Cohesiveness. Route Maps*. The Guild.

Smart City Mission Statement & Guidelines. (2015, June). Retrieved November 20, 2019, from Smart Cities Mission: shorturl.at/knIMP

Steger, M. (2013). *Globalization: A Very Short Introduction*. Oxford University Press. doi:10.1093/actrade/9780199662661.001.0001

Stiglitz, J. (2003). *Globalization and its Discontents*. W.W. Norton.

Tripathi, S. (2007, April). Bombay's Growth Gets Shanghaied. *Far Eastern Economic Review, 170*(3).

Wessner, G. (2011). Preface. In M. Shepard (Ed.), *Sentient City: Ubiquitous Computing, Architecture, and the Future of Urban Space* (pp. 8–9). Architectural League of New York, The MIT Press.

Chapter 10
Combinatorial Architecture:
Urban Causality, Creativity, and Socialisation

Caterina Tiazzoldi
Rhode Island School of Design, USA

Abeer Elshater
https://orcid.org/0000-0002-5061-6861
Ain Shams University, Egypt

ABSTRACT

At a time when laptops and Wi-Fi connections proliferate, it is possible to work and socialise anywhere. The pressing issue is how people should perceive their ambiance when released from physical interactions in public. This chapter addresses the possibility of designing a space that combines users' plurality with the coherence of the entire design. Argumentatively, the authors show professional and educational case studies with a mixture of physical and virtual environments, exploring their impact in the reconstruction of urban ambiance. These cases are organised into groups, moving from the most straightforward implementations of technology in order to create a new realm of interactions between the physical and virtual. A review of these cases showed collaborations between different institutions and artists. The concluding remarks call for the identification of a set of transdisciplinary design actions. These actions could respond to the transformations of public space with cooperation between academics and professionals.

INTRODUCTION

In recent decades, the notion of including smart technologies in urban places, and their influence on everyday life has become one of the broadly recognised and growing challenges in urban planning and design literature (Mossberger, Tolbert, & Franko, 2013; Sawyer & Tapia, 2005; Silva, 2018; Tiazzoldi, 2008). Researchers from diverse disciplinary backgrounds related the city design started to deliberate the profound implications of the use of new technologies, at different scales from small urban places to

DOI: 10.4018/978-1-7998-3856-2.ch010

smart cities, on the everyday urban environment (Abusaada, Vellguth, & Elshater, 2019; Salama, 2015; Thibaud, 2011). Some pieces of research provide opportunities to develop a new discourse that aims at understanding the ambiances of the public realm, especially when having technologies in users' hand (Abdel-Aziz, Abdel-Salam, & El-Sayad, 2016; Yamawaki & Filho, 2020).

Research on architecture and urban studies have a long tradition in how designers structure city through overlap cultural scenarios (Tiazzoldi, 2016; Holand, 1999). Only a few studies have shown the reflection on implemented projects of cultural overlaps between physical and virtual socialisation. One technique to stunned this challenge is to merge physical and virtual space by encouraging a new platform of urban causality and creativity for interaction in the city. Meanwhile, this chapter addresses several questions:

- How can smart technologies applied to art and architecture engender a new form of interaction in the public realm and promote new forms of socialisation?
- How is it possible to design a space combining users' plurality with the coherence of the whole design?
- How is it possible to smart technologies can impact the physical space and create new forms of ambiances?

The research questions go far toward the effects of smart technology on contemporary cities without removing the casualty and vivacity of the urban ambiances engendered by physical interaction. We, therefore, review projects, whether having technology may lead to creativity and socialisation by merging physical and digital interaction. The aim is to develop a method that can provide an outlook for integrating the presence of technology within the urban causality and creativity in everyday life. Furthermore, this chapter show sequence of case studies in collaboration between different institutions and artists.

STATE OF THE ART

In this section, scanning relevant literature recalls some of the visions McLuhan (1967) and Moos, (1997) and Levy and Bononno (1999) to synthesise state of the art in the field of smart device and transformation of the urban atmospheres through media. Followingly, the authors of the present chapter review and frame some case studies comprise three groups from the most straightforward implementations to the creation of the urban organisms and new form of interaction with the public institutions and the environment.

The first group of case studies adopt a methodology inspired by the French historian, Fernand Braudel. The review here focuses on analysing two case studies in which the coexistence between new techniques in architectural projects. A specific geographic situation and the copresence of low and high-income actors could transform the destiny and reinvent the urban atmosphere of a given area. The list of cases in this group includes Le Bateau Lavoir in Montmartre, Chelsea Hotel in Chelsea in New York during the 70s and 80s. In this group of cases, the authors compare the conditions of the historical context in each site, promoting and interaction between new techniques and the transformation of an urban realm.

The second group of case studies discuss how the relation between smart environment and physical environment can create new forms of urban organism with local and global socialisation. The socialisation development is seen in simple interactive devices or components (e.g., merging virtual and physical socialisation in the realm of educational projects and public art installations. The list of selected cases in this group includes smart components in the adaptive re-use of industrial building and how those can reactivate regenerate the urban ambiance as it was for the case of the Bateau Lavoir and the Chelsea Hotel in New York. The first of the three case studies is an installation exploring what. These two cases are Toolbox and Base Milano. Toolbox is one of them that is a coworking space within the campus of an industrial building released by the author in the city of Toronto following a concept developed by the author in 2002. The other case study in this group is Base Milano that show collaboration between Milan City Council and Private Partner and will quickly review the example of other intervention of urban regeneration created by the recreation

The last case study explores how the methodology implemented at the scale of the buildings expand at the landscape dimension contributing to the urban regeneration of an entire neighbourhood (Wong, 2016) and into the all New York Park System.

The third group of cases were redeveloped comparatively, by the first author as a visiting professor and practitioner in two different institutions in the United States and Italy. The product is a research-driven design project by graduate students in the first institution and from the first-year students in the other institution. Taking into consideration, we have a gap in levels of students' capability; the final pin-up shows the typological potentialities. New forms of communication drive these potentialities through inner and out spaces. The example of Social Cave was developed by the first author of this chapter at Columbia University for the Milano Furniture Fair in 2011. Another case is Mathematical Belonging, entry winner of the public art, competition for the starlight park in the Bronx, New York Park Service (NPS), with E. Duarte in 2018. Both last two cases reflect the smart technologies and new architectural constructions (organisms).

BACKGROUND

In the late 60s, Marshall McLuhan published a series of books establishing his reputation as a world-renowned communications theorist. Recognised as and the pre-eminent seer of the modern age. He coined the idea of "the medium is the message" and "the global village" and popularized other memorable terms "feedback" and "iconic." In the extension of McLuhan, proposes the potential of the men offer the vision, and the potentialities of the new media foresaw the development of personal computers at a time when computers were massive, unwieldy machines available only to institutions. McLuhan (1964) anticipated the wide-ranging effects of the Internet. Moreover, he provided a clear description of the transformations by digital technology that would shape our daily life—in particular, the rapid global communications. In 1994, two years after the public accessibility of the internet, the French anthropologist Pierre Levy in *Collective Intelligence* questioned how the communications revolution could affect our culture and society? Pierre Lévy shows how the unfettered exchange of ideas in cyberspace has the potential to liberate us from the social and political hierarchies that have stood in the way of mankind's advancement. The book was the premise of the development of software and commercial model such as Amazon created in 1993 having the capacity to re-use the amount of information that was inserted by user to create a new form of collective intelligence.

In the last fifty years, similarly, previous research showed how the growth of technology possessions have been impacting people daily life (Abusaada, 2020; Sepasgozar, Hawken, Sargolzaei, & Foroozanfa, 2019). Somehow, McLuhan (1964) automated lighting, sensors and automated systems that have changed our daily lives. Augmenting, he argued that access to education is the most regions of the world. McLuahn's visionary perspective in which the original ideal was eventually to release human form task. This task could have been easily developed by a machine, a robot, a sensor or a computer and to reduce the time spent to travel to work, to shop this somehow challenged the experience of the collection and the relation between the human. McLuhan's vision in which technology could release human from the task that could be quickly developed by computer, robots, sensors became more than real. Studies of computing architecture are well documented, it is also well acknowledged that the possibility of having laptops to allows people to work from home reduce the time of transportation for work, shopping, riding children at school (Jiang, 2020; Vanolo, 2014) as well as the Senseable City Lab at MIT (Claudel & Ratti, 2015).

Like Lévy's idea, pieces of research argue the fast growth of the social network and the smart device that can somehow support McLuhan's proposition about the rapid transformation due to smart devices on local societies. This fast development of social networks, smart context and technologies embedded the idea of automatisation between humans and physical tasks which supported McLuhan vision.

In the late 90s, the idea of the global village was started criticised earlier by Sasha Sassen (1999). Both scholars besides other researchers argue how human physical interaction is necessary to transform the city and to support human life (Abusaada, Vellguth, & Elshater, 2019; Hart, 2003; Lévy, 2011; Sassen, 1999).

On the one hand, it could be stated that the recent development of smart technologies positively transformed our daily life. On the other end, it is understandable that the task unusual of everyday human life, going to the supermarket, going to work, carry a child at school and interacting with friends and colleagues is a unique and powerful tool to develop society stability and innovation.

The literature on involving rapid changes in technology that affects our daily has highlighted several questioned the idea of the capacity of a fully automated system managed by smartness and to improve the quality of life of the citizen (Tiazzoldi, 2008; The United Nations, 2019). In 2002, the first author of the present chapter was requested to rethink a design program for a 20.000m^2. for an industrial area in Torino. The main challenge was designing a space combining users' plurality with the coherence of the whole digital design. She thought to mediate between a need for socialisation and privacy, relaxation and concentration.

This issue also extends to maintain the potential of the intangible components of daily urban life. These situations are defined by the need to move through the city, respecting coding, lining at the post office, sharing resources. Now, a question would be how to create a similar urban atmosphere. The question would be how the new typology of a creative hub can permit to connect to intangible heritage, creativity.

In the next paragraphs, the authors of the present chapter analyses how the coexistence of some given human and technological factor permitted to transform different areas of the globes during the past centuries. Because of this potential limitation, the reference to the idea of smart technology will be limited. They generally use the concept of innovative technologies such as movies, photography, new painting technique. The argument here exposes how the convergence of a set of conditions permitted to create an urban transformation and social transformation. The case studies are evidence of how other conditions promoted the booming of some specific area. In most case, the low rental of existing buildings, the wish of young artists and individual help their works and to share the cost of new technologies or assistants.

In the next paragraph, the authors also analyse how the coexistence of some given human and technological factor permitted to transform different areas of the globes during the past centuries (e.g., as Bateau Lavoir in Montmartre). However, in most case, the low rental of existing buildings, the wish of young artists and individuals promote their works and to share the cost of new technologies or assistants.

PUTTING CASES IN CONTEXT

In 2020, the full globe was affected by a virus the COVID 19. The main prescription from the worked health organisation was to avoid as much as possible human contacts and exchanges. The use of smart working, delivery systems implemented via computer application, medical assistance online was strong support to prevent further contaminations. Nevertheless, the experience of total social isolation, the impossibility to do some of the daily tasks such as going to work, interact with people in a casual way created one of the most globalised human crisis and the acknowledgements of the need of the casual human interaction qualifying and urban atmospheres. The following groups of projects paved the way for the review the urgent need for using technology in our daily. Table 1 shows the comparative review between these three groups of cases.

The First Group of Cases: Historical Case Studies

This section reviews cases where art and technological innovation contributed to the urban regeneration of Montmartre District in Paris and Chelsea in New York via the hubs Bateau Lavoir and Chelsea Hotel. The purpose of reviewing these cases is to highlight how decision-makers can define duties or task that could be replaced by machines and long-distance working. The reduction of causality in public life chance can also affect the quality of the atmosphere and urban life.

The case of Le Bateau Lavoir shows an example of special atmospheres in historical sites (Hewitt, 2001). Montmartre is a well know district in northern Paris, on a hill above the River Seine. This district is much frequently visited by artists in the late 19th and early 20th centuries when it was a separate village. Noteworthy, this a clustering of studios that attract artists in the early years of the 20th century till now. Picasso, as well as Modigliani Georges Braque, Max Jacob, Marie Laurencin, Guillaume Apollinaire, André Salmon, Maurice Raynal, Juan Gris, Fernand Léger, Robert Delaunay, were ones of these artists who lived there. The possibility was to share space, technology, camera, brushes, friends and to share or compete for art collectors. In this, the possible clients became strong catalyser of one of the most important art movement of the 20th century as described by Dan Frank book Bohemian Paris, Picasso, Modigliani, Matisse and the birth of Modern Art (Dan, 1998). The low income of the artist limited their capacity to travel. Therefore, Montmartre became one of the most represented locations of the art pieces. This contribution propagates the image of the area in all the world and to transform, along the 20th century. Therefore, this neighbourhood in one of the most romantic locations in the world.

Another case is Chelsea Hotel that was built as a city landmark in New York City between 1883 and 1885 (Claudel & Ratti, 2015). This hotel was featured out in films and novels (e.g., Chelsea Girls (Warhol, 1966; Davis, 2019), Romeo is Bleeding (Medak, 1993)) and published memoir by Patti Smith (Smith, 2010) Smith 2010—in *just kids*—writes about his relationship with artist Robert Mapplethorpe, were both born in 1946, at a time when “the iceman” and “the last of the horse-drawn wagons”. In such atmosphere Smith describes how she and Mapplethorpe were surrounded by artist like Andy Warhol, Lichtenstein, Bob Dylan, Nico, Tim Buckley, Janet Joplin and the Velvet Underground. They all could still be seen on city streets. The continuous cross-fertilisation derives the need to share support and technological resources permitted to create some of the most influential artists at this time.

Table 1. The comparison between the three groups of cases

	1st Group of Case Studies		2nd Group of Case Studies			Further Explorations
	Bateau Lavoir	**Chelsea Hotel**	**Social Cave**	**Toolbox Office Lab& Coworking**	**Base Milano**	**Mathematical Belonging**
Designer	Undefined	Undefined	Tiazzoldi with students at GSAPP	Tiazzoldi, C.	Onsite Studio and Tiazzoldi was consultant.	Tiazzoldi Duarte
Location	Montmartre Paris, France	New York, USA	Italy	Torino, Italy	Milano, Italy	Bronx, New York, USA
Date of construction	1905	1978-1985	2011	2007	2011	2018
Technology involvement	Painting photography lithography	Photography, Filmography, Electronic Music	Interactive Platform Kinetic camera	Smart managing, fab-lab Arduino Virtual meeting room	Virtual meeting room and music record	
Professional vs academia	Professional	Professional	Equal	Professional	Professional	Professional
Impact on the city	Montmartre	Transformed Chelsea	Testing installation no urban effect	Created a New Kind of Urban Institution	Amplified the area of zona Tortona	Urban regeneration of the park
Impact on the research	Very High	Very High	Low	Medium High	Medium	Medium
Impact on the economy of the sponsor	Undefined	Very High	Medium	Medium	High	Un-defined
Impact on context economy	High	Undefined	Undefined	High	Undefined	Undefined
Impact on social life in the area	High	High	Low	Medium	High	High
Virtual vs Physical	Physical	Physical	Mostly Virtual	Mostly Physical	Mostly Physical	Medium
Constructing ambiance in public space	High	High	low	Medium	High	High
Private vs Public investors	private	private	private	Mostly Private	Mostly Private	Public
Openness to Causality	Very High	Very High	Very Low	Medium	Very High	Medium

The historical atmospheres in these two places generate a distinction in the way of having the technology in users' hands. Pieces of research argue the crucial entrance of technology in historical sites which should consider the deep understanding of the local community, stakeholder in particular places and time and the authenticity of the place (Turner & Tomer, 2013; Araoz, 2013).

The Second Group of Cases: Combinatory Urbanism

In the previous two cases, the authors reviewed the coexistence between the coexistence of and artistic community exploring new technologies such as lithography, photography, filmography, electronic music in low incomes settlements. Artists and inventors motivated by their passion for art and their survival need radically change the destiny of specific districts, namely Montmartre, SoHo and Chelsea transforming them of some of the most vivid ambiances in the world.

The rapid transformation of contemporary cities implies an equally fast adaptive cycle. This process questions, on the one hand, the economic and environmental sustainability of cities while, on the other hand, requesting continuous spatial transformation. It is necessary to develop flexible tools that allow designers to change the uses of space without changing the physical structure and to use such tools as part of a coherent design concept capable of changing according to users' needs. In the next paragraph, the authors explore some case studies in which the smart environment when mixed with physical one has the capacity to create new unedited type of socialisation and to contribute to the recreation of urban ambiance.

The first case the Social Cave developed by the first author in Caterina Tiazzoldi as director of the research Lab Non-Linear-Solutions Unit at Columbia University with a group of 20 students is a straightforward implementation interactive platform challenging the boundaries between physical and physical socialisation.

The other cases recall the idea of creative hubs - explored in the case of the Bateau Lavoir and Chelsea Hotel- when supported by smart devices (digital fabrication, development of reactive sensor objects Arduino, the Internet of things and physical coexistence. This change permits to bring back life to industrial buildings and districts. The most relevant question will be how far it would be possible to create by design, a collaborative environment with the same qualities as it casually happened in the case of Le Bateau Lavoir and the Chelsea Hotel.

For this reason, the text explores the idea of combinatorial design methodologies. It will investigate the causality of socialisation and cross-fertilisation without the risk to create a standardised format.

The group of cases explore the classical architectural versus the combinatorial architecture as a movement of modern architecture. Like combinatory urbanism, combinatorial architecture attempted to mix between qualitative and quantitative attributes, the affects the relationship between measurable and non-measurable dimensions in composition and urban form (Abusaada & Elshater, 2020; Mayne, 2011; Tiazzoldi, 2016; Sobejano, 2011). Several pieces of research presented with case studies in the Second International Congress on Ambiances in Montréal in 2012. The concluded issue is to draw designers' attention toward what is experienced compared with what was designed (International Ambiances Network, 2012). Other types of research argue the effect of constructed materials and acoustics performance in creating particular atmospheres for specific users (Tiazzoldi & Tixier, 2008) like the project for a formal modulation for acoustic panels in a bridge and the urban playground Onion Pinch (Furuto, 2011).

The cases we present in the following paragraphs dwell on the principle that a combinatorial architecture concept may be applicable to new urban and architectural forms. This concept is responsive to issues inherent to existing urban form and proposes solutions for architecture practice in new urban settings.

The first idea these cases are following is 'creative reductionism' (Descartes, 1637; Inoue, Rodgers, Tennant, & Spencer, 2015). This concept consists of analysing some of the environmental conditions and translating them into elementary units and attributes that form the building blocks (re-usable certain parts). This system of construction can be adjusted during the architectural design process. Originally, developed in the field of cognitive science (Holand, 1999). It is possible to fragment a non-measurable intangible dimension of the environment, into a set of numeric data and to identify the logic to connect and transform them. In this concept, it is possible to transform the non-measurable into something measurable 'trees, buildings, automobiles, other humans, specific animals, and so on. This quick decomposition of complex visual scenes into familiar building blocks is something that we cannot yet mimic with computers' (Holand, 1999).

The creative reductionism method is conceived as a creative act permitting to approach a given problem unfolding new fields of the measurable. It could be said that in architectural design, the creative reductionism phase corresponds to the decomposing of a given reality into a set of elementary units. The example of these units are walls, windows, openings, slabs. The example also is given to their attributes (i.e., thickness, length, rotation, scale, reflectivity, transparency, porosity and sound absorbance) (Tiazzoldi, 2009; Descartes, 1637; Elshater, 2015; Munari, 1981). A group of units connected by a set of rules (e.g., rhythm, relative positions, geometry, proportion, alignments, relative size). This group of unites affects the perception of space (light, sound, visual dynamics) (Tiazzoldi & Tixier, 2008). Ultimately, the procedure of this construction system leads towards the creation of a tool permitting to manage qualitative and quantitative performance.

The review here of these cases consists of the 'reduction' or translation of a given problem into a series of parameters—through the formulation of specific design intentionality—a given space into a list of attributes that can be manipulated and edited (Figure 1).

Thus, the transition from traditional sciences to the sciences of complexity permitted to define new limits between what is measurable and what is non-measurable and to create new ways to integrate non-measurable data in scientific research protocols. Inspired for the work of Holland (1999) in the field of genetic algorithms and form the research on biological variation by Bateson (1979) in his research *Combinatorial Architecture* inherited from the science of complexity the capacity to produce design innovation while integrating quantitative and qualitative data in the design process.

Social Cave

In celebration of the 50th anniversary of the Furniture Fair of Milan, the Salone Satellite chief curator Marva Griffin invited the architecture and design Research Lab Non-linear Solutions Unit at GSAPP Columbia University to develop a small pavilion project. Growing virtual connectivity has undoubtedly created a network of unlimited communication pathways. Yet while our social reach has extended, relationships spawned by the web often remain confined to the digital space in which they were initiated. The protection of that dematerialisation is seductive; intimacy must no longer be defined by physicality. Emboldened by the blanket of physical anonymity, we may assert a redefined sense of privacy and closeness.

By merging physical and virtual space, how can design affect the changing vista of socialisation? Can design encourage a new platform for interaction? The project designers were invited to design a small, interactive installation that would be encountered by hundreds of thousands of people, most of whom are strangers to each other. In response, we have proposed the Social Cave, a typology that returns to the beginnings of our civilisation. The cave will blend the frenzied excitement of virtual connectivity's power and speed with the calm of its form and materiality, a parametric aggregation of 100% recycled and 100% recyclable foam cubes.

The barrier wall separating the two enclosures within the cave ensures that visitors inhabiting each of those spaces are initially concealed from each other. The presence of a visitor in the opposite space is revealed only through an abstracted projection capturing his movements. The physical anonymity created by the enclosure allows each visitor to feel comfortable engaging in a digital and visual conversation with the projected "shadow" or "ghost" of the visitor opposite him. Gestures and personalities are therefore made familiar to each other before the initial physical meeting. Thus, the Social Cave first hides and then exposes the proximity and identity of its visitors, allowing a conversation to begin that transcends traditional digital-physical boundaries.

The initiative attempted to provide a response to the question of what direction is design heading. The idea of the construction of this project is softening the 'traditional spatial conceptions, spawning unlimited potential for global communication networks' (Tiazzoldi, 2011). The construction system here tries to redefining privacy and closeness to exclude physical intimacy. Tiazzoldi (2011) raises question regarding the possibility of having a design that can inspire a new platform for interaction.

The production of forms proposes the interactive installation of Social Cave. While entering the 100% recycled foam cave, visitors are invited to interact with the projected 'shadow' of the visitor beyond the barrier wall, initiating a conversation that transcends traditional digital-physical boundaries (Ehmann, Borges, & Klanten, 2012; Kottas, 2014).

Figure 1. Photos for the social cave

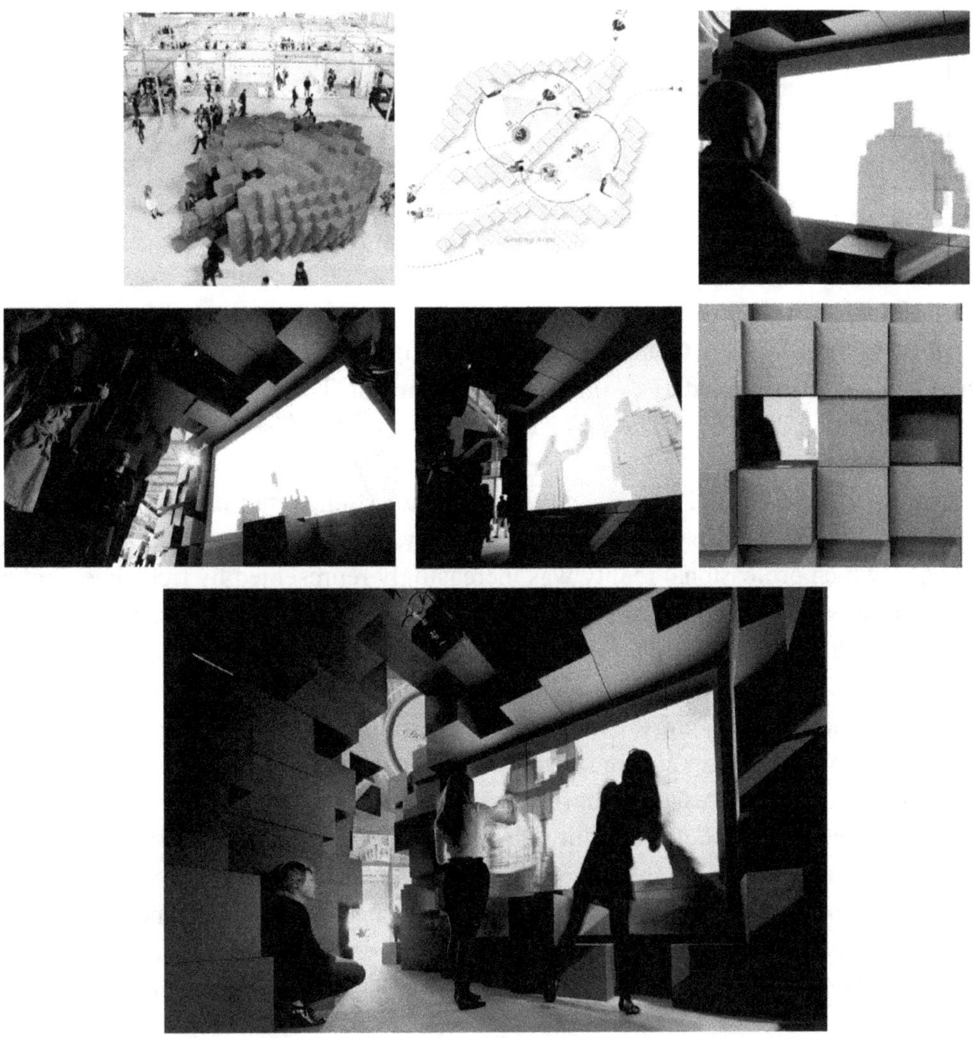

Social Cave combines atmospheres of excitement through virtual and connectivity between visitors. The digital technology has disintegrated traditional ideas of human connection inside spaces. Notably, socialisation once existed wholly within the physical realm; the virtual world has invited distinct rules and interaction within a previously atmospheres.

Previous research argues that the growing virtual connectivity has undoubtedly produced a network of unlimited communication pathways (Abdel-Aziz, Abdel-Salam, & El-Sayad, 2016) (Abusaada, Vellguth, & Elshater, 2019; Mossberger, Tolbert, & Franko, 2013; Tiazzoldi, Combinatorial Architecture Methods for the creation of ambiance in public space, 2016). However, while social has extended, relationships spawned by the web often remain confined to the digital space in which they were initiated. The strength of that dematerialisation is seductive; intimacy must no longer be defined by physicality. Emboldened by the blanket of physical anonymity, we may assert a redefined sense of privacy and closeness.

Rooms within the cave allow visitors inhabiting and concealing from each other. The presence of a visitor in the different area is shown only through an abstracted projection capturing movements. The dynamic anonymity created by the enclosure enables each visitor to feel ample engaging in digital and visual communication with the projected shades of the visitor on the opposite side. Gesticulations and characters are, therefore made familiar to each other before the beginning physical meeting. Hence, the idea of moving through the Social Cave started by hiding and followed by exposes the closeness and identity of its visitors, allowing a dialogue to begin that transforms conventional digital, concrete barriers.

Toolbox Office Lab

Toolbox is a coworking space conceived and designed by the first author of the current chapter in the city of Torino as an industrial building in the city of Torino in 2002. Toolbox is an entry winner of the Architecture Rivelate Award, a finalist to the Renzo Piano Award for a young Italian Architect with a special mention to the Bi-Ticino Contech award. The project attempts to investigate the relationship between smart technologies applied to a workspace environment. The client request was to envision the future transformation of a 20.000m^2. The first step of the project consists of developing 2000 m2 to be dedicated as a coworking place

By recalling the Chelsea Hotel and Bateau Lavoir Case studies, the designer intent for Toolbox was to propose a new possible use of the building in a city. This building was in a phase of profound transformation, in which professional reality was increasingly represented by freelance workers whose professional future was not yet well known. The aim was to define such new use for a 50,000m^2, three-story office building that is located in a transformation area of 220,000m^2, close to downtown Turin and the railroad station Toolbox project is located in the city of Turin in one of its many industrial buildings. Turin's history is characterised by the union of a historical downtown, developed in Roman, medieval and baroque periods, with surrounding industrial and post-industrial areas. Its history witnessed the continuous adaptation process necessitated by its attempt to survive against competing cities and regions. In 1861, Turin became the capital of Italy. But when the capital city was transferred to Rome in 1865,

Turin looked to reinvent itself. By that time, it had developed an industrial profile mostly centred on the mechanical field that gave birth to automobile factories such as Fiat and Lancia. At the beginning of the 20th century, the city witnessed a proliferation of concrete structures such as the Fiat car factory by Trucco, the Nervi Buildings of the Turin Exhibition, and the Palazzo del Lavoro. Between the

Eighties and the Nineties, most industrial buildings were converted into office spaces intended for the local branches of international firms. The recent economic recession forced most international firms to reduce the number of their secondary offices and their employees. Simultaneously, Turin's professional market experienced an increase in the name of freelance workers, likewise in response to the downturn.

For this reason, Toolbox was conceived of as a professional incubator. A kind of smart Bateau Lavoir was designer as a space in where the users contribute actively—even if indirectly—to the definition of a new professional identity for Turin.

From a functional standpoint, the intervention consisted of transforming a traditional office (divided into rooms) into a vast open space with 44 individual workstations, interspersed with communal areas including meeting rooms, printer rooms and informal meeting spaces. The goal of this transformation was to keep the original industrial concrete structure as intact and visible as possible. The industrial block's main building is divided into two parts: the side along the windows is used for coworking workstations, while on the opposite side, a corridor connects five enclosed boxes that contain the common functional services (kitchen, meeting rooms, mailboxes, bar, etc.). The service areas have been conceived of as "working tools." The secondary industrial building houses the lounge, bar, and a relaxation area. In this newly developed the area, the double concrete beams were left exposed while the service spaces (e.g. the bar) were enclosed in functional boxes (Figure 2).

Adaptation to users' needs and profiles Initially, the need for workspace itself was questioned as one can now work remotely from anywhere by using a laptop and a Wi-Fi network. Why then should people need a workspace? In effect, after the first stage in which working at home appeared as a real alternative to the traditional working environment, people rediscovered the need for spaces dedicated to work. Office space mostly is needed for three reasons: socialising, professionalism (in the eyes of other people but mostly in our own eyes), and the distinction between relaxed and professional environments. The Toolbox design targets a new generation of professional architects, web designers, artists, lawyers, accountants and independent contractors. It is derived from hybridisation of traditional European workspace (organised in small offices) and the post-war American open space model. Toolbox attempts to invent a new approach to work.

The concept of Toolbox is built on two main ideas: the coexistence of a plurality of users and a coherent architectural concept. How is it possible to create a rational structure with a multitude of users, needs, expectations and ways of working? What is the right level of severity, formality and solemnity for a workspace designed for independent workers, creative teams and dislocated corporate branches? The Toolbox project addresses these issues by modulating private and social levels, work and leisure gradients, and formality and informality ranges. In the last decade, most professionals have come to appreciate some of the positive aspects of working at home. The possibility of performing repetitive tasks on a couch with the computer on one's knee.

Some of the working-at home habits naturally became part of our working lives. For this reason, the project was developed by taking into account various kinds of settings and different ranges of intensity in the way we approach work. The spaces and their language reflect this concept - strict, almost austere for the coworking area, bright and colourful for the sofas and micro-meeting rooms and in areas where work and meetings are conducted in a more relaxed mode. The telephone pods are located some distance from the coworking areas to allow private phone calls or conference calls. In this way, everyone can follow his or her particular inclination or variation during the workday.

Figure 2. Plan and inner image of Toolbox office lab

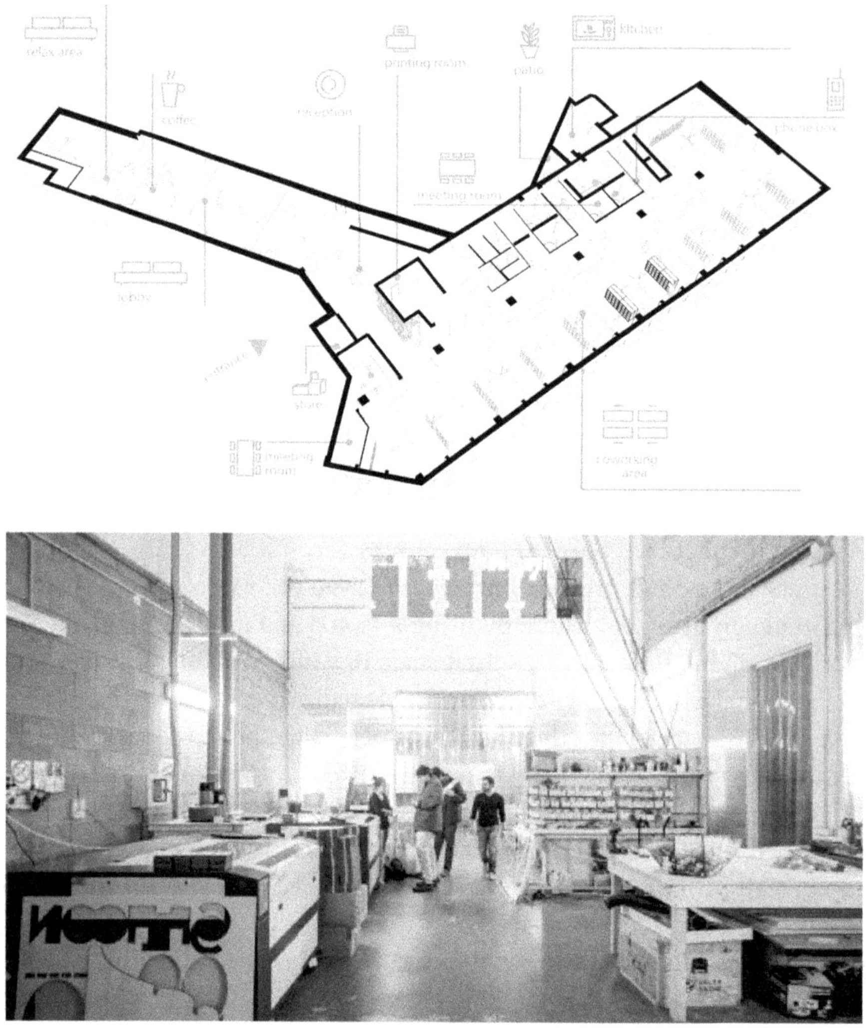

To provide different organisational structures for different users with their specific needs and profiles, Toolbox utilises an automated centralised system controlling access, the lighting system and services (printers, telephones). Each user is provided with a badge and can define her/his own profile in accordance with her/his needs. The management of the space is, therefore, related to an automated system that allows or denies access to the various functions in accordance with the user's choices. By varying the accessibility and use of the functions hosted in the space, such as coworking areas, meeting rooms, kitchen, patio, parking, it is possible to achieve an endless modality of use of the same space. As project transformation and evolution are fundamental for the success of the project, it was mandatory to define a strategy permitting constant adaption to different users' needs, without impacting the physical configuration of the structure.

Toolbox was one of the first coworking questioning the relation between privacy in compression to the shared space that is created from a multi-layered ambiance and user plurality and applying several automated systems to manage the different environments. For this reason, the project became an urban and regional attractor. Toolbox hosted several activities: The first Italian Fab Lab[1] in 2011, Casa Jasmina an artist in residences space for Makers and interaction designer in 2015, and in 2017 it included the new Arduino headquarter.

Base Milano

Base Milano is a collaboration between Milan City Council and Private Partner. This project is an urban regeneration project in which the first author of this chapter participated as a client consultant. This project was established within the boundaries of Savona-Tortona on an area of 12,000m^2. The idea behind this type of project into coworking space for generating insights from individuals and organisations. This project also engaged planners and innovators from creative and technical industries.

Cross-pollination between the arts, enterprises, technology and social innovation were also involved. The location of Base Milano is beside the ex-Ansaldo complex that is one of the most significant urban regeneration projects in Europe. The contribution of this project in the city is in exploring new models of collaboration between the public and private sector in the city. The state of Italy owned this building and is entrusted by Milan City Council to an individual, non-profit social enterprise. The self-financed outcome of a union between five different charities and businesses (Figure 3).

Figure 3. Collections of photos in Base Milano

Industry City, We Work and Talent Garden

Similar to Base Milano, this chapter reviewed the project of Industry City in New York City (Industry City, 2013). It started in the 1890s when Irving T. Bush began to build a sizable intermodal manufacturing, warehousing and distribution centre in Sunset Park, Brooklyn. Due to its prominent location, immense scale, and structure that supported a broad spectrum of businesses, Industry City (then known as Bush Terminal) flourished. It quickly became one of the most successful facilities of its type, employing nearly 25,000 workers per day and enabling Brooklyn to become a major international seaport.

By the 1960s, urban manufacturing had started its long decline. Most of Industry City's significant manufacturers – most notably the legendary Topps Baseball Card company – closed their doors or moved away, and Industry City suffered through a period of disinvestment and decay for 40 years.

In 2013, this all changed. A new ownership group, led by Belvedere Capital and Jamestown, began to redevelop Industry City. Over the past two years, Industry City has leased more than two million square feet of space and created over 2,000 jobs, capitalising on the rapidly emerging innovation economy. It encompasses the full arc of physical, digital and engineered product design and development, including initial research, engineering, design, manufacturing, and production. Once again, the property fulfils its purpose as a thriving centre of commerce and local employment (Figure 4).

Figure 4. Industry city

The Third Case: Further Explorations

There are growing appeals for Urbanism Next (Oregon, 2015) initiative that draws toward a centre the consequences of changes in technology on cities (Nunn, 2011; Winner, 1997). This initiative is fighting in the investigation of this field and is making a national channel of thought leaders from the private sector, public sector, and academia to address the challenge of having technology in our daily life.

Although there is a widespread scholarly effort of investigating technological characters, there is an essential need to review some of the implemented projects on their secondary effects on city ambiances. Worldwide, scholars discuss the interferences that facilitate our daily life (Abdel-Aziz, Abdel-Salam, & El-Sayad, 2016; Abusaada, Vellguth, & Elshater, 2019; Cudny, Comunian, & Wolaniuk, 2020). In this context, the challenges are in the assimilation of smart devices in the built environment and how they might to know whom to be used in future applications. The current work so far constitutes a relatively new area which has emerged from Urbanism Next.

Mathematical Belonging, Starlight Park

This project concerns with the future research application that addresses the implementation at the landscape level of what explored at the industrial scale. The last case study will approach the eight large-scale public art sculptures in the newly renovated Starlight Park, in the Bronx New York. The Bornx is one of the five NYC boroughs; it has the fourth-largest area, fourth-highest population, and third-highest population density. It is the only borough predominantly on the U.S. mainland. As per the Montmartre case study, it was initially classified as a dangerous district, recognisable by low-income inhabitants and the presence of Night Clubs like the Moulin Rouge (often represented by Toulouse Lautrec). The Bronx has been frequently associated with a critically social environment (abuse of drug, alcohol and prostitution). In the past twenty years, several projects have been developed to create a new based on education and art.

Turning now to another idea of sociability in public spaces, this section presents eight large-scale public art sculptures in the newly renovated Starlight Park, in the Bronx New York. Of the five boroughs, it has the fourth-largest area, fourth-highest population, and third-highest population density. It is the only borough predominantly on the U.S. mainland. If each borough were its own city, the Bronx would rank as the eighth-most populous in the U.S.

As per the Montmartre case at the beginning of the 20th century, with the night club Moulin Rouge often represented by Toulouse Lautrec and the lower Manhattan of the 60ies; the Bronx was associated with a critically social environment at the beginning of the 20th century (abuse of drug, alcohol and prostitution). In the past twenty years, governmental and private stakeholder has been investing in creating urban regeneration. By taking back the idea of Bateau Lavoir, the city promoted the creation of several institutions such as the Bronx River Art Center, conveying local artists, teenagers and public art.

It is in such a context that the project was established for the community to interact with and enjoy as modern pavilions. The sculptures artistically recreate the 1918 Bronx International Exposition of Science, Arts and Industries. This project was established for the community to interact with and enjoy as modern pavilions. The sculptures artistically recreate the 1918 Bronx International Exposition of Science, Arts and Industries.

The design of some selected cases here is about geometry, shapes, numbers and fitting elements together in a beautiful way that engages the viewer and inspires curiosity motivating user's appropriation. In the Starlight Park, the designer inspired from the idea of celebrating the scientific discoveries and optimism relating to the 1918 world fair. The idea of elements configurations proposes to design an environment in which the geometry of lines and layering of materiality are conceived to encourage a continuous experience and to solicit visitors' curiosity and playful reaction. The project is generated by a series of non-orientable loops having the capacity to twist the perception of space. This project also makes a framing of the surrounding landscape from inside and outside. This condition permits to bring in and out the user's views, the images of the nature of the park and remaining park landmarks now recaptured and merged in a visual game.

The inside out outside in curve permit to play with the feeling of belonging and transparency merging the Bronx community with science and the cityscape. Mathematical Belonging is an adaptive formal logic that can provide different functions and the environment from a counter desk for the BRAC Bronx River Art Center event or workshop as well as a small shelter and playground. The piece can be actively used for the BRAC event or used as a play space where people can enter and engage with the structure as a geometry learning platform.

The project is ultimately a spatial generator—a playful environment inviting visitors to engage with the semitransparency of the layers fanning from the landscape structures. The structure permits to define a feeling of belonging. Every slit is spaced within the acceptable regulations of the Parks Department. Children and adults may be seen freely from every perspective. The park visitors engage with an exhilarating non-orientable space, continuously revealing and recapturing while enhancing the presence of other people to form a community (Figure 5). This project was further expored in the Design Studio Re-Imagining City Island developed at the Intar Department Rhode Island School of Design.

CONCLUSION

The results of reviewing the presented cases from France, Italy and the United States show a different aspect of handling technology in constructing projects from scratch or in developing existing one to consider particle atmospheres. Unlike other pieces of research that handle the side effect of having technology in social-physical interaction, our cases show the possibility of having the significant contribution of technology in empowering the sociability (Nunn, 2011; Winner, 1997). As our review of the first group of cases figured out that technology might affect daily life, the other two groups of cases prove the importance of design-based atmospheres.

The review of these cases also shows the consequence of considering the impact of any project on several issues. The example of these issue is given to the city location where projects are implemented and its economy and economy of sponsors. The matter also extends to the project's impacts on research and social life. Scanning these projects in term on their impact on the social life show the significant role of historical buildings when having new adaption with modern technical elements. This notable appearance also was emerged in projects when it fits human needs in the community like Mathematical Belonging. The argument in the case of Social Cave extends to the impact of the project on raising social interaction between visitors once entered the site. However, it could not reach a broader appeal to the entire city. Involving student of Architecture in design process represented in Social Cave beside to professorial can provide creative ideas.

Figure 5. Collections of images show the design Mathematical Belonging, Starlight Park and Re-Imagining City Island

This chapter reviewed in the first group of cases to show how a specific urban typology – professional incubator, coworking and creative hubs, initially were generated by a form of causality (as per the case of the Bateau Lavoir and the Chelsea Hotel). These projects became a typological strategy to promote urban regeneration and the creative hub and coworking spaces, mixing creativity and intangible heritage. Such a phenomenon became one of the *unersco* (originally causally generated).

The second set of cases showed that a specific typology had been particularly affected by the development of smart technologies implemented in the coworking spaces engaging new forms of interaction in the semi-public places. This group of cases investigates the possibility to realised synthetically to define a strategic plan followed by creating smart and traditional technologies to regenerate urban environment without producing a standardised format. These case studies explore two different ways in which combinatorial architecture can transform public spaces by developing unprecedented forms. These urban forms made public spaces connect better to performative components and urban life.

To approach reconstructing urban ambiance in smart places, the authors illustrate the procedures of combinatorial architecture and its transposing to architecture and urban design that are applied in the sciences of complexity. How this to impact real life, two projects show this idea. The comparatives between cases show the importance of linking different disciplines such as geography, urban planning and sociology alongside the contribution of computer science could facilitate the application of responsive ambiances in implemented projects. This linkage should be at the level of academic and university scholarship. This study provided design implications for a transdisciplinary approach that link academia with professional practices.

ACKNOWLEDGMENT

The authors of the present work would love to accredit the following works to people who participate effectively in producing these projects, respectively.

Social Cave: Project Leader: C. Tiazzoldi - Interaction Design with: M. Arcese L. Biada

Research assistants D. Cashen, Roether V., Suite B., Zaratiegui J. Furthermore. Students: A. Robinson., Shim A., Kim A., Brichta B., Handal C., Pope C., Kaltis D., Kapompasopoulou E., Bauer E, Schmitt F., George Valdes G., Lalli G., Oh H.Y, Cho J.I, Dornic J., Thorn K, Lents K., Gonzales M., Kim Y., Ngy S.

With the Support of GSAPP Graduate School of Architecture Planning and Preservation, Columbia University: M. Wigley, DIPRADI Politecnico di Torino: L. Bazzanella, P. A. Croset, DIPRADI Politecnico di Torino. Sponsors Aipe, Cosmit, NSU. Griffin M., Chief Curator Salone Satellite

Toolbox: Concept: C.Tiazzoldi, A.Balestra, G.Milanese, L.Milanese. Project: C.Tiazzoldi

Base: Concept Base Milano, Project Onsite Studio, Consultant C. Tiazzoldi

Mathematical Belonging: Bronx River Art Center G. Nathan, KaN Landscape Design, C. Tiazzoldi, E. Benamor Duarte. With the support of Intar Department Rhodes Island School of Design.

Re-Imagining City Island Design Studio: Critic C. Tiazzoldi, Student Y., Intar Department Rhode Island School of Design

REFERENCES

Abdel-Aziz, A. A., Abdel-Salam, H., & El-Sayad, Z. (2016). The role of ICTs in creating the new social public place of the digital era. *Alexandria Engineering Journal, 55*(1), 487–493. doi:10.1016/j.aej.2015.12.019

Abusaada, H. (2020). Strengthening the affectivity of atmospheres in urban environments: The toolkit of multi-sensory experience. *Archnet-IJAR: International Journal of Architectural Research*. doi:10.1108/ARCH-03-2020-0039

Abusaada, H., Vellguth, C., & Elshater, A. (2019). *Handbook of Research on Digital Research Methods and Architectural Tools in Urban Planning and Design*. IGI Global. doi:10.4018/978-1-5225-9238-9

Araoz, G. (2013). Conservation Philosophy and its Development: Changing Understandings of Authenticity and Significance. *Heritage & Society, 6*(2), 144–154. doi:10.1179/2159032X13Z.00000000010

Bateson, G. (1979). *Mind and nature*. Dutton.

Cate, P. D., & Shaw, M. (1996). *The Spirit of Montmartre: Cabarets, Humor, and the Avant-Garde 1875–1905*. Rutgers University Press.

Claudel, M., & Ratti, C. (2015). *Open Source Architecture*. Thames and Hudson.

Columbia University. (2013). *Columbia Absreact.* Retrieved from None Linear Solutions Unit: http://abstract20122013.gsapp.org/non-linear-solutions-unit/

Cudny, W., Comunian, R., & Wolaniuk, A. (2020). Arts and creativity: A business and branding strategy for Lodz as a neoliberal city. *Cities (London, England), 100*, 100. doi:10.1016/j.cities.2020.102659

Dan, F. (1998). *Bohemian Paris, Picasso, Modigliani, Matisse and the birth of Modern Art.* Calmman Lévy.

Davis, F. (2019). *The Chelsea Girls.* Penguin Random House.

Descartes, R. (1637). *Discours de la méthode.* Leyde: Ian Maire.

Ehmann, S., Borges, S., & Klanten, R. (2012). The Social Cave. In S. Ehmann, S. Borges, & R. Klanten (Eds.), Learn for Life: New Architecture for New Learning (p. 236). Die Gestalten Verlag.

Elshater, A. (2015). The Principles of Gestalt Laws and Everyday Urbanism: A Visual Tactic of City Potentialities. *The International Journal of the Constructed Environment, 5*(3-4), 1–19. doi:10.18848/2154-8587/CGP/v05i3-4/37441

Furuto, A. (2011). *Onion Pinch & Napping Pod Projects at Cite' de l'architecture et du Patrimoine.* ArchDaily.

Hart, S. (2003, October 1). *Imagining the Future: How will we make buildings in 2030?* Retrieved May 2020, from Architectural Record: https://www.architecturalrecord.com/articles/12264-imagining-the-future

Hewitt, N. (2001). The Changing Landscape of Montmartre. *Journal of the Australasian Universities Language and Literature Association*, *96*(1), 1–18. doi:10.1179/aulla.2001.96.1.002

Holand, J. (1999). *Emergence: From Chaos to Order*. Perseus Books.

Industry City. (2013). *IC In Context*. Retrieved 2020, from Industry City: https://industrycity.com/

Inoue, S., Rodgers, P., Tennant, A., & Spencer, N. (2015). Creative Reductionism: How decreasing levels of information can stimulate designers imagination. In *International Conference on Engineering and Product Design Education*. Loughborough University.

International Ambiances Network. (2012, September 22). *2nd International Congress on Ambiances*. Retrieved May 2020, from Ambiances in action: https://ambiances2012.sciencesconf.org/

Jiang, D. (2020). The construction of smart city information system based on the Internet of Things and cloud computing. *Computer Communications*, *150*, 158–166. doi:10.1016/j.comcom.2019.10.035

Kottas, D. (2014). The Social Cave. In Digital Architecture: New Applications (pp. 250-257). Links International.

Lévy, P. (2011). The Semantic Sphere 1: Computation, Cognition and Information Economy 1st Edition. London: Wiley-IST.

Levy, P., & Bononno, R. (1999). *Collective Intelligence: Mankind's Emerging World in Cyberspace*. Basic Books.

Leydecker, S. (2016). Toolbox. In *Adaptive Reuse: Extending the Lives of Buildings* (p. 87). Birkhauser.

Massachusetts Institute of Technology (MIT). (2020). *Urban imagination and social innovation through design & science*. Retrieved from Senseable City Lab: http://senseable.mit.edu/

Mayne, T. (2011). *Combinatory Urbanism A Realignment of Complex Behavior and Collective Form*. Stray Dog Cafe.

McLuhan, M. (1964). *Understanding media: The extensions of man*. McGraw-Hill.

McLuhan, M. (2004). *Understanding Me: Lectures and Interviews*. MIT Press.

McLuhan, M., & Moos, M. (1997). *Media Research: Technology, Art and Communication (Critical Voices in Art, Theory and Culture)*. Routledge.

Medak, P. (Director). (1993). *Romeo Is Bleeding (1993)* [Motion Picture].

Mossberger, K., Tolbert, C. J., & Franko, W. W. (2013). *Digital Cities: The internet and the geography of opportunity*. Oxford University Press.

Munari, B. (1981). *Da Cosa Nasce Cosa*. Laterza.

Nunn, S. (2011). Police technology in cities: Changes and challenges. *Technology in Society*, *23*(1), 11–27. doi:10.1016/S0160-791X(00)00033-6

Oregon, U. o. (2015). *Sustainable Cities Institute*. Retrieved 2020, from Urbanism Next: https://sci.uoregon.edu/urbanism-next-0

Oxford Reference. (2020). *Overview: Bateau-Lavoir*. Retrieved May 2020, from https://www.oxfordreference.com/view/10.1093/oi/authority.20110803095451214#showlesscontent

Salama, A. M. (2015). *Spatial Design Education: New Directions for Pedagogy in Architecture and Beyond*. Routledge.

Sassen, S. (1999). *Globalization and its Discontents*. New Press.

Sawyer, S., & Tapia, A. (2005). The sociotechnical nature of mobile computing work: Evidence from a study of policing in the United States. *International Journal of Technology and Human Interaction, 1*(3), 1–14. doi:10.4018/jthi.2005070101

Sepasgozar, S. M., Hawken, S., Sargolzaei, S., & Foroozanfa, M. (2019). Implementing citizen centric technology in developing smart cities: A model for predicting the acceptance of urban technologies. *Technological Forecasting and Social Change*, *142*, 105–116. doi:10.1016/j.techfore.2018.09.012

Silva, C. N. (2018). *New Approaches, Methods, and Tools in Urban E-Planning*. IGI Global. doi:10.4018/978-1-5225-5999-3

Smith, P. (2010). Just Kids. London: Bloomsbury.

Sobejano, E. (2011). Combinatorial Architecture. In C. Gengnagel (Ed.), *Computational Design Modelling* (pp. 55–61). Springer. doi:10.1007/978-3-642-23435-4_7

The United Nations. (2019). The Impact of Rapid Technological Changes. In *United Nations Conference on Trade and Development*. Geneva: United Nations Publications.

Thibaud, J.-P. (2011). The Sensory Fabric of Urban Ambiances. *The Senses and Society*, *6*(2), 203–215. doi:10.2752/174589311X12961584845846

Tiazzoldi, C. (2008). Applied responsive devices. In M. Voyatzaki (Ed.), Emerging possibilities of testing and simulation methods and techniques (pp. 265-272). Thessaloniki: Charis ltd.

Tiazzoldi, C. (2009). Prototyping the city. In E. Lupo & C. Campagnaro (Eds.), *Designing connected places* (pp. 130–150). Compositori Editori.

Tiazzoldi, C. (2011). *Archilovers*. Retrieved from The Socail Cave: https://www.archilovers.com/projects/107616/the-social-cave.html

Tiazzoldi, C. (2016). Combinatorial Architecture Methods for the creation of ambiance in public space. In *Proceedings of 3rd International Congress on Ambiances* (pp. 865 - 872). Volos, Greece: HAL.

Tiazzoldi, C., & Tixier, N. (2008). Formal modulation for acoustic performances of a bridge. *ARCC Journal*, *5*(2), 104–115. doi:10.17831/enq:arcc.v5i2.20

Turner, M., & Tomer, T. (2013). Community Participation and the Tangible and Intangible Values of Urban Heritage. *Heritage & Society*, *6*(2), 185–198. doi:10.1179/2159032X13Z.00000000013

Vanolo, A. (2014). Smartmentality: The smart city as disciplinary strategy. *Urban Studies (Edinburgh, Scotland)*, *51*(5), 883–898. doi:10.1177/0042098013494427

Warhol, A. (Producer), Warhol, A., & Morrissey, P. (Directors). (1966). *Chelsea Girls* [Motion Picture]. The United States.

Weisberg, G. (2001). *Montmartre and the Making of Mass Culture*. Rutgers University Press.

Winner, L. (1997). Technology today: Utopia or dystopia? *Social Research*, *66*(3), 998–1017.

Wong, L. (2016). Toolbox. In Adaptive Reuse: Extending the Lives of Buildings (p. 87). Birkhauser.

Yamawaki, Y., & Filho, F. M. (2020). Do cultural issues matter on the adoption of technologies on public spaces? The case of Brazil through Black Mirror series. *Cities (London, England)*, 98.

ENDNOTE

1 The Italian Fab Lab was established in 2012 by Officine Arduino, Fablab Torino. It is a workshop open to everyone, equipped with 3D printers, numeric control milling and laser cutting equipment for the creation of prototypes and small-scale productions, experimenting with open source programming and the Arduino board, approaching new production processes and new business models, comparing ideas with other digital fabrication and open source technology enthusiasts and experts.

Chapter 11
Situational Atmospheres:
Filmmaking Approaches for Conveying Urban Design

Hisham Abusaada
https://orcid.org/0000-0001-6530-7714
Housing and Building National Research Center (HBRC), Egypt

Abeer Elshater
https://orcid.org/0000-0002-5061-6861
Ain Shams University, Egypt

ABSTRACT

This chapter examines how placemaking and filmmaking can improve teaching urban design-based situations of atmospheres. The qualitative-deductive form of content analysis addresses atmospheres in four Egyptian and American films and thus creates a conceptual framework for design strategy in studio of urban design. In this chapter, the authors discuss this framework with regard to the works of the Egyptian novelist Naguib Mahfouz. They show that filmmaking and placemaking are similar in the intellectual discourse because both produce ideas, themes, and events together with architectural elements and experience of everyday life. The findings highlight the impact of local architecture on the quantitative and nonquantitative factors that create affective atmospheres. The theoretical argument proved that the process of filmmaking could constitute a method for teaching in urban design studios and urban design history. The conclusion gives an action plan to use cinematic works as a tool for enhancing teaching affective atmospheres in urban-design studios.

INTRODUCTION

For decades, the concept of atmospheres has been one of the prevailing issues in many fields of humanities, such as physical sciences (Beggs, Šikoparija, & Smith, 2017; Reichler, 2009), artworks (Egan, 2014; Hsu, 2017), as well as architecture (Ford, 2012; Jenner, 2013; Semper, 2004, 2011). However, many of the theories of environmental design have devoted considerable space to the atmospheres from a purely

DOI: 10.4018/978-1-7998-3856-2.ch011

technical perspective, whether natural or physical (Lenzhölzer, 2010; Pelsmakers, 2012). Besides, recent philosophers, psychologists, sociologists, and anthropologists see atmospheres as a cognitive phenomenon, which relates to sensation and corporeality (Li et al., 2016; Merleau-Ponty, 2002; Schmitz, 1967, 2017; Schmitz, Owen, & Slaby, 2011).

According to them, atmospheres are perceived as an invisible thing that changes an individual's mood after influencing his/her feelings through the notion of the felt body experience (Abusaada, 2020; Schmitz, 2017; Schmitz et al., 2011). One of the crucial topic researchers have investigated in this field is perceptual qualities, which are related to the visual-aesthetic dimensions (Böhme, 1998; Li et al., 2016; Schmitz et al., 2011). Some architectural branches have addressed urban atmospheres as a human phenomenon, especially those which are associated with senses and behaviour, such as landscape architecture (Weidinger, 2016; Zhang, Cha, & Constable, 2014) and urban planning and design (Abusaada & Elshater, 2019; Buser, 2014; Stefansdottir, 2017). They, along with others, have rated the atmospheres as an aesthetic phenomenon in everyday life situations (Böhme, 1998; Borch, 2015; Griffero, 2014; Labuhn, 2016). Some of them have observed that atmospheres in the artworks (including novels, cinema, and architecture) are a purely satisfying emotional experience effect on the perceptual qualities of outdoor places in the city (Abusaada, 2019; Ford, 2012; Jenner, 2013; Semper, 2004).

Only a few previous types of architectural research have used the concepts of urban atmospheres (Augoyard, 2007; Stefansdottir, 2017), affective atmospheres (Bissell, 2010; Buser, 2014; Michels & Steyaert, 2017), and urban ambiance (Stefansdottir, 2017; Thibaud, 2002) as widespread phenomena, including human, natural, physical, psychological, and behavioural aspects of design (Abusaada, 2019; Böhme, 2017; Labuhn, 2016; Pallasmaa, 2014; Zumthor, 1998, 2006). Over time, many thinkers have revealed that urban environments have situations that can change the characteristics of the built environment, and, by changing these characteristics, the indications of people's impressions towards these characteristics change, which are evident in people's response and reactions in the body and faces (Abusaada & Elshater, 2019; Dewey, 2005, 1984; Pallasmaa, 2014).

One primary problem with these concepts is latent in the absence of its applications in some educational activity in the colleges of architecture, particularly in the departments of urban planning and design. However, the study of affective atmospheres has not found yet a significant foothold in urban planning and design, mainly about the issues of what emotions affect and what their role in place experience and placemaking change (Abusaada, 2019; Buser, 2014).

Accordingly, a well-known problem with urban design is that it does not consider the applications of affective atmospheres and relevant situations, mainly in some educational activities in the college of architecture, particularly in design studios, in the disciplines of urban planning and design. In this chapter, this dilemma concerns the understanding of the strong relationship between cinema and the history of thought, in order to use the output of the analysis of this relationship in urban design education. Other researchers studied this problem before and found many viable solutions (Andjelković, 2015; Egan, 2014; Krause & Petro, 2003).

A solution to this problem is latent in the design process, mainly in two stages. The first is data collections and analysis of place characteristics and indicators. The second is philosophy and design based on (criteria, principles, and guidelines) (Moughtin, Cuesta, Signoretta, & Sarris, 1999). It is necessary to know the urban characteristics of each locus regarding the most relevant events in each everyday experience and in each situation (Dewey, 1981; Pallasmaa, 2014; Stefansdottir, 2017). It is also important to identify the impacts on people's sensations, which correspond to the five senses: Taste, smell, sight, hearing, and touch (Pallasmaa, 2011).

Finally, there are growing appeals to recognise the indicators that are related to people's reactions (Reichler, 2009), which impacts the physical and nonphysical components of each locus by using the newest methods, such as sensors (Abusaada, Vellguth, & Elshater, 2019; Zeng & Zhao, 2011). In addition, it is interesting to explore the potential effects of affective atmospheres more carefully, which depends on the nonquantitative analysis in the design process concerning intangible yet vital factors that measure the interactions of sensations and feelings. For instance, the geographic information system (GIS) technique allows to collect and analyse data that are associated with statistical and quantitative spatial analysis (Pavlovskaya, 2009). Nevertheless, the primary problem with this technique is that it is now used outside its traditional technical fields and is being rapidly integrated with the latest multimedia and web-based technologies (Pavlovskaya, 2009, p. 5).

The objective of this chapter is to provide an action plan and some of the lessons learned in the field of affective atmospheres. Specifically, the chapter aims to analyse some of the cinematic works that have achieved different affective atmospheres in diverse human and urban situations and places. This refers to the idea of linking characteristics and indicators. The authors of the current chapter try to investigate the change of urban situations, which depends on the outputs of the architectural history of thought. They propose that the change in emotions is based on the intended situations of dramatic event (scenario of events.

The contribution of this chapter has widespread applicability, mainly in the studios of urban design in the colleges of architecture. This application has significant benefits in terms of improving the quality of the urban design project, not only related to the idea and concepts, but also at the level of the final product (master plan). Again, these contributions take into consideration two folds: The urban situations based on the events of daily life experiences and the people's impressions on the impacts of the characteristics of each situation in outdoor public places.

The question remains whether the urban designer can connect the history of urban design thought with affective atmospheres. Some of the interesting questions in this context are:

- Similar to filmmaking, can the urban designer make *staging* of affective atmospheres through situations of the built environment?
- How can the designer use the concept of affective atmospheres and its relevant situations in the academic discipline?

This work responds to these questions and is organised in six sections. The first section (*Background*) addresses the emergence of affective atmospheres in literature and how cinematic works contribute to screening our cities. The second section (*State of the Art*) discusses the methods used by reviews of four Western and Eastern films to determine the methods of emerging case of Naguib Mahfouz's. Thought of using the movies, the stage the traditional *Hara* ("alley" in Arabic) can create affective atmospheres. This selection reflects the nature of Egyptian life in a specific period of time. The third section (*Drawing on Research of Affective Atmospheres in Filmic Narratives*) discusses cinematic works vs. urban design teaching by an analysis of Mahfouz's works. The fourth section (*An Eye Opener from Moving Images to Urban Design Studio*) illustrates a two-steps conceptual framework of the common ties between place and time. The fifth section (*Processing Situation-Based Atmospheres*) provides a discussion of the results of this study. The sixth and last section provides concluding remarks and future research directions, and presents an action plan and some lessons learned from the application of cinematic methods for examining the history of architecture and using its tools in teaching urban design.

BACKGROUND

Several pieces of research tackle the screening cities from films (AlSayyad, 2006; Bordwell, 1985; Lima, 2008). Few studies have investigated the creation of affective atmospheres in urban places as cinematic moving images do (Abusaada, 2019; Anderson, 1994; Andjelković, 2015; Dorrian & Rose, 2003; Georgiadou, 2016; Lucas, 2008; Pallasmaa, 2008; Troiani & Carless, 2014; Weidinger, 2018).

Previous studies as well as the current work focus on the moving images and architecture of the space as essential tools in filmmaking and architectural design (Shiel & Fitzmaurice, 2001). The common elements in cinema and making the city image support visually and mentally the goals and subjectivity of creative design, such as "the constructed spaces" and "functions as a concentrated experience" (Georgiadou, 2016, p. 2). Films help to understand the meaning in architectural design and the city life (Narkunas, 2001). The atmospheres and mood are not vague categories in cinematic works (Tawa, 2017). They can produce through "the concilient discrepancy" between "kinematic and narrative in film, tectonic, and materials in architecture" (Tawa, 2017, p. 1). The "tectonic practice" is a cinematic method that creates the atmospheres due to the technological paradigm. This method enhances the design and experience of architecture. The production of affective atmospheres in cinematic work is based on the idea of "consilient discrepancy" and the case of "spatiality" that promotes the narrative, the "tectonic crafting of form, materials, and light" (Tawa, 2017, p. 6).

Additionally, two primary speeches in which filmmakers and architects are involved based on the concept of communication process. The first is the intellectual discourse, which includes the theme, idea, narrative structure, and scenario of events (Emmott, 1997; Werth, 1999). The second speech encompasses film techniques (Rabiger, 2008) or the audiovisual discourse (Janney, 2012), which includes physical elements and characteristics, everyday life experience, audio-visual effects (i.e., image, light, and sound), and vision (i.e., montage, which is piecing separate sections together to form a continuous whole).

Research concerning affective atmospheres in cinematic works has a long tradition in the areas of production design and cinematographic set (Egan, 2014; Tashiro, 2004). Concerning space and time, both terms mediate the use of the idea of *staging* (placing on stage), which is known in the French expression *mise en scène* (Mallet-Stevens, 1924). Some elements make the placing on stages, such as decor, lighting, depth of space, and costumes and makeup. This expression means "placing on stage" literally in English. It refers to what we see onscreen in a film. It means everything in the frame can carry meaning and it explains the visual presentation and the story it tells (McGrail, 2018). This term is also used "for culminating in a singular visual impression and represents an opportunity to how to tell any story visually (McGrail, 2018).

In one way or another, the authors of this work believe that this is what occurs during the production of the works of urban design. The designer reveals his/her themes based upon an audiovisual discourse (theme), such as utopia or dystopia, cities for people, making places, contemporary city, smart city, livable city, class inequality, and loss of hope. Afterwards, based on the design team, each designer achieves the requirements of each theme based on what his/her field of specialisation requires. Urban designers are cooperating in every detail in the project, namely in the steps of analysis (appreciating the context), a proper architectural style, construction, suitable building materials (colour and textures), and artefacts, as well as the elements of landscape architecture based on the visual and audio effects. However, most theories and paradigms of urban design are focused on meeting the requirements of function, beauty, recreation, movement, sustainability, and technology, as well as innovation and metaphor (Elshater, 2015; Elshater & Abusaada, 2016).

Continually, the matter of reconstructing affective atmospheres in the design process remains in the background of the scene. Meanwhile, the study of affect and its role in place experience in planning does not meet the needs of reconstructing different affective atmospheres (Buser, 2014). In any case, the designer in this field rarely believes in the idea of offering the proper design of events that match with the user's daily life, regarding sadness or happiness, excitement or boredom, and day or night.

This matter requires more explicit disclosure of the effect of affective atmospheres on changing people's feelings, through studying the visual sensations of architectural styles, colour, texture, scale, lighting, and other sensory elements (Jaaniste, 2010; Pallasmaa, 2008; Zumthor, 2006). This is what the field of environmental design can help the designer to achieve. The technical effects come in the forefront of sound effects. This is what the field of landscape architecture can perform (Dorrian & Rose, 2003; Weidinger, 2016).

One of the major topics to be investigated in these fields is exploring the impact of architecture in creating an affective atmosphere—wherein the incredible cinematic experience is produced from "the mixture between history and cinema" (Egan, 2014, p. 5). Therefore, the authors investigated whether cinematic works are analogous to urban designs, and how distinctive affective atmospheres can be accessed by applying filmmaking methods to generate effective solutions in this regard. In order to explore this undiscovered area, the authors reviewed many cinematic works that rely on architecture to create distinct affective atmospheres. They also examined whether this effect is possible in the field of urban design education or whether this is far-fetched.

STATE OF THE ART

This study hypothesises that the characteristics of the situations of the daily experiences of each place can create different affective atmospheres. In order to discuss the validity of this hypothesis, the authors selected a research strategy based on the deductive content analysis (Krippendorff, 2004; Neuendrof, 2002; Pandey, 2019), which they divided into two phases (Figure 1). In these two phases, the content analysis focused on the following issues: a) How people and places appear in these movies can alter affective atmospheres; b) how the diversity of the urban situations of the built environment can touch the feeling and emotions in affective atmospheres.

In the first phase, the authors analysed the content by scanning four American and Egyptian films for devising a framework. Going deep into developing this framework, in the second phase, they also analysed the content of the cinematic works produced from Mahfouz's novels to display the validity of the hypothesis on the Egyptian context. The authors' hypothesis proposes that filmmakers have plans to create atmospheres. Methodologically, in the two phases of content analysis, the researchers scanned a selection of films to identify the theme of the scripts, and the idea and settings of events, in the first phase. Then, in the second phase, they tackled how the filmmakers manage the production process for the physical elements and characteristics, everyday life experience, audio-visual effects, and image assembly. Also, in the second phase, the authors' analysis focused on what alters the situations of affective atmospheres in the Egyptian context.

In the first phase, the authors analysed four American and Egyptian films. The American films were *Blade Runner*, by director Ridley Scott (Fancher & Peoples, 1982), and *The Great Gatsby*, by director Baz Luhrmann (2013), by F. Scott Fitzgerald (2013). The Egyptian films were *Palace of Longing* (*Qasr El-Shouq* in Arabic), by director Hassan al-Imam (1967), and *Miramar*, by director Kamal El Sheikh in

Figure 1. The two phases of content analysis

1st phase — Tow Egyptian movies — Two American movies

Contents

- *Miramar* (1960), turned to film by director Kamal El Sheikh (1969).
- *Qasr El-Shouq* (1947), turned to film by the director Hasan Al-Emam (1961)
- *Blade Runner* Hampton Fancher and David Peoples by the director Ridley Scott (1982).
- *The Great Gatsby* F. Scott Fitzgerald (1925), Baz Luhrmann is the director and Catherine Martin as a production designer (2013)

Investigations

Intellectual Discourse:

- the theme of the scripts (rational discourse),
- the idea, and
- the settings of events.

2nd phase — Four Egyptian movies

Contents

- *Khan Al-Khalili,* written in 1946 turned to a film by the director Atef Salem in 1967
- *Midaq Alley (Zuqaq al-Midaq),* written in 1947 turned to a film by the directors Hassan Al-Emam in 1963
- *The Beginning and the End* (Bidaya wa Nihaya) written in 1949 turned to a film by the director Salah Abouseif in 1960
- *Palaces Walk (Bayn al-Qasrayn)* written in 1956 and turned to a film by the directors Hassan Al-Emam and Said Tantawi in 1964

Investigations

Intellectual Discourse:

- the theme,
- the idea, and
- the social characteristics.

Mise en scène (staging the settings) through:

- The situations of the atmospheres,
- the arrangement of the physical elements and their characteristics,
- The nonphysical experiences (everyday life experiences), and
- The technical works: Audio-visual effects and vision (montage) audio-visual effects and photos assembly.

1969, written by the Egyptian writer Naguib Mahfouz (1911-2006). These four films allowed to verify the validity of the authors' hypothesis that the architecture of the place can create differentiated or possibly affective atmospheres. This phase presented the initial theoretical framework.

In the second phase, the authors discussed the impacts of affective atmospheres in four movies produced based on Naguib Mahfouz's novels. These works by Mahfouz are *Khan Al-Khalili*, written in 1946 (Hameed, 2011; Salem, 1967), *Zuqaq al-Mida*q (Midaq Alley in English), written in 1947 (Battaoui, 2014; Al-Emam, Hassan Al-Emama, 1963), *The Beginning and the End* (*Bidaya wa Nihaya* in Arabic), written in 1949 (Ali, 2019; Abouseif, 1960), and *Palaces Walk* (*Bayn al-qasrayn*), written in 1956 (Alkaaby, 2017; Al-Emam & Tantawi, 1964; Makhail, 1992). The researchers explored the potential link between Mahfouz's works, urban morphology, and affective atmospheres in the four films. This subsection presents an Egyptian alley (*Al-Hara*) as an urban situation and staging in Mahfouz's works. This situation focuses on the physical and nonphysical elements of urban morphology, while staging (*mise en scène)* shows the complementary and mutual elements.

The authors chose the American and Egyptian films in the first phase because they have together a great history in the motion picture industry. The American movies also have a broad audience in most of the countries of the world. Likewise, Egyptian films have a significant spread in Arabic speakers. The authors selected the movies produced from Naguib Mahfouz' works not only because he had won the Nobel Prize in literature, and his novels have been subsequently globally recognised, but also because these novels were produced as movies in the Egyptian cinema. In addition, the authors selected these films because they personalise Egyptian everyday life. In this second phase, the content analysis linked

Mahfouz's works, urban morphology, and affective atmospheres as an urban situation and their staging in Mahfouz's works. These situations tackle the physical and nonphysical of the urban morphology, while staging (*mise en scène*) shows the complementary and mutual elements.

This phase confronted the effect of architecture in creating affective atmospheres between the personification of Egyptian public life in the 1940s and 1960s. On another vein, the authors of this chapter have more than twenty years of experience in Egyptian urban design studios based on the performance of public life. In this regard, cinema represents the most effective method to create and control virtually affective atmospheres (Bader, 2018).

DRAWING ON RESEARCH OF AFFECTIVE ATMOSPHERES IN FILMIC NARRATIVES

Previous studies show that the levels of studying atmospheres are diverse and focus on the evaluation and description of qualitative phenomena (Weidinger, 2018, p. 13). These phenomena linked with precise concepts in the fields of humanities ranges between behavioural events, which related to the reactions of people according to their mood, aesthetics, which connected to the visual effects, and socio-cultural elements tied to stress, hope, and desire (Anderson, 2009; Böhme, 1993; Griffero, 2014; Merleau-Ponty, 2002; Michels & Steyaert, 2017; Oakley, 1970; Pallasmaa, 2011).

However, today, it can be said that the phenomena of atmospheres have been associated with quantitative, measurable phenomena, which are linked with the natural phenomena according to the geographical location, climatic change, and street warming, with technical phenomena associated with place techniques, and with attributes phenomena concerning audio and visual characteristics, sound, and light (Lenzhölzer, 2010; Weidinger, 2016; Zumthor, 2006).

Moreover, a series of contemporary studies have indicated that the phenomenon of qualitative atmosphere appeared in the literature on spatial planning, landscape architecture, and environmental design. The first is concerned with cognitive research (Dorrian & Rose, 2003; Sidanin, 2007). The second deals with natural and technological influences (Gissen, 2009; Weidinger, 2016). Previous studies seem to focus on demystifying the concept of atmospheres (or ambiances) in the areas of architecture that are interested in humanism, on the one hand, and technology, on the other (Weidinger, 2018). This chapter focusses on demystifying these concepts through tracking the situations in the both cinematic work and history of architecture.

Finding Intellectual Discourse in American and Egyptian Films

Architecture and cinema are widespread phenomena (Oakley, 1970; Salingaros & Mehaffy, 2006); they include combined atmospheres and inspired by the natural, physical, sensory, and behavioural aspects to create an impactful scene on people's lives. Based on understanding the wider meaning of the phenomena of affective atmospheres, this subsection focusses on the similarities between cinema and architecture, particularly in the urban situations, and what follows of time, space, and multimedia (Georgiadou, 2016; Jacobs, 2007; Pallasmaa, 2008). Both "represent a vision of the fusion of space and time" (Vidler, 1993, p. 46), wherein the film classifies as a "spatial art" (p. 46).

A mutual intersection of these two spatial arts, the architects build the construction sets in the film, while the production designer constructs its architecture based on the impacts of light and shade, and scale, and movement. The filmmakers used the idea of affective atmospheres to shape the environment in the cinematic work, and these environments have been constructed mainly by using the architectural styles and elements, as well as time. Time helps to identify not only the historical moments, but also to send messages about the spirit of the place. While the architectural style improves the dramatic sensation in time, many movies create affective atmospheres through the emergence of the reality of the city.

In cinematic works, the impressions of the one scene do not take minutes, yet have lasting effects on the feelings of the viewers, altering their mood and experience. The consequences of the atmospheres of any scene still affect who lives in any place. Nevertheless, filmmakers are cautious in creating the different atmospheres, as each stage captures the viewers' mind and makes them believe everything they see.

The data analysis of the four films of the first phase revealed significant differences in the places and time of their events. The events of the first film were located in the West, in the city of Los Angeles, in the year 2019. The second film was situated in Long Island, New York, in 1922, in the height of the roaring twenties during the Jazz era (Fisher et al., 2013). In Egypt, the third locations are Al-Gamaliya and Al-Abbassia, which are respectively traditional and old districts in Cairo and the new Cairo suburb during the events of 1919. Finally, the fourth film was located in a fabulous motel (pension) to show the impacts of the foreigner architects in Alexandria.

Concerning the first phase, the authors' results showed that each of these films has a central theme. These are dystopia, the falsity of the upper class, the neo-rich portrayed as vulgar and disgraceful, epic realistic literature, and the popular struggle. The results also revealed that the ideas behind each film were several, namely: A terrible beauty, the post-disaster affective atmospheres, the visual influences dominated in the machine age, the core Long Island jewels, patriarchy and the masculine duality between the claim values of family, misguided abroad, female fight, and the small lines conflicting through the life of a seven retired.

The authors classified the events of affective atmospheres into the following five categories: class differences, a microcosm of the Off-World/Earth, and the high towers of commerce and power, which are accessible only to individual flying devices, in the sole possession of the police, and by controlled-access elevators (Williams, 1988). The upper-class characters were in immense palaces with soaring ceilings and cold marble floors, while the lower class lived in lower-quality housing and apartments. Internal courtyards created the privacy, and Al-Mashrabiya covered the windows to avoid pedestrians on the street from being able to see the female that sits behind it. The differences between the social classes: the lives of the lower level and the nobility. Inside and outside the building are dominated by feminized atmospheres in the third and fourth films.

Concerning the Western movies in the first phase, the filmmakers reconstructed affective atmospheres based on the characteristics of the place and the physical and social elements in each film, starting from splendour and mystery in *Blade Runner* (Williams, 1988). In this, the atmospheres emerged in the city of postmodernity and the postmodern urban landscape. It is also developed based on the post-industrial decays, hybrid architectural design, deindustrialization, dark shadows, and dim lighting. In another vein, in *The Great Gatsby*, the filmmakers revive the style of visual arts and France architecture style.

In the Egyptian films, the date refers to the atmospheres that emerged based on the daily life of the Egyptian families in private homes in the early 20th century. Besides, finding atmospheres occurred due to belligerent occupation and street revolutions, obscenity and prostitutes' lives. In the 1920s, atmospheres change due to the romanticism, differentiation between classes, impacts of an aristocratic era, and experience of the places. In the 1960s, atmospheres are changeable due to the Egyptian political situation, the violence against women, and urban and rural violence.

Regarding the second phase, the results of the analysis of the four Egyptian films revealed differences in the characteristics of affective atmospheres, which ranged between acid rain and choking on pollution (Milner, 2004). Accordingly, the architectural elements emerged in the dead-ended streets, similar buildings as a robot, plate glass windows, and flashes of neon in all directions (Dempsey, 1982). In another vein, the facades of the buildings emerged in the simple cottage and in immense palaces. Besides, the traditional courtyards in Islamic building appeared with small narrow streets for pedestrians. The residential neighborhood unit, the individual buildings, and the Islamic elements consisted of the narrow streets of Al-Gamaliya as a traditional form of Al-Hara, while the vast roads in Al-Abbassia are found there. Besides, scene showed the English-style garden, kiosks and the beautiful landscape. The building in *Miramar* film was built by the Italian architect Giacomo Alessandro Luria in 1926 brought together the Islamic and Italian styles.

The content analysis of the first phase showed also the relationships between urban form and everyday life in the built environment to explore the situations and staging. These situations clarify that everything aims to have different affective atmospheres. For instance, in the early 20th century, the first situation emerged in the film *Blade Runner* as consumer capitalism. It appeared through the large corporate towers that cluster at the heart of the metropoles (Milner, 2004). The following situation is in the film of *The Great Gatsby*, with prosperity and grandeur, which the film expressed by using the Art Deco style as consisting of bold geometric shapes, luxurious decorations, and technical aspects.

The situations in these films emerges in the class discrimination among the Egyptian people through the following two scene: 1) The body, style, and construction of architecture between the building with courtyards and solid facades with Islamic architecture vs. the significant palaces with the European style; 2) everyday life experiences in the street cafe vs. the significant garden of the palace.

The situations in the film *Miramar* recognised in the building, which were constructed by the Italian architect Giacomo Alessandro Luria in 1926, and which brought together the Islamic and Italian styles. The filmmakers chose the building of the *Miramar* for its aesthetic qualities. The film also showed this building and its architectural splendor in many scenes. The authors think this link the nature of the story, as well as the atmosphere achieved by the structure of *Miramar*, reflects how it is possible to build a different atmosphere through the artifacts.

The authors' analysis now can show a primary step toward an in-depth understanding of the role of cinema in reconstructing affective atmospheres in the status quo. Presently, an important challenge in this domain is how urban design instructors can teach how to recognise affective atmospheres as cinematic works did.

Retrieving Situational Atmospheres from Mahfouz's Works

In the selected movies, the authors tracked the change in time to provide a sensation of the places, the transformation of physical architectural elements, people, and social life. The period that these selected works tackled is the beginning of the end of the Ottoman Era, in the period of English occupation in the early 19th century, which was known as the period of transformation from the colonial to the independence era.

Regarding the urban morphology, Al-Hara occupied a vast space from the making of the place in many films. Al-Hara "as in life itself, there is much gaiety, colour, and excitement to enliven the passing scene" (Bhuiya, 2014, p. 142). The central theme of most Mahfouz's novels is the consequence of the rapid "change in the social values and radical alteration of the perspective of reality on the common people" (Bhuiya, 2014, p. 142). The Al-Hara represents the pulsating heart of the residential neighbourhood unit (Husseinya & Husseiny, 2012; Raymond, 1984; Saad, 2018).

Four movies focus on the social, morphological, and temporal dimensions that create affective atmospheres through the concept of the Egyptian Al-Hara. Arguably, Al-Hara is the smallest planning unit in the structural hierarchy of the traffic routes in the ancient Egyptian cities. Al-Hara is similar to the closed-end roads in Western thought.

Figure 2 shows the analyses of the alley in the four films in two steps. The first step includes the theme, the ideas, and the social characters. The second step interprets the film atmospheres in the alley through the situations and staging the physical elements.

The film directors, with the production designers and decorations designers, created the context of the place based on the concept of Al-Hara. They built affective atmospheres based on the vocabulary of the places, namely the physical architectural elements and nonphysical social aspects. The physical elements are the context components, such as the style of building facades, lined walls with arabesque finishes, perfumery and spice shops, the bakery, the barber, the coffee shop, and the alley paved with stones. The nonphysical elements include the transformations of events (scenarios) and the audio-visual effects, such as light and shadow, sound, and the climatic effects (rainfall, winds, and temperatures). Perhaps the characters of the films are one of the most critical impacts on affective atmospheres.

Since the 1920s, Egyptian architecture has derived its inspiration from European styles. The impacts of aesthetic dimension in the new residential buildings has appeared along with remnants of the Islamic architectural heritage that were built in the eras of the Mamluk, Fatimid, and Ottoman Caliphate, which ended before the 1920s. The architectural design of that period was dominated by affective atmospheres in the literary works of Mahfouz. Starting in the 1920s, the influence of European thought began to change distinctively, the French Haussmann school of construction appeared, and the work of architectural models committed to the Egyptian style dominated. From 1920 to 1930, there was a continuation of the movement of modernity. After the year 1925, the international style of architecture appeared, mainly in residential buildings. The change in thinking began between the 1930s and 1950s. During this period, many different European architects were coming from different countries, such as Italy, France, Belgium, and England, to stay in Egypt. Mahfouz's literary works, most of which were turned into cinematic works, show how to blend this architectural diversity with the atmosphere of the places where the events took place.

Followers of Mahfouz' novels and based films can discover a strong bond between the historical, traditional, and old areas (Makhail, 1992; al-Tawati, 2006). Each story focuses on creating the affective atmospheres released from people, places, and the surrounding buildings. In the selected novels, the ethnicity of Egyptian architecture helps in creating affective atmospheres. In cinematic works based on Mahfouz's novels, the interest is in making affective atmospheres that seem clear based on the changes in the architectural patterns. Affective atmospheres have been ranging between being sympathetic, in the traditional Egyptian areas by the use of the preferable type to Egyptian people for Al-Hara, to dictatorial and aggressive affective atmospheres, related to the European architecture in the new suburban districts. From the short review above, key findings emerge as follows:

Figure 2. The second phase tackles the content analysis of the Egyptian alley in four of Mahfouz's works

		Khan Al-Khalili	Midaq Alley	The Beginning and the End	Palace Walk
Date of novel publishing		1945	1947	1950	1956
Date of film production		1967	1963	1960	1964
Time of events		1942	1940	1930	1919
Step 1: Intellectual Discourse	1. The aim	How the Egyptian Hara (alley) reflects the reality of the public life of people in Egypt			
	2. The theme	The change in the social values lead to the alteration of the reality of people			
		A philosophical approach to the idea of the life and death	Poverty and destitution. Inadequate education. Neighbours relations. The transformation to modernity.	Human compassion	Sovereignty and female servitude
	3. The ideas	Life is a tragedy and the world are a boring theatre The misery resulting from emotional deprivation	Transformation from tradition to modernity	A widowed mother's hellish struggle against poverty The torments of the human soul.	The contradiction between the visible and the hidden Injustices towards women. The subjugation of women
Step 2: Mise en scène (staging the settings) of Al-Hara: The arrangement of everything that appears in the framing – actors, lighting, décor, props and costume	4. The situations of the atmospheres (a distinction between inside and outside places)	Inside events: The life in the apartment and the traditional market.	Inside events: The life in the traditional room that looking the alley, the barber and brothel.	Inside events: The life in the basement residence, the night club and the prison.	Inside events: Family life in the traditional building with courtyards and home of prostitutes.
		In inside and outside events: neighbours' relations in the lobby of apartments.	In inside and outside events: Windows overlooking the alley.	In inside and outside events: Not appeared	In inside and outside events: Overlooking from Mashrabiya to the alleys
		Outside events: the movement in the streets and the traditional bazars.	Outside events: the sittings of the traditional alley and the activities of people in the popular café and surrounding buildings and shops	Outside events: the events in the street, casino bridge on the Nile It focuses on the world of cannabis, disguised unemployment, mutilation, and begging	Outside events: Life in the alleys, fragrant shop and café. Friends' meetings in the house luxurious garden
	5. The physical elements (urban form)	The commercial shops of treasures and antiques. The materials: the warm stone tiles of basalt black metallic color.	The traditional building facades, the traditional market, and the popular coffee.	The alley here is devoid of characteristics, it is a space conducive to a house has no characteristics and its heroes live in the basement. Poor small shops Buildings without styles	The residential quarters of middle-class families Traditional form and style of houses.
	6. The nonphysical experiences (everyday life experiences)	The ability to a) make carnival of clothes and b) follow everyday life experience in the street that sells the antiques	The ability to follow the social life through the Kirsha's café. The men spent most of the night drinking tea and coffee, smoking shisha, playing card games and dice, and most of all, they chatted about everything.	The conflict between the modernity and classical/traditional. The characters of the film are one of the most influential influences on the atmospheres.	Blind ambition of a self-centred person who destroys his family for his benefit. From respect and chastity inside home to deceit and vulgarity outside home.
	7. The technical works: Audio-visual effects and vision (montage)	The ability to a) mix of heritage goods, b) make aromatherapy perfumes, c) the lighting lamps suspended in sling between houses above the alley and d) wooden ceilings allow daylight access.	The centre of the process of packing and manufacturing boxes for the bride, along with incense, spices, and perfumes, in addition to natural fat (not artificial margarine)	The ability to feel a) the effects of the poverty related to the furniture, cloths and foods, b) c) the woman's fear of her brother through the lighting effects on the river.	The ability to feel the effects of sound, odours and texture inside and outside all places in the urban environments.

1. The first impression from watching the selected films and reading the novels is that the designer can consider the impacts of types of urban tissue, open spaces, and buildings to create different affective atmospheres.
2. The second impression is that Mahfouz and the directors of the selected movies made unprecedented changes in patterns of place scenes to explain the impacts of social life on affective atmospheres. In *Cairo Trilogy*, these atmospheres reflect the domination of the aristocrats against the bourgeoises.
3. The third impression is that Mahfouz was playing with the influence of the private residence related to the structure and construction of building blocks. These building blocks figured outhouses that have heavenly courtyards, representing the Islamic style recognized in Al-Hara alongside a palace with a Western-style in modern suburbs. Mahfouz and all the film directors show that people who live in Al-Hara have ethics, and that their roots belong to the homeland. Instead, peoples who leave it are losing their ethics and values, which needs to be reconsidered.

As in the making of atmospheres in literature and cinematic art, the question however is: Can the urban designer draw the characteristics that create urban atmosphere for reality?

AN EYE OPENER FROM MOVING IMAGES TO URBAN DESIGN STUDIO

Academic programmes have initiatives to screen cities and public life from films (Bader, 2018; Lima, 2008; Lucas, 2008; Rowe, 2020). Linking these initiatives to the authors' outcomes provides evidence to the urban designer by considering the design-based atmospheres from the beginning similar to what novelists or directors do in films. At the beginning of the design process, urban designers should mention that designed architectural elements will create atmospheres for a unique experience. Thus, working during the design stages can list the indented atmospheres and emotions that the users should have. Moreover, the commitment to the modalities, indicators, and signs of design is essential, so that the recipient can feel atmospheres in various parts of the city that have the same activities. These findings support the notion that the cinematic language is one of the closest technical arts to the realm of urban design.

In this work, the authors' suggested two-steps in the suggested conceptual framework illustrates the common ties between place and time. The first phase is the intellectual discourse and the second phase is the *mise en scène* based on the situations of the atmospheres. Each phase consists of two components, namely the actions and the actors. It ends with a statement about the common ties, which includes the urban form and experiences of daily life and the impact of positioning in a situation. Figure 3 summarises the authors' conceptual framework for reading and analysing cinematic works.

Indeed, the filmmakers create the atmospheres for the viewer to live within them, whether these atmospheres expressed a reality or were inspired by the imagination of the writers of fantasy films. In the same way, an architect creates buildings and spaces to provide an environment for living, work or recreation, and each life an architect makes is different thanks to the affective atmospheres of the realm of architecture and the art of the city. This matter could be possible and available in the field of architecture, so all these feelings can be transferred to the user wherever he/she lives.

Urban designers should determine the meanings in the design process at the outset. Then, they could review the program of activities, events, relationships, behaviour, buildings, spaces, and vocabularies to stimulate a scenario that creates feelings and emotions. In this vein, the next question revolves around whether it is possible to create festive, repulsive or charming atmospheres in a place without the principles

Figure 3. A conceptual framework

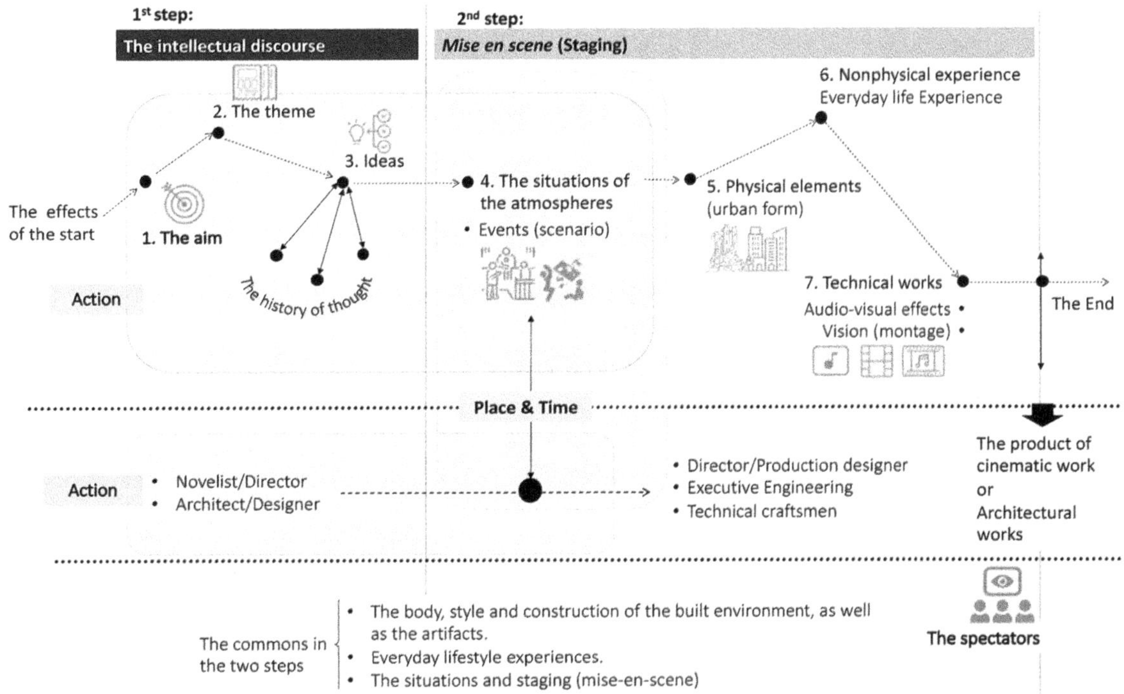

that affect people's feelings. The answer is that all these kinds of atmospheres depend on the designer's imagination and his/her ability to create the convenient built environment that generates different feelings. Consequently, at the beginning of designing a place, designers should think about the possible materials and elements of staging the scene that creates affective atmospheres.

This review now provides evidence to the matter of urban design that should exceed the uses of the social, economic, functional, visual, perceptual, morphological, temporal, and environmental dimensions. In this regard, one can consider how an urban designer deals with design process as narratives and scenarios. The authors' content analysis showed that the city is a real tale; it includes real people who live their actual events and scene in moving images. Hence, their buildings and their situations should represent authentic architecture in real-time. In this section, the authors present the answer to how to use the analysis of the cinematic works in teaching urban design in design studios.

The results in this chapter match the methods of the state of the art. Below, the authors compare the results of the proposed method with the atmospheres produced in other 16 films. Figure 4 includes examples of how personalised atmospheres effect on outdoor places in many cities around the world. These movies, together with the movies the authors selected, demonstrated that atmospheres emerge as "Paris city image," "Italy real space," and "Egyptian street markets" (Dodds, 2014; Knapton, 2012; Leaman, 2003; Project for public places, 2012). The last example shows different atmospheres in many public places from the film *Mission Impossible: Ghost Protocol* (Bird, 2011; Movie-Locations.com, 2019).

The present readings of literature and analysing the random selection of cinematic works in both Eastern and Western thought can represent a methodical addition to create affective atmospheres during the rehabilitation of traditional areas, as well as the design of the new cities. Concerning learning to produce affective atmospheres, in the realm of architecture, particularly in urban design at academic

Figure 4. Examples to reconstructing the atmospheres in cinematic works

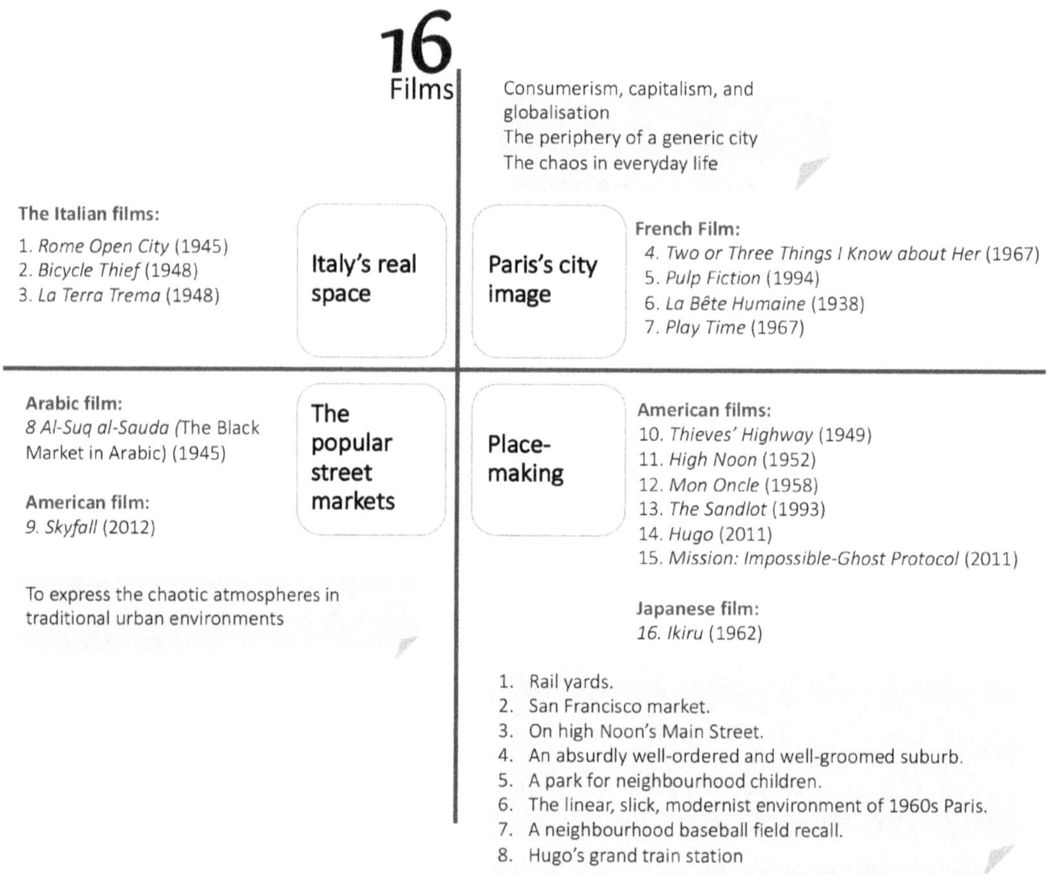

level, two courses could be considered to include teaching the issues which are studying the effects of affective atmospheres on urban forms based on the cinematic works. The first is the history of urban design thought, and the second is urban design studios.

The authors confirm the results of scanning relevant literature and cinematic works, namely that designing based on atmospheres can link urban design history of thought through reviewing cinematic documentation of cities. These archives include the results of the analysis of the atmospheres that are related to the history of urban form. Furthermore, designers of today's cities can go through the same process of designing and controlling unreal movie images and situations that might affect the emotions. These findings support the notion that designing the real situation in everyday life of cities means using a process based on four stepladders (Figure 5).

The first step studies the history of thought in discrete periods. It aims at analysing the general context of knowledge based on the relationship between urban form and societal transformations. This step focuses on knowing what happens today regarding the general context of culture. It includes a deep understanding of the present by analysing historical events through time: patterns and meanings. Concerning preparing a document of cinematic works, this knowledge is acquired based on two aspects of an experience. The first explores types of affective atmospheres for one city of a bygone period, and the second examines affective atmospheres in one period for many cities.

Figure 5. Four stepladders to teach the history of urban design thought and history of cities

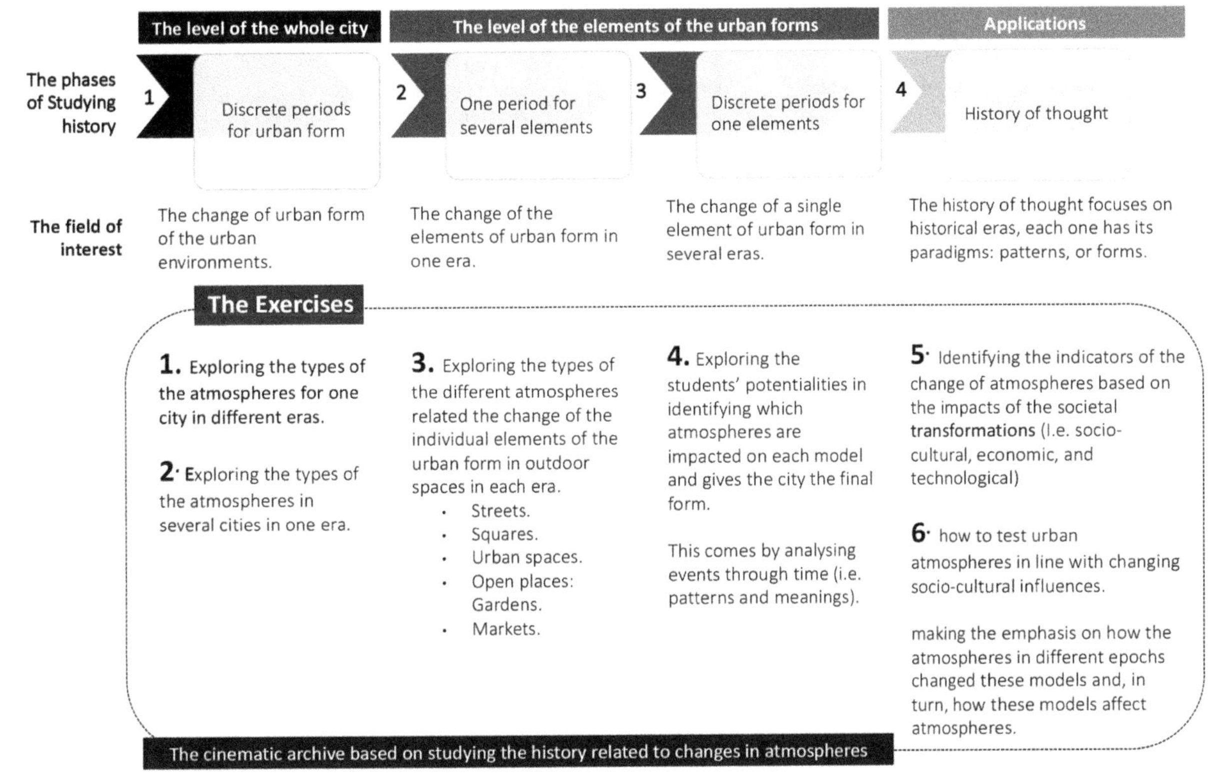

The second step explores the effects of affective atmospheres on the main elements of urban form in diverse eras. The third step focuses on examining the impact of the affective atmospheres related to a single element of urban form, such as Al-Hara as a primary planning unit in the old Egyptian cities.

The fourth step shows the reasons behind the changing of the urban form, which is reviewing the societal transformations and how they synchronise with historical events. As such, this step develops the cinematic archives to include a cognitive connotation related to the relationship between the change of atmospheres combined with the change of urban forms and societal transformations. This step concerns linking the shift in the atmospheres with the societal change via different historical periods, such as the Mamluk, Ottoman, and modern eras. It also searches for the ideas that enable one to solve the problems in the past to address the current issues: the lessons learned. Moreover, this reviews how the pioneers of urban design created the ideas of designing the cities not only to solve the emerging problems, but also to teach the researchers of today to explore innovations, theories, and principles.

It is by now generally accepted that, in urban design studios, the design process of projects should start from the development of the theme and scenarios of events related to the perception of specific atmospheres as in the cinematic works. The process of the study begins with the identification of time and place for each project. Afterwards, students should analyse the atmospheres of their projects based on the situations of urban design in three phases and five steps (Figure 6).

Figure 6. Three phases to teach affective atmospheres in urban design studios

The First phase: Drama of thought		The second phase: Concept		The third phase: Design
1st step: The nonphysical elements	**2nd step: The physical elements**	**3rd step: the atmospheres of the project**	**4th step: The urban form and architectural styles**	**5th step: the product**
1) The theme: the intellectual discourse 2) The ideas 3) The events (scenario) 4) Visions (placing on stage)	5) The program (the components): buildings, open and urban spaces, and the circulation movement and pathways. considering the components of landscape architecture, such as landform, water, vegetation, and wildlife. 6) The spatial organisation: bubble diagram Understanding the determinants of spatial relations.	7) The situations and staging atmospheres. 8) The characteristics of the places.	9) The design of the atmosphere is a thoughtful and philosophical presenting, touches the spirit of the spectators; it assumes that architectural products should be distinctive based on the urban form and architectural styles.	10) Producing the projects depends on selected atmospheres and based on (a) the intellectual discourse: themes, ideas, visions, (b) the physical and non-physical architectural elements, (c) situations, staging and characteristics and the audio-visual effects and (d) peoples. 11) Evaluation process: the viewpoints of the instructors and students.

PROCESSING SITUATION-BASED ATMOSPHERES

This study confirmed that the findings of the analysis of the artworks addressed affective atmospheres as human, aesthetic, and purely satisfying emotional experiences. They are overwhelming cinematic experiences. These experiences are associated with senses and behaviour in everyday situations and effect on the perceptual qualities of places in the city. An essential finding focuses on the understanding that the realm of architecture plays a significant role in constructing atmospheres similar to cinematic works. Therefore, by recognising the factors that enable filmmakers to create the different atmospheres, the designers may reconstruct affective atmospheres in urban design studios.

In line with many previous studies, the analyses of atmospheres focus on quantitative, measurable, and qualitative phenomena in two aspects. The first focuses on the cognitive phenomenon and relates to people's reactions based on sensation, behavioural and socio-cultural events, and audio and visual-aesthetic characteristics and effects. The second includes the technical and attributes phenomena, such as climatic change, street warming, place techniques, sound and light audio and visual characteristics, spatial planning, landscape architecture, and environmental design.

Many theorists have demonstrated that experience in cinematic and architectural works is an integrated emotional and technical experience. An experience represents an essential factor that influences atmospheres in filmmaking and urban design projects. Both depend on the image of architecture. It is necessary to highlight the fact that the tectonic practice is a cinematic method to create atmospheres. It depends on the knowledge of architecture, production techniques, materials, and new technologies, which promotes the narrative through the case of spatiality (Tawa, 2017).

Moreover, the results highlight that two factors effect on cinematic and architecture work, namely intellectual discourse and techniques. They include the theme, narrative structure, and scenario of events, and audio-visual aspects (image, light, and sound). The data of the content analysis regarding the four American and Egyptian films provide a shred of evidence about the effects of changing affective atmospheres on filmmaking. These results show that similar elements occur in cinema and architecture: The subject, the idea, the time of narration, events, induction faces, characteristics, and architectural elements. These data in the results of the content analysis of the prior films enabled the author to construct a framework to explore the possible relationship between Mahfouz's works and the atmospheres based on the idea of the traditional Egyptian alley, which emerges as a remarkable situation in the cinematic works and as a fundamental morphological unit in urban design.

The outcomes show that Mahfouz's works relied on presenting the alley as a hero in his novels, as the main characters, which represents an excellent opportunity to show how a locus can create a situation. Wherein he focused on his writing on the dramatic scenes that describe the nature of the characters accurately in line with the time, theme, and events of the story. Mahfouz illustrated the selection of locus was based on changes in the characteristics of the architectural situation. This situation included two types of architecture: One in the old Cairo Downton (the birthplace of the author), side by side with the second place located in in the new districts with modern architecture. Besides, it arises in each novel to show the impact of Eastern and Western styles on the scenario of events. The findings confirm that the time (the temporal dimension) plays a useful role in many of Mahfouz's works.

A further novel finding uses the idea of traditional Al-Hara as a way for identifying atmospheres. Another novel finding uses the idea of traditional Al-Hara—as a social phenomenon in Egyptian cinema—as a way for showing atmospheres in urban design studios. The authors' results show how their implications help raise students' creativity during their architectural education, which occurs mainly in urban design studios, to create affective atmospheres that change according to social, cultural, economic, and political conditions. Moreover, these implications present a way to reconstruct affective atmospheres in designing public places, depending on people's reactions in the face of the changing of events, characteristics, and situations.

This approach has two limitations. The first is due to the traditional way of presenting the concept of Al-Hara in cinematic works, even though this limitation does not exist in the status quo. The ideas of contemporary urban design that may use many themes applies the concept of Al-Hara, such as cities for people, cities for pedestrian, and car-free and sustainable design. The second limitation of this approach is that the urban designer cannot control people's reactions as the director can in the movies. Anyhow, for each situation, affective atmospheres in the four cases of Mahfouz's movies are based on the architectural elements, social aspects, events transformation, audio-visual effects, and the characters of the films.

The leading conclusion that can be drawn is that the artworks, such as the novel and cinema, can explain the history of urban design, on one side, and provide an approach to create the urban atmospheres of the places based on the element of urban design, on the other hand. More generally, these basic findings are consistent with research showing that architecture plays a pivotal role in creating an affective atmosphere in literature and cinematic works. The atmospheres in literature and cinema are made up of the architecture of the place, the people, activities, and events (AlSayyad, 2006; Georgiadou, 2016; Shiel & Fitzmaurice, 2001). Nevertheless, the designer does not take this emotional dimension seriously at the beginning of the process of urban design practice to create the realm of real life. The concept of atmospheres in history and literature provides a reliable and comprehensive theoretical basis based on the beliefs of Western thinkers.

CONCLUSION AND FUTURE RESEARCH DIRECTIONS

The authors conclude by arguing the findings of the cinematic works to be the nearest to the field of urban design. This discussion refers to the aims of this research, which began by studying the possibility of creating urban atmospheres in reality—as in art and literature—and primarily in cinema as a field that is very similar to reality. This aspect of the research suggested that it can change the way of thinking for teaching architecture, particularly in the projects of urban design studios. Many issues based on the idea of how to create affective atmospheres are an essential subject of this path to analyse the paradigms of architecture throughout history, such as styles, trends, movements, and theories.

Besides, the authors' leading conclusion is two-fold. The first is what artworks can explain concerning the history of architecture. The second provides an approach to create affective atmospheres of the public places based on the factors that effect on urban design. This aspect of the research suggested that it is possible to change the way of thinking in teaching urban design, particularly in two courses, which are the history of urban design thought and some exercises in the projects of design studios. This way of thinking can be considered a way to teach the styles, trends, movements, and theories of the history of thought based on the idea of how to create urban affective atmospheres.

As the authors argued, architecture plays a pivotal role in creating affective atmospheres in cinematic works. This creation may be considered a promising aspect to take this emotional dimension seriously at the beginning of the process of urban design practice to create the realm of real life. This conclusion follows the fact that the affective atmospheres in literature and cinema are made up of the characteristics of the place, surrounding situations, events in daily experiences. It also follows the indicators of people's reactions toward all these elements—which is very similar to what happens in the urban design process. The current work concluded with some lessons learned to determine how different affective atmospheres can create in ways close to those used by filmmakers. Moreover, these lessons provide suggestions which focus on how to take advantage from the history of architectural thought to generate well-defined affective atmospheres.

The contribution of this work has broad applicability, if included as an educational goal in urban design studios. Because of the lack of availability of these data above, the authors decided to investigate the strong relationship between cinema and the history of architecture to use the output of analysing this relationship in urban design education. This analysis focuses on the events of daily life experiences and the indicators of people's interactions relevant to the impacts of the characteristics of each situation in public places. Considering a selected film, regarding one or more situations, different events, and architectural characteristics, the manifestations of the impacts of the affective atmospheres on people's reactions and applying in design project occur through five educational stages. First, the instructor can identify the theme, the ideas, and the social, economic, and political characteristics. Second, students should identify the situations for the scenarios, events of everyday life experiences, and architectural characteristics. In the third stage, both instructors and students should together classify the people's reactions, bodily movements, and facial reactions, identify the types of affective atmospheres related to these reactions, and people's reactions that should be presented in intended actions

The authors' results require further investigation for developing the methodology that exploits the GIS as tool to explore the change in the situations spatially. This investigation may form an introduction to link the alleys that have distinctive affective atmospheres to a database the public places in the city. Moreover, this future study may explore the tools of spatial analysis for supporting the formulation of urban planning and design in terms of affective atmospheres based on the available data of the GIS in public places. The GIS can help the designer to identify the characteristics of situations based on the styles of architecture, events, and daily experiences. This method can perform very well in constructing an adequate database related to the relationships between the atmospheres and the characteristics of the place. A further novel finding is that the GIS technique for collecting and analysing the data takes into consideration nonquantitative analysis, such as the interactions of sensations and feelings. The applicability of these new results can be tested on urban design studios, notably in the phases of data collections, analysis, philosophy, and design. In urban design studios, regarding situations, the GIS technique can be used to identify the urban characteristics, events, everyday experience, and the indicators that are related to people's reactions. Future research may consider the potential effects of the possibility of how GIS techniques can strengthen the teaching of affective atmospheres in urban design studios.

Questions for future research can derive from how the GIS can strengthen the teaching of affective atmospheres in urban design studios more carefully. For example, how the GIS plays a role to link the alleys that have different atmospheres to a database for two purposes. The first is to be able to define the situations, characteristics, and styles of architecture data that are related to affective atmospheres, which creates distinctive people's reactions. The second is to be able to link events and daily experience in the public places that change people's attitudes related to the changes of the atmospheres.

REFERENCES

Abouseif, S. (Director). (1960). *The Beginning and the End* [Motion Picture].

Abusaada, H. (2019). Revisiting the Word "Atmospheres" in the Urban Design Academic Field Based on the Artworks. *International Journal of Sustainable Entrepreneurship and Corporate Social Responsibility, 3*(2), 22–42. doi:10.4018/IJPPPHCE.2019070102

Abusaada, H. (2020). Strengthening the affectivity of atmospheres in urban environments: The toolkit of multi-sensory experience. *Archnet-IJAR: International Journal of Architectural Research.* doi:10.1108/ARCH-03-2020-0039

Abusaada, H., & Elshater, A. (2019). Effects of urban atmospheres on changing attitudes of crowded public places: An action plan. *International Journal of Community Well-Being*, 3(2), 109-159. doi:10.100742413-019-00042-w

Abusaada, H., Vellguth, C., & Elshater, A. (Eds.). (2019). *Handbook of Research on Digital Research Methods and Architectural Tools in Urban Planning and Design*. IGI Global. doi:10.4018/978-1-5225-9238-9

Al-Emam, H. (Director). (1963). *Hassan Al-Emama* [Motion Picture].

Al-Emam, H., & Tantawi, S. (Directors). (1964). Palaces Walk [Motion Picture]

al-Tawati, M. (2006). Place in three Novels by Mahfouz. In T. L. Gassick (Ed.), *Critical Perspectives on Naguib Mahfouz* (pp. 71–84). Three Continent Press.

AliN. (2019, March 1). *YouTube*. Retrieved from https://www.youtube.com/watch?v=0LU-vIoH0Zk

AlkaabyK. (2017, January 17). *YouTube*. Retrieved from https://www.youtube.com/watch?v=0X4O1AugiWA

AlSayyad, N. (2006). *Cinematic Urbanism: A History of the Modern from Reel to Real*. Routledge.

Anderson, B. (2009). Affective Atmospheres. *Emotion, Space and Society*, *2*(2), 77–81. doi:10.1016/j.emospa.2009.08.005

Anderson, L. (1994). Film Theory and Architectural Design. *ACADIA Conference Proceedings* (pp. 219-227). Saint Louis, MO: Washington University.

Andjelković, K. (2015). The Spatial Context of the Cinematic Aspects of Architecture. *Architecture and Civil Engineering*, *13*(2), 123–135. doi:10.2298/FUACE1502123A

Augoyard, J.-F. (2007). *Step by Step: Everyday Walks in a French Urban Housing Project*. University of Minnesota Press.

Bader, A. P. (2018). Museums and urban life in the cinema: On the ordinary and extraordinary architectural experiences. *Emotion, Space and Society*, *29*, 22–31. doi:10.1016/j.emospa.2018.07.011

BattaouiM. (2014, March 4). *YouTube*. Retrieved from https://www.youtube.com/watch?v=yqlYVru9q5A

Beggs, P. J., Šikoparija, B., & Smith, M. (2017). Aerobiology in the International Journal of Biometeorology, 1957–2017. *International Journal of Biometeorology*, *61*(S1), 51–58. doi:10.100700484-017-1374-5 PMID:28607999

Bhuiya, H. A. (2014). Khan Al-Khalili: A Masterpiece of Naguib Mahfouz. *Pratidhwani the Echo*, *III*(II), 142–145.

Bird, B. (Director). (2011). *Mission Impossible-Ghost Protocol* [Motion Picture].

Bissell, D. (2010). Passenger Mobilities: Affective Atmospheres and the Sociality of Public Transport. *Environment and Planning. D, Society & Space*, *28*(2), 270–289. doi:10.1068/d3909

Böhme, G. (1993). Atmosphere as the Fundamental Concept of a New Aesthetics. *Thesis Eleven*, *36*, 113-126.

Böhme, G. (1998). Atmosphere as an aesthetic concept. In G. Confurius (Ed.), *Constructing Atmospheres* (Vol. 68, pp. 112–115). Daidalos.

Böhme, G. (2014). Urban Atmospheres: Charting New Directions for Architecture and Urban Planning. In C. Borch (Ed.), *Architecture, Architectural Atmospheres – On the Experience and Politics of Architecture* (pp. 42–59). Basel: Birkhäuser.

Borch, C. (2015). Introduction: Why Atmospheres? In C. Borch (Ed.), *Architectural Atmospheres: On The Experience and Politics of Architecture* (pp. 6–17). Birkhauser.

Bordwell, D. (1985). *Narration in the Fiction Film*. Routledge.

Buser, M. (2014). Thinking through non-representational and affective atmospheres in planning theory and practice. *Planning Theory*, *13*(3), 227–243. doi:10.1177/1473095213491744

Dempsey, M. (1982). Review: Blade Runner by Ridley Scott, Michael Deeley. *Film Quarterly*, *36*(2), 33–38. doi:10.2307/3696992

Dewey, J. (1981). *The Later Works of John Dewey, 1925-1953: 1925, Experience and Nature* (J. A. Boydston, Ed., Vol. 1). Southern Illinois University Press.

Dewey, J. (2005). *Art as experience* (1934 ed.). Penguin.

Dodds, K. (2014). Shaking and Stirring James Bond: Age, Gender, and Resiliencein Skyfall (2012). *The Journal of Popular Film and Television*, *42*(2), 116–130. doi:10.1080/01956051.2013.858026

Dorrian, M., & Rose, G. (Eds.). (2003). *Deterritorialisations... Revisioning: Landscapes and Politics*. Black Dog Publishing.

Egan, K. (2014). Film Production Design: Case Study of The Great Gatsby. *Elon Journal of Undergraduate Research in Communications*, *5*(1), 5–14.

Elshater, A. (2015). Urban design redux: Redefining a professional practice of specialization. *Ain Shams Engineering Journal*, *6*(1), 25–39. doi:10.1016/j.asej.2014.08.004

Elshater, A., & Abusaada, H. (2016). *What Is Urban Design?: Learning and Thought*. Partridge Publishing.

Emmott, C. (1997). *Narrative Comprehension: A Discourse Perspective*. OxfordUniversity Press.

Engels-Schwarzpaul, T. (Ed.). (2017). *Atmospheric Architectures: The Aesthetics of Felt Spaces*T. Engels-Schwarzpaul, Trans.). Bloomsbury Academic.

Fancher, H., & Peoples, D. W. (Writers), & Scott, R. (Director). (1982). *Blade Runner* [Motion Picture]. Retrieved from https://www.imdb.com/title/tt0083658/

Fitzgerald, F. S. (2013). The Great Gatsby. Newbury, Berkshire.

Ford, T. (2012, December 6). *Nineteenth-Century Climate Change: Atmosphere, Culture and Romanticism*. Retrieved from rsha.cass.anu.edu.au: http://rsha.cass.anu.edu.au/nineteenth-century-climate-change-atmosphere-culture-and-romanticism

Georgiadou, Z. (2016). Cinema and architectural design education. *artciencia.com*, 1-21.

Gissen, D. (2009). *Subnature: Architecture's Other Environments*. Princeton Architectural Press.

Griffero, T. (2014). *Atmospheres: Aesthetics of Emotional Spaces* (S. D. Sanctis, Trans.). Ashgate Publishing Limited.

HameedA. (2011, November 22). *YouTube*. Retrieved from https://www.youtube.com/watch?v=aYA8084o9EE

Hsu, H. L. (2017). Literary Atmospherics. *Literary Geographies*, *3*(1), 1–5.

Husseinya, A. M., & Husseiny, A. A. (2012). Integration of Static and Dynamic Spaces in Historic Cairo: A community spirit that may vanish. *AicE-Bs 2012 Cairo, ASIA Pacific International Conference on Environment-Behaviour Studies*, 68, 687 – 699. doi:10.1016/j.sbspro.2012.12.259

Jaaniste, L. O. (2010). The Ambience of Ambience. *A Journal of Media and Culture, 13*(2). Retrieved from http://www.journal.media-culture.org.au/index.php/mcjournal/article/view/238

Jacobs, S. (2007). *The Wrong House: The Architecture of Alfred Hitchcock*. Rotterdam: 010 Publishers.

Janney, R. W. (2012). Pragmatics and Cinematic Discourse. *Lodz Papers in Pragmatics*, *8*(1), 85–113. doi:10.1515/lpp-2012-0006

Jenner, R. (2013). Ambient atmospheres: Exhibiting the immaterial in works by Italian Rationalists Edoardo Persico and Franco Albini. *Interstices: Journal of Architecture and Related Arts*, 13-24.

Knapton, S. (2012). *Istanbul for James Bond fans*. Retrieved from www.telegraph.co.uk: https://www.telegraph.co.uk/travel/destinations/europe/turkey/istanbul/articles/Skyfall-Istanbul-for-James-Bond-fans/

Krause, L., & Petro, P. (Eds.). (2003). *Global Cities: Cinema, Architecture, and Urbanism in a Digital Age*. Rutgers University Press.

Krippendorff, K. (2004). *Content Analysis: An Introduction to Its Methodology*. Sage.

Labuhn, B. (2016). Conceptualisations of Atmosphere in Text and in Image in Architectural Journalism (1991 - 2013). In *Ambiances, tomorrow. Proceedings of 3rd International Congress on Ambiances* (pp. 289-294). Volos, Greece: International Network Ambiances; University of Thessaly.

Leaman, O. (2003). *Companion Encyclopedia of Middle Eastern and North African Film*. Routledge. doi:10.4324/9780203426494

Lenzhölzer, S. (2010). Designing Atmospheres, Research and Design for Thermal Comfort in Dutch Urban Squares. Wageningen, The Netherlands: S.n.

Li, X., Hijazi, I., Reinhard Koenig, Z. L., Zhong, C., & Schmitt, G. (2016). Assessing Essential Qualities of Urban Space with Emotional and Visual Data Based on GIS Technique. *International Journal of Geo-Information*, *5*(11), 1–18. doi:10.3390/ijgi5110218

Lima, C. (2008). Filmic Narratives of the City. Cities in Film: Architecture, Urban Space and the Moving Image (pp. 142-148). University of Liverpool and Arts and Humanities Research Council (AHRC).

Lucas, R. (2008). Acousmêtric Architecture: Filmic Sound Design and its Lessons for Architects. In Cities in Film: Architecture, Urban Space and the Moving Image (pp. 157-164). University of Liverpool and the Arts and Humanities Research Council (AHRC).

Luhrmann, B. (Director). (2013). *The Great Gatsby* [Motion Picture].

Makhail, M. N. (1992). Studies in the Short Fiction of Mahfouz and Idris. New York University Press.

Mallet-Stevens, R. (1924). *Le Cinéma et les arts: L'Architecture*. L'Herbier.

McGrail, L. (2018). *5 Essential Elements of Successful Mise en Scène in Film*. Retrieved from www.lightsfilmschool.com: https://www.lightsfilmschool.com/blog/mise-en-scene-in-film-afk

Merleau-Ponty, M. (2002). *Phenomenology of Perception*. Psychology Press. doi:10.4324/9780203994610

Michels, C., & Steyaert, C. (2017). By accident and by design: Composing affective atmospheres in an urban art intervention. *Organization*, *24*(1), 79–104. doi:10.1177/1350508416668190

Milner, A. (2004). Darker Cities: Urban Dystopia and Science Fiction Cinema. *International Journal of Cultural Studies*, *7*(3), 259–279. doi:10.1177/1367877904046303

Moughtin, C., Cuesta, R., Signoretta, P., & Sarris, C. (1999). *Urban Design: Method and Techniques*. Architectural Press. Retrieved from www.qualitestgroup.comhttps://www.qualitestgroup.com/white-papers/advantages-disadvantages-crowdsourcing/

Movie-Locations.com. (2019). *Mission Impossible– Ghost Protocol*. Retrieved from movie-locations.com: http://www.movie-locations.com/movies/m/Mission-Impossible-Ghost-Protocol.php

Narkunas, J. P. (2001). Streetwalking in the Cinema of the City: Capital Flows Through Saigon. In M. Shiel & T. Fitzmaurice (Eds.), Cinema and the City (pp. 147-157). Hoboken, NJ: Blackwell Publishing Ltd.

Neuendrof, K. A. (2002). *The content analysis*. Sage Publications.

Oakley, D. (1970). *The Phenomenon of Architecture in Cultures in Change*. Pergamon.

Pallasmaa, J. (2008). *The Architecture of Image: Existential Space in Cinema*. Rakennustieto Publishing.

Pallasmaa, J. (2011). An Architecture of the Seven Senses. In L. Weinthal (Ed.), *Theory, Toward a New Interior. An Anthology of Interior Design* (pp. 40–49). Princeton Architectural Press.

Pallasmaa, J. (2014). Space, Place, and Atmosphere: Peripheral Perception in Existential Experience. In C. Borch (Ed.), *Architectural atmospheres: On the experience and politics of architecture* (pp. 18–41). Birkhauser. doi:10.1515/9783038211785.18

Pandey, J. (2019). Analysis, Deductive Approach to Content. In M. Gupta, M. Shaheen, & K. P. Reddy (Eds.), *Qualitative Techniques for Workplace Data Analysis* (pp. 145–169). IGI Global., doi:10.4018/978-1-5225-5366-3.ch007

Pavlovskaya, M. (2009). Non-Quantitative GIS. In M. Cope & S. Elwood (Eds.), *Qualitative GIS: A Mixed Methods Approach* (pp. 13–38). SAGE Publications Ltd. doi:10.4135/9780857024541.n2

Pelsmakers, S. (2012). *The Environmental Design Pocketbook* (J. Thompson, Ed.). Riba Publishing.

Project for Public Places. (2012). *The Great Movies for Placemajers*. Retrieved from pps.org: https://www.pps.org/article/ten-great-movies-for-placemakers

Rabiger, M. (2008). *Directing: Film Techniques and Aesthetics*. Focal Press.

Raymond, A. (1984). *The Great Arab Cities in the 16th-18th centuries, An Introduction*. New York University Press.

Reichler, T. (2009). Changes in the Atmospheric Circulation as Indicator of. In T. M. Letcher (Ed.), *Climate Change: Observed* (pp. 145–165). Elsevier. doi:10.1016/B978-0-444-53301-2.00007-5

Rowe, J. (2020). *Urban Lab Films*. Retrieved from University College London: https://www.ucl.ac.uk/urban-lab/events/urban-lab-films

Saad, S. A. (2018). The Spatial Identity of Historic Cairo. Part two: Must Prominent Buildings be Enframed. In *Proceedings of the conference titled: Cities' Identity through Architecture and Arts 2* (pp. 183-194). Cairo: IEREK Press.

Salem, A. (Director). (1967). *Khan Al-Khalili* [Motion Picture].

Salingaros, N. A., & Mehaffy, M. W. (2006). *A Theory of Architecture*. UMBAU-VERLAG Harald Püschel.

Schmitz, H. (1967). *System der Philosophie* (Vol. 3, Part 1). H. Bouvier.

Schmitz, H. (2017). The Felt Body and Embodied Communication. *Yearbook for Eastern and Western Philosophy*, 9-19. https://doi.org/ doi:10.1515/yewph-2017-0004

Schmitz, H., Owen, R., & Slaby, M. (2011). Emotions outside the box—The new phenomenology of feeling and corporeality. *Phenomenology and the Cognitive Sciences, 10*(2), 241–259. doi:10.100711097-011-9195-1

Semper, G. (2004). *Style in the Technical and Tectonic Arts, Or, Practical Aesthetics*. Getty Research Institute.

Semper, G. (2011). *Four Elements of Architecture* (1851 ed.). (H. F. Mallgrave & W. Herrmann, Trans.). Cambridge University Press.

Shiel, M., & Fitzmaurice, T. (2001). *Cinema and the City: Film and Urban Societies in a Global Context*. City of Hoboken: Blackwell Publishers Ltd. doi:10.1002/9780470712948

Sidanin, P. (2007). On Lynch's and Post-Lynchians Theories. *Architecture and Civil Engineering*, *5*(1), 61–69.

Stefansdottir, H. (2017). The role of urban atmosphere for non-work activity locations. *Journal of Urban Design*, 319–335. doi:10.1080/13574809.2017.1383150

Tashiro, C. (2004). Passing for the Past: Production Design and the Historical Film. *Cinéaste (New York, N.Y.)*, *29*(2), 40–44.

Tawa, M. (2017). Consilient Discrepancy: Porosity and Atmosphere in Cinema and Architecture. *Architecture_media_politics_society, 11*(13), 1-17. doi:10.14324/111.444.amps.2017v11i3.001

Thibaud, J. (2002). From Situated Perception to Urban Ambiences. In *First International Workshop on Architectural and Urban Ambient Environment* (pp. 1-11). Nantes, France: Centre de recherche m'ethodologique d'architecture; Ecole d'architecture de Nantes.

Troiani, I., & Carless, T. (2014). Architectural Design Research through Cinematic Collage. *Architecture & Education Journal*, 255-278.

Vidler, A. (1993). The explosion of Space: Architecture and the Filmic Imaginary. *Assemblage*, *21*(21), 44–59. doi:10.2307/3171214

Weidinger, J. (2016). Atmosphere—an Emerging Phenomenon in Landscape Architecture. *Wolkenkuckucksheim*, 147.

Weidinger, J. (Ed.). (2018). *Designing Atmospheres*. Technical University Berlin.

Werth, P. T. (1999). *Text Worlds: Representing Conceptual Space in Discourse*. Longman.

Williams, D. E. (1988). Ideology as Dystopia: An Interpretation of "Blade Runner". *International Political Science Review / Revue internationale de science politique, 9*(4), 381-394.

Zeng, H., & Zhao, Y. (2011). Sensing Movement: Microsensors for Body Motion Measurement. *Sensors (Basel), 11*(1), 638–660. doi:10.3390110100638 PMID:22346595

Zhang, X., Cha, K. L., & Constable, M. (2014). Atmospheric Perspective Effect Enhancement of Landscape Photographs Through Depth-Aware Contrast Manipulation. *IEEE Transactions on Multimedia, 16*(3), 653–667. doi:10.1109/TMM.2014.2299511

Zumthor, P. (1998). *Thinking Architecture.* Lars Müller.

Zumthor, P. (2006). *Atmospheres: Architectural Environments - Surrounding Objects.* Birkhauser.

Compilation of References

Abaya Gomez, J. E. (2008). Waterfront design without policy? The actual uses of Manila's Baywalk. *Cities (London, England), 25*(2), 86–106. doi:10.1016/j.cities.2008.03.001

Abdel-Aziz, A., Abdel-Salam, H., & El-Sayad, Z. (2015). The Role of ICTs in Creating the New Social Public Place of the Digital Era. *Alexandria Engineering Journal, 55*(1), 487–493. doi:10.1016/j.aej.2015.12.019

Abdel-Galil, R. E. S. (2012). Desert reclamation, a management system for sustainable urban expansion. *Progress in Planning, 78*(4), 151–206. doi:10.1016/j.progress.2012.04.003

Abdulkarim, D., & Nasar, J. (2014). Do Seats, Food Vendors, and Sculptures Improve Plaza Visitability? *Environment and Behavior, 46*(7), 805–825. doi:10.1177/0013916512475299

Abel, C. (1982). Architecture as Identity, I: the Essence of Architecture. In M. Herzfeld & M. D. Lenhart (Eds.), *Semiotiqs 1980* (pp. 1–12). Plenum Press. doi:10.1007/978-1-4684-9137-1_1

Abouseif, S. (Director). (1960). *The Beginning and the End* [Motion Picture].

Abrams, M., & Harpham, G. (2011). *A Glossary of Literary Terms.* Cengage Learning.

Abrego, L. J., & Schmalzbauer, L. (2018). Illegality, motherhood, and place: Undocumented Latinas making meaning and negotiating daily life. *Women's Studies International Forum, 67*(March), 10–17. doi:10.1016/j.wsif.2017.12.004

Abusaada, H. (2017). Architecture Evermore: The World of Architecture via the Western Thought an Extended Term (A. Elshater, Ed.). Partridge Africa, A Penguin Random House Company.

Abusaada, H. (2019). Revisiting the Word "Atmospheres" in the Urban Design Academic Field Based on the Artworks. *International Journal of Sustainable Entrepreneurship and Corporate Social Responsibility, 4*(2), 22–43. doi:10.4018/IJPPPHCE.2019070102

Abusaada, H. (2020). Strengthening the affectivity of atmospheres in urban environments: The toolkit of multi-sensory experience. *Archnet-IJAR: International Journal of Architectural Research.* doi:10.1108/ARCH-03-2020-0039

Abusaada, H., & Elshater, A. (2018). *The Fifth Dimension: Urban design - The atmospheres of the City.* Academic Bookshop. (in Arabic)

Abusaada, H., & Elshater, A. (2019a). Internet of Things Interpreting a Myth: Internally Displaced Persons Transform Cities of Hardship into Chaos Cities. In H. Abusaada, C. Voulguth, & A. Elshater (Eds.), *Handbook of Research on Digital Research Methods and Architectural Tools in Urban Planning and Design* (pp. 47–81). IGI Global. doi:10.4018/978-1-5225-9238-9.ch004

Abusaada, H., & Elshater, A. (2019b). A Crowdsourcing Approach in Urban Design: A bibliographic review of cities of singularity. In H. Abusaada, C. Voulguth, & A. Elshater (Eds.), *Handbook of Research on Digital Research Methods and Architectural Tools in Urban Planning and Design* (pp. 1–24). IGI Global. doi:10.4018/978-1-5225-9238-9.ch001

Abusaada, H., & Elshater, A. (2020). Effects of urban atmospheres on changing attitudes of crowded public places: An action plan. *International Journal of Community Well-Being*, *3*(2), 109–159. doi:10.100742413-019-00042-w

Abusaada, H., Vellguth, C., & Elshater, A. (2019). *Handbook of Research on Digital Research Methods and Architectural Tools in Urban Planning and Design*. IGI Global. doi:10.4018/978-1-5225-9238-9

Acuto, M., Steenmans, K., Iwaszuk, E., & Ortega-Garza, L. (2019). Informing urban governance? Boundary-spanning organisations and the ecosystem of urban data. *Area*, *51*(1), 94–103. doi:10.1111/area.12430

Adajania, N. (2008). *Anant Joshi*. Chemould Prescott Road.

Adams, P., Hoelscher, S., & Till, K. (Eds.). (2001). Texture of place. Univ of Minnesota Press.

Adams, D., Smith, M., Larkham, P., & Abidin, J. (2019). Encounters with a future past: Navigating the shifting urban atmospheres of place. Journal of Urban Design. *Journal of Urban Design*, *25*(3), 1–20. doi:10.1080/13574809.2019.1592668

Adarkar, N., & Menon, M. (2004). *One Hundred Years, One Hundred Voices: The Millworkers of Girangaon: An Oral History*. Seagull Books.

Adler, J. (2009). Goethe's use of chemical theory in his *Elective Affinities*. In A. Cunningham & N. Jardine (Eds.), *Romanticism and the sciences* (pp. 263–279). Cambridge University Press.

Ahuja, A. (2016). Integration of Nature and Technology for Smart Cities (3rd ed.). Springer. doi:10.1007/978-3-319-25715-0

Akl, M. H. (2015). *Towards Sustainable Internet and ICT for Sustainability Egypt*. Mansoura University.

Albertsen, N. (1999). Urban Atmospheres. *Sociologi I Dag*, *29*(4), 5-29.

Al-Bishawi, M., & Ghadban, S. (2011). A Methodological Approach for Reading Urban Open Space. *Archnet-IJAR*, *5*(1), 73–85.

Al-Emam, H. (Director). (1963). *Hassan Al-Emama* [Motion Picture].

Al-Emam, H., & Tantawi, S. (Directors). (1964). Palaces Walk [Motion Picture]

Algeo, J. (2010). The Origins and Development of the English Language (6th ed.). Wadsworth, Cengage Learning.

AliN. (2019, March 1). *YouTube*. Retrieved from https://www.youtube.com/watch?v=0LU-vIoH0Zk

AlkaabyK. (2017, January 17). *YouTube*. Retrieved from https://www.youtube.com/watch?v=0X4O1AugiWA

Allaire, J. (1960). *Neighborhood Boundaries*. American Society of Planning Officials.

Allam, Z., & Dhunny, Z. A. (2019). On big data, artificial intelligence and smart cities. *Cities (London, England)*, *89*, 80–91. doi:10.1016/j.cities.2019.01.032

AlSayyad, N. (2006). *Cinematic Urbanism: A History of the Modern from Reel to Real*. Routledge.

al-Tawati, M. (2006). Place in three Novels by Mahfouz. In T. L. Gassick (Ed.), *Critical Perspectives on Naguib Mahfouz* (pp. 71–84). Three Continent Press.

Al-Thani, S. K., Skelhorn, C. P., Amato, A., Koc, M., & Al-Ghamdi, S. G. (2018). Smart Technology Impact on Neighborhood Form for a Sustainable Doha. *Sustainability, 10*(4764), 17. doi:10.3390u10124764

Alvarez, L., Borsi, K., & Rodrigues, L. (2017). The role of social network analysis on participation and placemaking. *Sustainable Cities and Society, 28*, 118–126. doi:10.1016/j.scs.2016.06.017

Ambiances. (n.d.). *The Ambiances Network.* Retrieved from https://www.ambiances.net/network/about-the-international-ambiances-network.html

Ambience. (2018). In *Meriam Webester's dictionary.* Retrieved from https://www.merriam-webster.com/dictionary/ambience

Anciaes, P. R., Nascimento, J., & Silva, S. (2017). The distribution of walkability in an African city: Praia, Cabo Verde. *Cities (London, England), 67*(April), 9–20. doi:10.1016/j.cities.2017.04.008

Andersen, L., Gustat, J., & Becker, A. B. (2015). The Relationship Between the Social Environment and Lifestyle-Related Physical Activity in a Low-Income African American Inner-City Southern Neighborhood. *Journal of Community Health, 40*(5), 967–974. doi:10.100710900-015-0019-z PMID:25808676

Anderson, B. (2009). Affective atmospheres. *Emotion, Space and Society, 2*(2), 77–81. doi:10.1016/j.emospa.2009.08.005

Anderson, J., Ruggeri, K., Steemers, K., & Huppert, F. (2017). Lively Social Space, Well-Being Activity, and Urban Design: Findings from a Low-Cost Community-Led Public Space Intervention. *Environment and Behavior, 49*(6), 685–716. doi:10.1177/0013916516659108

Anderson, L. (1994). Film Theory and Architectural Design. *ACADIA Conference Proceedings* (pp. 219-227). Saint Louis, MO: Washington University.

Andjelković, K. (2015). The Spatial Context of the Cinematic Aspects of Architecture. *Architecture and Civil Engineering, 13*(2), 123–135. doi:10.2298/FUACE1502123A

Andrews, G. F. (1974). *Maya Cities: Placemaking and Urbanization* (1st ed.). Univ of Oklahoma Pr.

Angelidou, M., Karachaliou, E., Angelidou, T., & Stylianidis, E. (2017). Cultural heritage in smart city environments. *ISPRS - The International Archives of the Photogrammetry, Remote Sensing and Spatial Information Sciences, 42*(W5), 27–32. doi:10.5194/isprs-archives-XLII-2-W5-27-2017

Angioni, L. (2014). Definition and Essence in Aristotle's Metaphysics vii 4. *Ancient Philosophy (Garland Publishing), 34*(1), 75–100. doi:10.5840/ancientphil20143416

Araoz, G. (2013). Conservation Philosophy and its Development: Changing Understandings of Authenticity and Significance. *Heritage & Society, 6*(2), 144–154. doi:10.1179/2159032X13Z.00000000010

Artificial Intelligence and Life in 2030. (2016), *One Hundred Year Study on Artificial Intelligence: Report of the 2015-2016 Study Panel.* Stanford, CA: Stanford University. Retrieved from https://ai100.stanford.edu/2016-report

Ash, J. (2013). Rethinking affective atmospheres: Technology, perturbation and space times of the non-human. *Geoforum, 49*, 20–28. doi:10.1016/j.geoforum.2013.05.006

Atmosphere. (n.d.). Retrieved from http://www.literarydevices.com/atmosphere/

Audirac, I., & Fitzgerald, J. (2003). Information technology (IT) and urban form: An annotated bibliography of the urban deconcentration and economic restructuring literatures. *Journal of Planning Literature, 17*(4), 480–511. doi:10.1177/0885412203017004002

Augoyard, J.-F. (2007). *Step by Step: Everyday Walks in a French Urban Housing Project.* University of Minnesota Press.

Aurigi, A. (2017). Space Is Not a Platform – Foregrounding Place in Smart Urban Design. In *Enhancing Places Through Technology* (pp. 69–81). Edições Universitárias Lusófonas.

Australia ICOMOS & International Council on Monuments and Sites. (2013). *The Burra Charter: The Australia ICOMOS charter for places of cultural significance 2013.* http://australia.icomos.org/wp-content/uploads/The-Burra-Charter-2013-Adopted-31.10.2013.pdf

Babina, B. J., & Attaway, J. S. (2000). Atmospheric affect as a tool for creating value and gaining share of customer. *Journal of Business Research, 49*(2), 91–99.

Backer, M. D. (2019). "Being different together" in public space: Young people, everyday cosmopolitanism and parochial atmospheres. *Social & Cultural Geography*, 1–20. doi:10.1080/14649365.2019.1594352

Bader, A. P. (2018). Museums and urban life in the cinema: On the ordinary and extraordinary architectural experiences. *Emotion, Space and Society, 29*, 22–31. doi:10.1016/j.emospa.2018.07.011

Bakker, I., van der Voordt, T., Vink, P., & de Boon, J. (2014a). Pleasure, Arousal, Dominance: Mehrabian and Russell revisited. *Current Psychology (New Brunswick, N.J.), 33*(3), 405–421. doi:10.100712144-014-9219-4

Balzac, H., & Saintsbury, G. (1901). *The Works of Honoré de Balzac.* Retrieved November 26, 2019, from Google Books: shorturl.at/joCNT

Banerjee, T. (2001). The Future of Public Space. *Journal of the American Planning Association, 67*(1), 9–24. doi:10.1080/01944360108976352

Barrie, T., Bermudez, J., & Bermudez, J. (Eds.). (2015). *Architecture, Culture & Spirituality.* Ashgate Publisher.

Bašová, S., & Štefancová, L. (2017). Creative and Smart Public Spaces. *International Journal of Liberal Arts and Social Science, 5*(1), 17–33.

Bateson, G. (1979). *Mind and nature.* Dutton.

BattaouiM. (2014, March 4). *YouTube.* Retrieved from https://www.youtube.com/watch?v=yqlYVru9q5A

Batty, M. (2017). *The Age of the Smart City.* London: UCL Working Paper Series.

Bauer, C., Recreation, S. C., Fletcher, R., Travieso, C., & Woodward, D. (2010). High performance landscape guidelines: 21st century parks for NYC. In *Design Trust for Public Space : City of New York. Parks & Recreation.*

Beggs, P. J., Šikoparija, B., & Smith, M. (2017). Aerobiology in the International Journal of Biometeorology, 1957–2017. *International Journal of Biometeorology, 61*(S1), 51–58. doi:10.100700484-017-1374-5 PMID:28607999

BELBIM. (2019a). *Istanbulkart Mobile.* Retrieved from https://www.istanbulkart.istanbul/istanbulkart-mobil

BELBIM. (2019b). *Our smartcards.* Retrieved from https://www.istanbulkart.istanbul/kartlarimiz

Bell, D. (1973). *The Coming of Postindustrial Society.* Basic Books.

Benjamin, W. (2006). *The Writer of Modern Life: Essays on Charles Baudelaire* (M. Jennings, Ed.). Harvard University Press.

Berman, B., & Evans, J. R. (2013). *Retail management: A strategic approach* (12th ed.). Pearson.

Bhuiya, H. A. (2014). Khan Al-Khalili: A Masterpiece of Naguib Mahfouz. *Pratidhwani the Echo, III*(II), 142–145.

Bird, B. (Director). (2011). *Mission Impossible-Ghost Protocol* [Motion Picture].

Bissell, D. (2010). Passenger mobilities: Affective atmospheres and the sociality of public transport. *Environment and Planning. D, Society & Space*, *28*(2), 270–289. doi:10.1068/d3909

Bitner, M. J., & Obermiller, C. (1984). *The elaboration likelihood model: Limitations and extensions in marketing*. Academic Press.

Bitner, M. J. (1992). The Impact of Physical Surroundings on Customers and Employees. *Journal of Marketing*, *56*(2), 57–71. Advance online publication. doi:10.1177/002224299205600205

Böhme, G. (1993). Atmosphere as the Fundamental Concept of a New Aesthetics. *Thesis Eleven, 36*, 113-126.

Böhme, G. (1993). Atmosphere as the Fundamental Concept of a New Aesthetics. *Thesis Eleven, 36*, 113-126. doi:10.1177/072551369303600107

Böhme, G. (1995). *Atmosphare. Essays zur Neuen Asthetik*. Frankfurt a. M: Suhrkamp Verlag.

Böhme, G. (2005). Atmosphere as the subject matter of architecture. In H. Meuron & P. Ursprung (Eds.), Natural History (pp. 398–407). London: Springer Science & Business Media.

Böhme, G. (2013a). Atmosphere as Mindful Physical Presence in Space. *OASE Journal for Architecture*, 21-32.

Böhme, G. (2017a). Atmospheric Architectures: The Aesthetics of Felt Spaces. London: Bloomsbury.

Böhme, G., & Farrell, J. (1992). An Aesthetic Theory of Nature: An Interim Report. *Thesis Elev., 32*, 90-102, Doi:10.1177/072551369203200107

Böhme, G. (1995). *Atmosphäre: Essays zur neuen Ästhetik* [Atmosphere: Essays on a new aesthetics]. Suhrkamp.

Böhme, G. (1998). Atmosphere as an aesthetic concept. In G. Confurius (Ed.), *Constructing Atmospheres* (Vol. 68, pp. 112–115). Daidalos.

Böhme, G. (2006). *Architektur und Atmosphäre* [Architecture and Atmosphere]. Wilhelm Fink Verlag.

Böhme, G. (2006). Atmosphere as the subject matter of architecture. In P. Ursprung (Ed.), *Herzog and Meuron: Natural History* (pp. 398–407). Lars Muller Publishers.

Böhme, G. (2013). Atmosphere as Mindful Physical Presence in Space. *OASE Journal for Architecture*, *91*, 21–32.

Böhme, G. (2013b). Encountering atmospheres: A reflection on the concept of atmosphere in the work of Juhani Pallasmaa and Peter Zumthor. *Building atmosphere. OASE*, *91*, 93–100.

Böhme, G. (2014). Urban atmospheres: Charting new directions for architecture and urban planning. In C. Borch (Ed.), *Architecture, architectural atmospheres – On the experience and politics of architecture* (pp. 42–59). doi:10.1515/9783038211785.42

Böhme, G. (2014). Urban Atmospheres: Charting New Directions for Architecture and Urban Planning. In C. Borch (Ed.), *Architecture, Architectural Atmospheres – On the Experience and Politics of Architecture* (pp. 42–59). Basel: Birkhäuser.

Böhme, G. (2017). *Atmospheric architectures: The aesthetics of felt spaces* (T. Engels-Schwarzpaul Ed. & Trans.). Bloomsbury Academic.

Böhme, G. (2017b). *Critique of Aesthetic Capitalism*. Mimesis.

Böhme, G. (2017c). *The Aesthetics of Atmospheres* (J.-P. Thibaud, Ed.). Routledge.

Bonanno, P., Pierdicca, R., Franco, F., Martinez, A., Marcheggiani, E., & Malinverni, E. (2017). A Connectivist Approach to Smart City Learning: Valletta City Case-Study. In *Enhancing Places Through Technology* (pp. 69–81). Edições Universitárias Lusófonas.

Bonn, M. A., Joseph-Mathews, S. M., & Dai, M. (2005). *International versus Domestic Visitors: An Examination of Destination Image Perceptions*. doi:10.1177/0047287504272033

Bonn, M. A., Joseph-Mathews, S. M., Dai, M., Hayes, S., & Cave, J. (2007). Heritage/Cultural Attraction Atmospherics: Creating the Right Environment for the Heritage/Cultural Visitor. *Journal of Travel Research*, *45*(3), 345–354. doi:10.1177/0047287506295947

Boon, B. (2017). Bibliographic research. In M. Allen (Ed.), *The SAGE Encyclopedia of communication research methods*. SAGE Publications, Inc., doi:10.4135/9781483381411.n37

Borch, C. (2015). Introduction: Why Atmospheres? In C. Borch (Ed.), *Architectural Atmospheres: On The Experience and Politics of Architecture* (pp. 6–17). Birkhauser.

Bordwell, D. (1985). *Narration in the Fiction Film*. Routledge.

Botteldooren, D. (2017). Cyber Enhancing the Urban Soundscape. In *Enhancing Places Through Technology* (pp. 181–187). Edições Universitárias Lusófonas.

Boutin, A. (2012). Rethinking the Flâneur: Flânerie and the Senses. *Dix-Neuf, 16*(2).

Bouvier, L. (2005, September). A Taste for Words: Gastronomy and the Writing of Loss in Brillat-Savarin's Physiologie du Goût. *Mosaic*, *38*(3).

Boyd, S. (2002). Cultural and Heritage Tourism in Canada: Opportunities, Principles and Challenges. *Tourism and Hospitality Research*, *3*(3), 211–233. doi:10.1177/146735840200300303

Bratton, B. H. (2015). *The Stack: On Software and Sovereignty*. MIT Press.

Brisky, B. A. (2014). *Physical environment and women's physical activity behaviors*. ProQuest Dissertations and Theses.

BTK. (2018). *Annual statistics bulletin on provincial basis for the electronic communication sector* [Elektronik haberleşme sektörüne ilişkin il bazında yıllık istatistik bülteni]. BTK.

Burgess, E. W. (2008). The Growth of the City: An Introduction to a Research Project. In J. M. Marzluff, E. Shulenberger, W. Endlicher, M. Alberti, G. Bradley, C. Ryan, & C. Zumbrunnen (Eds.), *Urban Ecology: An International Perspective on the Interaction Between Humans and Nature* (pp. 71–78). Springer US. doi:10.1007/978-0-387-73412-5_5

Buser, M. (2014). Thinking through non-representational and affective atmospheres in planning theory and practice. *Planning Theory*, *13*(3), 227–243. doi:10.1177/1473095213491744

Büyüksalih, G., Bayburt, S., & Jacobsen, K. (2018). Analysis of height models based on KOMPSAT-3 images. *International Archives of the Photogrammetry, Remote Sensing and Spatial Information Sciences-ISPRS Archives*, *42*(3W4), 115-119.

Çakmak, B., & Topçu, M. (2018). An Evaluation of Urban Open Spaces in Historical City Center of Konya in the Context of Pedestrian Mobility. *International Journal of Human Sciences*, *15*(4), 1827–1846. doi:10.14687/jhs.v15i4.5506

Carmona, M., Heath, T., Oc, T., & Tiesdell, S. (2003). *Public places-urban spaces: The dimensions of urban design*. Architectural Press.

Carmona, M., Heath, T., Oc, T., & Tiesde, S. (2003). *Public places, urban spaces: The dimensions of urban design*. Architectural Press.

Carr, S. (1992). Public spaces. Cambridge, UK: Cambridge University Press.

Casey, E. S. (1997). *The fate of place: A philosophical history*. University of California Press.

Cate, P. D., & Shaw, M. (1996). *The Spirit of Montmartre: Cabarets, Humor, and the Avant-Garde 1875–1905*. Rutgers University Press.

Cerwén, G., Wingren, C., & Qviström, M. (2017). Soundscape actions: A tool for noise treatment based on three workshops in landscape architecture. *Frontiers of Architectural Research*, *6*(4), 505–518. doi:10.1016/j.foar.2017.10.002

Chang, T.-Y., & Horng, S.-C. (2010). *Conceptualizing and measuring experience quality: The customer's perspective*. doi:10.1080/02642060802629919

Cheliotis, K. (2016). Capturing Real-Time Public Space Activity Using Publicly Available Digital Traces. In *The Workshops of the Tenth International AAAI Conference on Web and Social Media CityLab, Technical Report WS-16-16* (pp. 8-13). Association for the Advancement of Artificial Intelligence.

Claudel, M., & Ratti, C. (2015). *Open Source Architecture*. Thames and Hudson.

Columbia University. (2013). *Columbia Absreact*. Retrieved from None Linear Solutions Unit: http://abstract20122013.gsapp.org/non-linear-solutions-unit/

Community toolbox. (2019). *Section 7. Changing the Physical and Social Environment*. Implementing Promising Community Interventions. https://ctb.ku.edu/en/table-of-contents/implement/physical-social-environment

Cooper, C. (2014). *Wiley Encyclopedia of Management*. John Wiley & Sons, Ltd.

Cozens, P., & Van Der Linde, T. (2015). Perceptions of Crime Prevention Through Environmental Design (CPTED) at Australian Railway Stations. *Journal of Public Transportation*, *18*(4), 73–92. doi:10.5038/2375-0901.18.4.5

Credits, P. (n.d.). *Safe Cities for Women and Girls*. Academic Press.

Cresswell, T. (2009). Place. In N. Thrift & R. Kitchen (Eds.), *International Encyclopedia of Human Geography* (Vol. 8, pp. 169–177). Elsevier. doi:10.1016/B978-008044910-4.00310-2

Crystal, D. (2014). *Words in Time and Place: Exploring Language Through the Historical Thesaurus of the Oxford English Dictionary*. OUP Oxford.

Cudny, W., Comunian, R., & Wolaniuk, A. (2020). Arts and creativity: A business and branding strategy for Lodz as a neoliberal city. *Cities (London, England)*, *100*, 100. doi:10.1016/j.cities.2020.102659

Dahl, N. O. (1997). Two kinds of essence in Aristotle: A pale man is not the same as his essence. *The Philosophical Review*, *106*(2), 233-265. doi:phr1997106215

Dahlberg, K. (2006). The essence of essences – the search for meaning structures in phenomenological analysis of lifeworld phenomena. *International Journal of Qualitative Studies on Health and Well-being*, *1*(1), 11–19. doi:10.1080/17482620500478405

Dajian, Z., & Peter, R. (2006). 2010 World Expo and Urban Life Quality in Shanghai in Terms of Sustainable Development. *Zhongguo Renkou Ziyuan Yu Huanjing*, *4*(1), 15–22. doi:10.1080/10042857.2006.10677445

Dan, F. (1998). *Bohemian Paris, Picasso, Modigliani, Matisse and the birth of Modern Art*. Calmman Lévy.

Darmawan, R. (2011). Web 2.0 and idiosyncrasy of cultural heritage: a perspective from indonesia. In G. Styliaras, D. Koukopoulos, & F. Lazaniris (Eds.), *Handbook of Research on Technologies and Cultural Heritage: Applications and Environments* (pp. 481–494). IGI Global. doi:10.4018/978-1-60960-044-0.ch024

Davis, F. (2019). *The Chelsea Girls*. Penguin Random House.

DCMS. (2010). *Principles of selection for listed buildings*. The Department for Digital, Culture, Media and Sport is a Department of the United Kingdom Government.

Deakin, M. (2011). From the City of Bits to E-Topia: Space, Citizenship and Community as Global Strategy in the Governance of the Digitally-Inclusive Regeneration Thesis. In D. Piaggesi, K. J. Sund, & W. Castelnovo (Eds), *Global strategy and practice of e-governance: examples from around the world* (pp. 124-141). Hershey, PA: IGI Global.

Deakin, M. (2014). From the city of bits to e-topia: Space, citizenship and community as global strategy. *International Journal of E-Adoption*, *6*(1), 16–33. doi:10.4018/ijea.2014010102

Deleuze, G. (1990). *The logic of sense*. Columbia University Press.

Deleuze, G. (1990). *The Logic of Sense*. Columbia University Press.

Dempsey, M. (1982). Review: Blade Runner by Ridley Scott, Michael Deeley. *Film Quarterly*, *36*(2), 33–38. doi:10.2307/3696992

Department of e-Government Services. (2016). *2016-2019 National e-Government Strategy and Action Plan*. Author.

Derr, V., & Tarantini, E. (2016). 'Because We Are All People': Outcomes and Reflections from Young People's Participation. *Local Environment: international Journal of Sustainability and Justice, 21*(12), 1534-1556.

Descartes, R. (1637). *Discours de la méthode*. Leyde: Ian Maire.

Dewey, J. (1981). *The Later Works of John Dewey, 1925-1953: 1925, Experience and Nature* (J. A. Boydston, Ed., Vol. 1). Southern Illinois University Press.

Dewey, J. (2005). *Art as experience* (1934 ed.). Penguin.

Dewey, J. (2008). *The later works of John Dewey, 1925-1953: 1934, Art as experience* (A. Boydston, Ed., Vol. 10). Southern Illinois University Press.

Dhage, V. (2013). Transnationalism and the Transformation of Architecture of Textile Mill-lands in Mumbai. *History Today: Journal of History and Historical Archaeology,* (14), 110-113.

Dickens, C., & Shruckburg, J. (1949). *A Tale of Two Cities (Oxford illustrated Dickens)*. Oxford University Press.

Dickens, P. (2001). *Global Shifts: Mapping the Changing Contours of the World Economy* (6th ed.). Sage Publications Ltd.

Digital Transformation Office. (2019). *e-Government Gateway*. Retrieved from https://www.turkiye.gov.tr

Dixit, S., Mandal, S. N., Thanikal, J. V, & Saurabh, K. (2019). *Study of Significant Factors Affecting Construction Productivity Using Relative Importance Index in Indian Construction Industry*. Academic Press.

Dixit, S., & Sharma, K. (2020). An Empirical Study of Major Factors Affecting Productivity of Construction Projects. In K. G. Babu, H. S. Rao, & Y. Amarnath (Eds.), *Emerging Trends in Civil Engineering* (pp. 121–129). Springer Singapore.

Dixit, S., Sharma, K., & Singh, S. (2020). Identifying and Analysing Key Factors Associated with Risks in Construction Projects. In K. G. Babu, H. S. Rao, & Y. Amarnath (Eds.), *Emerging Trends in Civil Engineering* (pp. 25–32). Springer Singapore.

Djukic, A., Vukmirovic, M., Jokovic, J., & Dinkic, N. (2017). Tweeting in Open Public space – Case Study Belgrade. In *Enhancing Places Through Technology* (pp. 191–203). Edições Universitárias Lusófonas.

Dodds, K. (2014). Shaking and Stirring James Bond: Age, Gender, and Resiliencein Skyfall (2012). *The Journal of Popular Film and Television*, *42*(2), 116–130. doi:10.1080/01956051.2013.858026

Dodman, D. (2007). Post-independence optimism and the legacy of waterfront redevelopment in Kingston, Jamaica. *Cities (London, England)*, *24*(4), 273–284. doi:10.1016/j.cities.2006.09.001

Dorrian, M., & Rose, G. (Eds.). (2003). *Deterritorialisations... Revisioning: Landscapes and Politics*. Black Dog Publishing.

Douglas, O., Lennon, M., & Scott, M. (2017). Green space benefits for health and well-being: A life-course approach for urban planning, design and management. *Cities (London, England)*, *66*, 53–62. doi:10.1016/j.cities.2017.03.011

Dovey, K. (2010). *Becoming places, urbanism/architecture/identity/power*. Routledge.

Dovey, K. (2010). *Becoming Places, Urbanism/Architecture/Identity/Power*. Routledge.

Dragulanescu, N. G. (2002). Social impact of the "digital divide" in a Central–Eastern European country. *The International Information & Library Review*, *34*(2), 139–151. doi:10.1080/10572317.2002.10762570

Duarte, T., & Mateus, D. (2017). The Contribution of ICT in Planning of Public Open Spaces Reflections on the City of Lisbon. In *Enhancing Places Through Technology* (pp. 41–56). Edições Universitárias Lusófonas.

Dubey, H., Yang, J., Constant, N., Amiri, A. M., Yang, Q., & Makodiya, K. (2015). Fog Data: Enhancing Telehealth Big Data Through Fog Computing. In *Proceedings of the ASE bigdata & socialinformatics* (p. 6). ACM.

Dufrenne, M. (1973). *The phenomenology of aesthetic experience*. Northwestern University Press.

Dufrenne, M. (1973). *The Phenomenology of Aesthetic Experience*. Northwestern University Press.

Duke, D. (2003). Alzira Rufino's a casa de cultara de Mulher Negra as a form of female empowerment: A look at the dynamics of a black women's organizations in Brazil today. *Women's Studies International Forum*, *26*(4), 357–368. doi:10.1016/S0277-5395(03)00080-3

Durisch, T., & Zumthor, P. (Eds.). (2014). *Peter Zumthor: Buildings and projects, 1985-2013*. Scheidegger and Spiess.

Eberhard-Kaechele, M. (2012). Memory, metaphor, and mirroring in movement therapy with trauma patients. In S. C. Koch, T. Fuchs, M. Summa, & C. Müller (Eds.), *Body Memory, Metaphor and Movement* (pp. 267–287). John Benjamins Publishing Company.

Eccleston, E. (2016). *Embodied: The psychology of physical sensation*. Oxford University Press.

Egan, K. (2014). Film Production Design: Case Study of The Great Gatsby. *Elon Journal of Undergraduate Research in Communications*, *5*(1), 5–14.

Ehmann, S., Borges, S., & Klanten, R. (2012). The Social Cave. In S. Ehmann, S. Borges, & R. Klanten (Eds.), Learn for Life: New Architecture for New Learning (p. 236). Die Gestalten Verlag.

Ekkekakis, P. (2012). Affect, mood and emotion. In G. Tenenbaum, R. C. Eklund, & A. Kamata (Eds.), Measurment in sports and exercise psychology (pp. 321-332). Human Kinetics.

Elgobashi, A., & Elsemary, Y. (2019). *Towards Social Quality of life Through Cairene Public Parks*. Architecture and Urban Planning Departments.

El-khateeb, S. M. (2019). Sustainable Neighborhoods in Arab world; a Path Towards Sustainability The Case of the Sustainable City in Dubai. *The Academic Research Community Publication*, *3*(437), 14.

Ellis, C. (2002). The {New} {Urbanism}: {Critiques} and {Rebuttals}. *Journal of Urban Design*, *7*(3), 261–291. doi:10.1080/135748002200003933

Elshater, A. (2012). The New Urbanism Principles vs. Urban Design Dimensions towards Behavioral Performance in Egyptian Neighborhood. In *Future Communities: Socio-Cultural & Environmental Challenges* (pp. 826-843). Giza: AicE-Bs- ASIA Pacific International Conference on Environment-Behavior Studies.

Elshater, A. (2013, March). Towards Refined a New Concept of New Urbanism in Egypt. *Asian Journal of Environment-Behavior Studies*, *4*(12).

Elshater, A. (2015). The Principles of Gestalt Laws and Everyday Urbanism: A Visual Tactic of City Potentialities. *The International Journal of the Constructed Environment*, *5*(3-4), 1–19. doi:10.18848/2154-8587/CGP/v05i3-4/37441

Elshater, A. (2015). Urban design redux: Redefining a professional practice of specialization. *Ain Shams Engineering Journal*, *6*(1), 25–39. doi:10.1016/j.asej.2014.08.004

Elshater, A. (2016). The Ten-Minute Neighbourhood Is[not] a Basic Planning Unit for Happiness in Egypt. *Archnet-IJAR*, *10*(1), 344–357. doi:10.26687/archnet-ijar.v10i1.878

Elshater, A., & Abusaada, H. (2016). *What Is Urban Design?: Learning and Thought*. Partridge Publishing.

Emmott, C. (1997). *Narrative Comprehension: A Discourse Perspective*. OxfordUniversity Press.

Engels-Schwarzpaul, T. (Ed.). (2017). *Atmospheric Architectures: The Aesthetics of Felt Spaces*T. Engels-Schwarzpaul, Trans.). Bloomsbury Academic.

Ercan, M. A. (2011). Developing sustainable communities in historic neighbourhoods of istanbul. *Tijdschrift voor Economische en Sociale Geografie*, *102*(2), 205–219. doi:10.1111/j.1467-9663.2010.00625.x

Erkip, F. (2005). The rise of the shopping mall in Turkey: The use and appeal of a mall in Ankara. *Cities (London, England)*, *22*(2), 89–108. doi:10.1016/j.cities.2004.10.001

Escoubas, E. (2003). La Bildung des Couleurs (M. Elie, Trans.). In *Goethe, Matériaux pour l'histoire de la théorie des couleurs* (pp. 9–23). Presses Universitaires du Mirail.

Fancher, H., & Peoples, D. W. (Writers), & Scott, R. (Director). (1982). *Blade Runner* [Motion Picture]. Retrieved from https://www.imdb.com/title/tt0083658/

Faris Hmood, K. (2019). Introductory Chapter: Heritage Conservation - Rehabilitation of Architectural and Urban Heritage. In K. Hmood (Ed.), *Urban and Architectural Heritage Conservation within Sustainability*. IntechOpen. doi:10.5772/intechopen.86670

Farrar, M. E. (2011). Amnesia, Nostalgia, and the Politics of Place Memory. *Political Research Quarterly*, *64*(4), 723–735. doi:10.1177/1065912910373553

Fatih Municipality. (2019a). *e-Municipality*. Retrieved from http://www.fatih.bel.tr/ebelediye/

Fatih Municipality. (2019b). *Fatih Municipality city information system*. Retrieved from https://gis.fatih.bel.tr/webgis/

Feigenbaum, R. (2015). Toward a Nonanthropocentric Vision of Nature: Goethe's Discovery of the Intermaxillary Bone. *Goethe Yearbook*, *22*, 73–93. doi:10.1353/gyr.2015.0012

Feilden, B. M. (2005). Conservation of historic buildings (3rd ed.). Elsevier.

Ferdous, F. (2013). Examining the Relationship between Key Visual Characteristics of Urban Plazas and Aesthetic Response. *SAGE Open*, *3*(2), 1–10. doi:10.1177/2158244013485581

Ferreira, M. I. (2018). *Cognitive architectures*. Springer.

Fitzgerald, F. S. (2013). The Great Gatsby. Newbury, Berkshire.

Floridi, L. (2014). *The 4th Revolution: How the infosphere is reshaping human reality*. Oxford University Press.

Ford, T. (2012, December 6). *Nineteenth-Century Climate Change: Atmosphere, Culture and Romanticism*. Retrieved from http://rsha.cass.anu.edu.au: http://rsha.cass.anu.edu.au/nineteenth-century-climate-change-atmosphere-culture-and-romanticism

Ford, T. (2012, December 6). *Nineteenth-Century Climate Change: Atmosphere, Culture and Romanticism*. Retrieved from rsha.cass.anu.edu.au: http://rsha.cass.anu.edu.au/nineteenth-century-climate-change-atmosphere-culture-and-romanticism

Foster, S., & Giles-Corti, B. (2008). The built environment, neighborhood crime and constrained physical activity: An exploration of inconsistent findings. *Preventive Medicine*, *47*(3), 241–251. doi:10.1016/j.ypmed.2008.03.017 PMID:18499242

Foster, S., Wood, L., Christian, H., Knuiman, M., & Giles-Corti, B. (2013). Planning safer suburbs: Do changes in the built environment influence residents' perceptions of crime risk? *Social Science & Medicine*, *97*, 87–94. doi:10.1016/j.socscimed.2013.08.010 PMID:24161093

Freestone, R. (2014). Progress in Australian planning history: Traditions, themes and transformations. *Progress in Planning*, *91*, 1–29. doi:10.1016/j.progress.2013.03.005

Frost & Sullivan. (2015). *Strategic Analysis of Smart Parking Market in Europe and North America*. Author.

Fuchs, T. (2012). The phenomenology of body memory. In S. C. Koch, T. Fuchs, M. Summa, & C. Müller (Eds.), *Body Memory, Metaphor and Movement* (pp. 9–22). John Benjamins Publishing Company.

Fuchs, T. (2019). The Uncanny as Atmosphere. In G. Francesetti & T. Griffero (Eds.), *Psychopathology and Atmospheres: Neither Inside nor Outside* (pp. 101–118). Cambridge Scholars Publishing.

Furuto, A. (2011). *Onion Pinch & Napping Pod Projects at Cite' de l'architecture et du Patrimoine*. ArchDaily.

Gandy, M. (2017). Urban Atmospheres. *Cultural Geographies*, *24*(3), 353–374. doi:10.1177/1474474017712995 PMID:29278257

Gehl, J. (2011). *Life between Buildings*. Island Press.

Georgiadou, Z. (2016). Cinema and architectural design education. *artciencia.com*, 1-21.

Giddings, B., Charlton, J., & Horne, M. (2011). Public Squares in European City Centres. *URBAN DESIGN International*, *16*(3), 202–212. doi:10.1057/udi.2011.6

Gilling, D., & Barton, A. (2007). Crime prevention and community safety. In Critical Social Policy (Vol. 17). doi:10.1177/026101839701705204

Ginzarly, M., Pereira Roders, A., & Teller, J. (2019). Mapping historic urban landscape values through social media. *Journal of Cultural Heritage*, *36*, 1–11. doi:10.1016/j.culher.2018.10.002

GIS Academy. (n.d.). *Istanbul Fire Department automation project*. Retrieved from https://cbsakademi.ibb.istanbul/proje/istanbul-itfaiyesi-otomasyon-projesi/

Gissen, D. (2009). *Subnature: Architecture's Other Environments*. Princeton Architectural Press.

Graham, S., & Marvin, S. (1999). Planning cybercities: Integrating telecommunications into urban planning. *The Town Planning Review*, *70*(1), 89–114. doi:10.3828/tpr.70.1.w34454x3475g2858

Greed, C. (2005). An investigation of the effectiveness of gender mainstreaming as a means of integrating the needs of women and men into spatial planning in the United Kingdom. *Progress in Planning*, *64*(4), 241–321. doi:10.1016/j.progress.2005.08.003

Griffero, T. (2016). Atmospheres and felt-bodily resonances. *Studi di estetica, 49*(4), 1-49.

Griffero, T. (2014). *Atmospheres: Aesthetics of Emotional Spaces* (S. D. Sanctis, Trans.). Ashgate Publishing Limited.

Griffero, T. (2016). Atmospheres and felt-bodily resonances. *Studi di Estetica*, *XLIX*(IV), 1–49.

Griffero, T. (2017). *Quasi-things: the paradigm of atmospheres* (S. De Sanctis, Trans.). State University of New York.

Griffero, T. (2019). The invasion of felt-bodily atmospheres: Between pathic aesthetics and psychopathology. In G. Francesetti & T. Griffero (Eds.), *Psychopathology and Atmospheres: Neither Inside nor Outside* (pp. 6–34). Cambridge Scholars Publishing.

Gulersoy-Zeren, N., Tezer, A., Yigiter, R., Koramaz, T. K., & Gunay, Z. (2008). Istanbul Project: Istanbul Historic Peninsula Conservation Study; Zeyrek, Süleymaniye and Yenikapı Historic Districts: Vol. 1. *Conservation of Cultural Assets in Turkey*. Istanbul Technical University.

Gullberg, J. (2016). Voids and bodies: August Schmarsow, Bruno Zevi and space as a historiographical theme. *Journal of Art Historiography*, 1-20.

Guthey, G. T., Whiteman, G., & Elmes, M. (2014). Place and Sense of Place: Implications for Organizational Studies of Sustainability. *Journal of Management Inquiry*, *23*(3), 254–265. doi:10.1177/1056492613517511

Haarstad, H., & Wathne, M. W. (2019). Are smart city projects catalyzing urban energy sustainability? *Energy Policy*, *129*, 918–925. doi:10.1016/j.enpol.2019.03.001

Haberturk. (2018). *"ATAK" model for Istanbul traffic: Green light will be on according to intensity.* Retrieved from https://www.haberturk.com/tv/gundem/video/son-dakika-haberi-istanbul-trafigine-atak-modeli-yogunluga-gore-yesil-isik-yanacak/505204

Hadot, P. (1995). *Philosophy as a way of life: Spiritual exercises from Socrates to Foucault.* Wiley-Blackwell.

HameedA. (2011, November 22). *YouTube.* Retrieved from https://www.youtube.com/watch?v=aYA8084o9EE

Hampton, K., Goulet, L., & Albanesius, G. (2015). Change in the Social Life of Urban Public Spaces: The Rise of Mobile Phones and Women, and the Decline of Aloneness over 30 Years. *Urban Studies (Edinburgh, Scotland)*, *52*(8), 1489–1504. doi:10.1177/0042098014534905

Harris, A. (2008). From London to Mumbai and Back Again: Gentrification and Public Policy in Comparative Perspective. *Urban Studies (Edinburgh, Scotland)*, *45*(12), 2408. doi:10.1177/0042098008097100

Hart, S. (2003, October 1). *Imagining the Future: How will we make buildings in 2030?* Retrieved May 2020, from Architectural Record: https://www.architecturalrecord.com/articles/12264-imagining-the-future

Hasse, J. (2014). Atmosphere as expressions of medial power, Understanding atmospheres in urban governance and under self-guidance. *Lebenswelt*, *1*(4), 214–229.

Hasse, J. (2014). Atmosphere as Expressions of Medial Power, Understanding Atmospheres in Urban Governance and Under Self-Guidance. *Lebenswelt*, *1*(4), 214–229.

Haughton, G. (2000). Environmental Justice and the Sustainable City. In M. Miles, T. Hall, & I. Borden (Eds.), *The City Cultures Reader*. Routledge.

Hay, R. B. (1998). Sense of place in a developmental context. *Journal of Environmental Psychology*, *18*(1), 5–29. doi:10.1006/jevp.1997.0060

Healey, P. (2010). *Making Better Places*. Palgrave Macmillan. doi:10.1007/978-1-137-01379-8

Hedayati Marzbali, M., Abdullah, A., Ignatius, J., & Maghsoodi Tilaki, M. J. (2016). Examining the effects of crime prevention through environmental design (CPTED) on Residential Burglary. *International Journal of Law, Crime and Justice*, *46*, 86–102. doi:10.1016/j.ijlcj.2016.04.001

Heintzelman, M. D., & Altieri, J. A. (2013). Historic Preservation: Preserving Value? *The Journal of Real Estate Finance and Economics*, *46*(3), 543–563. doi:10.100711146-011-9338-8

Hekmat, S., Morgan, K., Soltani, M., & Gough, R. (2015). Sensory evaluation of locally-grown fruit purees and inulin fibre on probiotic yogurt in Mwanza, Tanzania and the microbial analysis of probiotic yogurt fortified with Moringa oleifera. *Journal of Health, Population and Nutrition*, *33*(1), 60–67. doi:10.1016/j.jenvp.2011.09.004 PMID:25995722

Hendarti, R., Tomasowa, R., & Katarina, W. (2016, December). Study on Smart City Concept of Energy Source and Distribution for A Special Economic Zone in Jakarta. In *2016 IEEE 18th International Conference on High Performance Computing and Communications; IEEE 14th International Conference on Smart City; IEEE 2nd International Conference on Data Science and Systems (HPCC/SmartCity/DSS)* (pp. 1312-1316). IEEE. 10.1109/HPCC-SmartCity-DSS.2016.0186

Herzfeld, M., & Lenhart, M. D. (1982). *Semiotiqs 1980*. Plenum Press. doi:10.1007/978-1-4684-9137-1

Hes, D., & Hernandez-Santin, C. (Eds.). (2019). Placemaking Fundamentals for the Built Environment. Academic Press.

Hewitt, N. (2001). The Changing Landscape of Montmartre. *Journal of the Australasian Universities Language and Literature Association*, *96*(1), 1–18. doi:10.1179/aulla.2001.96.1.002

Höjer, M., Gullberg, A., & Pettersson, R. (2011). *Images of the Future City. Time and Space for Sustainable Development*. Springer. doi:10.1007/978-94-007-0653-8

Holand, J. (1999). *Emergence: From Chaos to Order*. Perseus Books.

Hollands, R. G. (2008). Will the real smart city please stand up? Intelligent, progressive or entrepreneurial? *City*, *12*(3), 303–320. doi:10.1080/13604810802479126

Holt-Jensen, A. (2009). *Geography: History and Concepts* (4th ed.). SAGE.

Hough, M. (1990). *Out of place: Restoring identity to the regional landscape*. Yale University Press.

Hough, M. (1990). *Out of Place: Restoring Identity to the Regional Landscape*. Yale University Press.

Ho, W. C., & Cheung, C. K. (2011). Social sustainability for mothers in Hong Kong's low-income communities. *Habitat International*, *35*(4), 529–536. doi:10.1016/j.habitatint.2011.03.001

Hsu, H. L. (2017). Literary Atmospherics. *Literary Geographies*, *3*(1), 1–5.

Hurriyet. (2018). *Here are the new taxis of Istanbul! Turquoise, black...* Retrieved from https://www.hurriyet.com.tr/gundem/iste-istanbulun-yeni-taksileri-turkuaz-siyah-40902239

Husseinya, A. M., & Husseiny, A. A. (2012). Integration of Static and Dynamic Spaces in Historic Cairo: A community spirit that may vanish. *AicE-Bs 2012 Cairo, ASIA Pacific International Conference on Environment-Behaviour Studies*, 68, 687 – 699. doi:10.1016/j.sbspro.2012.12.259

Husserl, E. (1913). Ideas Pertaining to a Pure Phenomenology and to a Phenomenological Philosophy – First Book: General Introduction to a Pure Phenomenology (F. Kersten, Trans.). The Hague: Nijhoff.

Hyden, C. T., & Sheckels, T. F. (2016). *Public Places: Sites of Political Communication*. London: Lexington Books.

ICOMOS, (2017). *UNESCO-WHC Committee report and Conclusion by ICOMOS Advisory Bodies for Historic Areas of Istanbul (Turkey) No 356 Bis*. Retrieved from file:///C:/Users/ITU/Downloads/356bis-ICOMOS-2255-en.pdf

ICOMOS. (2005). *Xi'an declaration on the conservation of the setting of heritage structures, sites and areas*. International Council on Monuments and Sites(ICOMOS).

Iliyasu, I. I. (2014). *Challenges of preservation of cultural landscapes in traditional cities: Case study of Kano Ancient City*. Academic Press.

Industry City. (2013). *IC In Context*. Retrieved 2020, from Industry City: https://industrycity.com/

Information Society Department. (2018). *2015-2018 Information Society Strategy and Action Plan*. Author.

Inoue, S., Rodgers, P., Tennant, A., & Spencer, N. (2015). Creative Reductionism: How decreasing levels of information can stimulate designers imagination. In *International Conference on Engineering and Product Design Education*. Loughborough University.

International Ambiances Network. (2012, September 22). *2nd International Congress on Ambiances*. Retrieved May 2020, from Ambiances in action: https://ambiances2012.sciencesconf.org/

Ionescu, V. (2016). Architectural symbolism: Body and space in Heinrich Wölfflin and Wilhelm Worringer. *Architectural Histories*, *4*(1), 1–9. doi:10.5334/ah.213

ISBAK. (2019). *iTaksi*. Retrieved from https://itaksi.com/en/#Welcome

ISKI. (2016). *ISKI 2017 budget approved*. Retrieved from https://www.iski.istanbul/web/tr-TR/kurumsal/haberler1/haberler-detay/iski-2017-butcesi-onaylandi

ISKI. (2017). *2017 Annual Report*. ISKI.

ISKI. (n.d.a). *Scada technology*. Retrieved from https://www.iski.istanbul/web/en-US/kurumsal/iski-hakkinda/scada-technology

ISKI. (n.d.b). *Smart meter application*. Retrieved from https://www.iski.istanbul/web/tr-TR/musteri-hizmetleri/kartli-saya-uygulamasi

ISPARK. (2017). *Smart parking systems*. Retrieved from https://ispark.istanbul/ispark-ile-akilli-otopark-sistemleri/

ISPARK. (2019a). *Isbike*. Retrieved from https://www.isbike.istanbul

ISPARK. (2019b). *Isbike: How does it work?* Retrieved from https://www.isbike.istanbul/nasilcalisir

Istanbul Metropolitan Municipality. (2015). *Traffic Management for a Smarter City*. Author.

Istanbul Metropolitan Municipality. (2016). *2016 Annual Report*. Author.

Istanbul Metropolitan Municipality. (2017). *IBB YolGosteren: Your New Guide on the Road*. Author.

Istanbul Metropolitan Municipality. (2018). *"ATAK" model from IMM to Istanbul traffic*. Retrieved from https://www.ibb.istanbul/News/Detail/34842

Istanbul Metropolitan Municipality. (2019a). *Air quality monitoring centre*. Retrieved from https://havakalitesi.ibb.istanbul/Pages/AirQuality

Istanbul Metropolitan Municipality. (2019b). *IMM White Desk*. Retrieved from http://alo153.ibb.gov.tr/

Istanbul Metropolitan Municipality. (2019c). *Istanbul city guide portal.* Retrieved from https://sehirharitasi.ibb.gov.tr/

Istanbul Metropolitan Municipality. (2019d). *Watch Istanbul: Touristic cameras.* Retrieved from http://www.istanbuluseyret.com/TuristikKameralar/

ISTTELKOM. (2017). *Common communications infrastructure is coming to Istanbul.* Retrieved from https://isttelkom.istanbul/en/common-communications-infrastructure-is-coming-to-istanbul/

Ja'afar, N., Sulaiman, A., & Shamsuddin, S. (2012). Article. *Social and Behavioral Sciences, 50*, 643–656.

Jaaniste, L. O. (2010). The Ambience of Ambience. *A Journal of Media and Culture, 13*(2). Retrieved from http://www.journal.media-culture.org.au/index.php/mcjournal/article/view/238

Jackson, J. B. (1994). *A sense of place, a sense of time.* Walker Art Center.

Jackson, J. B. (1994). *A Sense of Place, a Sense of Time.* Walker Art Center.

Jacobs, S. (2007). *The Wrong House: The Architecture of Alfred Hitchcock.* Rotterdam: 010 Publishers.

Jacobs, J. (1961). *The Death and Life of Great American Cities.* Random House.

Janney, R. W. (2012). Pragmatics and Cinematic Discourse. *Lodz Papers in Pragmatics, 8*(1), 85–113. doi:10.1515/lpp-2012-0006

Jarass, J., & Heinrichs, D. (2014). New Urban Living and Mobility. *Transportation Research Procedia, 1*(1), 142–153. doi:10.1016/j.trpro.2014.07.015

Jayasuriya, K. (2009, January). Asian Regional Governance: Crisis and Change. *ASEAN Economic Bulletin, 26*(2), 227. doi:10.1355/ae26-2g

Jencks, C. (2013). *Architecture becomes music.* Retrieved from The architectural review: https://www.architectural-review.com/archive/viewpoints/architecture-becomes-music/8647050.article

Jenner, R. (2013). Ambient atmospheres: Exhibiting the immaterial in works by Italian Rationalists Edoardo Persico and Franco Albini. *Interstices: Journal of Architecture and Related Arts*, 13-24.

Jiang, D. (2020). The construction of smart city information system based on the Internet of Things and cloud computing. *Computer Communications, 150*, 158–166. doi:10.1016/j.comcom.2019.10.035

Jorgensen, B., & Stedman, R. (2001). Sense of Place as an Attitude: Lakeshore Owners' Attitudes Toward Their Properties. *Journal of Environmental Psychology, 21*(3), 233–248. doi:10.1006/jevp.2001.0226

Kahn, L. (1957). Architecture is the thoughtful making of spaces. *Perspecta*, (4): 2–3.

Karimi, K. (2012). A Configurational Approach to Analytical Urban Design: 'Space Syntax' Methodology. *Urban Design International, 0*(0), 1–22. doi:10.1057/udi.2012.19

Kazig, R. (2016, April 27). *Presentation of Hermann Schmitz' paper, "Atmospheric spaces".* Retrieved April 30, 2017, from Ambiances [En ligne]: http://ambiances.revues.org/709

Kazig, R. (2016, April 27). *Presentation of Hermann Schmitz' paper, "Atmospheric Spaces".* Retrieved April 30, 2017, from Ambiances [En ligne]: http://ambiances.revues.org/709

Kellner, D. (2002, November). Theorizing Globalization. *Sociological Theory, 20*(3), 285–305. doi:10.1111/0735-2751.00165

Kemmis, D. (1995). *The Good City and the Good Life.* Houghton Mifflin Co.

Kersten, T. P., Büyüksalih, G., Baz, I., & Jacobsen, K. (2009). Documentation of Istanbul historic peninsula by kinematic terrestrial laser scanning. *The Photogrammetric Record, 24*(126), 122–138. doi:10.1111/j.1477-9730.2009.00528.x

Khakpour, A., Sabouri, S., & Harirchian, M. (2017). Analysis of Effective Factors on Presence of Citizens in Urban Spaces, Case Study: Towhid Square in Tehran. *Journal of Sustainable Development, 10*(3), 35–45. doi:10.5539/jsd.v10n3p35

Kitchin, R. (2014). The real-time city? Big data and smart urbanism. *GeoJournal, 79*(1), 1–14. doi:10.100710708-013-9516-8

Knapton, S. (2012). *Istanbul for James Bond fans.* Retrieved from www.telegraph.co.uk: https://www.telegraph.co.uk/travel/destinations/europe/turkey/istanbul/articles/Skyfall-Istanbul-for-James-Bond-fans/

Knight, D. (2009). Romanticism and the sciences. In A. Cunningham & N. Jardine (Eds.), *Romanticism and the sciences* (pp. 13–24). Cambridge University Press.

Koch, S. C. (2012). Testing Fuchs' taxonomy of body memory: A content analysis of interview data. In S. C. Koch, T. Fuchs, M. Summa, & C. Müller (Eds.), *Body Memory, Metaphor and Movement* (pp. 171–186). John Benjamins Publishing Company.

Koramaz, T. K. (2017) Information - Communication Technologies and Management Plans for Heritage Sites. In *Abstract Book of 30th Annual AESOP 2017 Congress, SPACES OF DIALOG FOR PLACES OF DIGNITY: Fostering the European Dimension of Planning* (pp. 599-600). Lisbon, Portugal: Técnico Lisboa, The Institute of Geography and Spatial Planning, Faculty of Architecture.

Koramaz, T. K. (2018b). Housing Renewal Sites and Spatial Features of Deterioration and Deprivation in Istanbul's Historic Peninsula. *Journal of Urban Planning and Development, 144*(1), 1-10.

Koramaz, T. K., Nasrollahzade, S., & Ozdemir, Z. (2019) Spatial Pattern of Digital Divide in Turkey. In *Proceedings of REAL CORP 2019.* Karlsruhe, Germany: Karlsruher Institut für Technologie.

Koramaz, T. K. (2018a). Information and Communication Technologies in Cultural Heritage Management. In L. Reynolds, A. Campelo, A. Lindgreen, & M. B. Beverland (Eds.), *Cultural Heritage: A Research Anthology* (pp. 155–165). Routledge. doi:10.4324/9781315107264-12

Korupp, S. E. (2006). Causes and trends of the digital divide: a European perspective. In *Proceedings of Deutsche Gesellschaft für Soziologie (DGS), Soziale Ungleichheit, kulturelle Unterschiede: Verhandlungen des 32. Kongresses der Deutschen Gesellschaft für Soziologie.* Campus Verl.

Koskela, H., & Pain, R. (2000). Revisiting fear and place: Women's fear of attack and the built environment. *Geoforum, 31*(2), 269–280. doi:10.1016/S0016-7185(99)00033-0

Kotler, P. (1974). *Atmospherics as a Marketing Tool.* Academic Press.

Kottas, D. (2014). The Social Cave. In Digital Architecture: New Applications (pp. 250-257). Links International.

Kramers, A., Höjer, M., Lövehagen, N., & Wangel, J. (2014). Smart sustainable cities–Exploring ICT solutions for reduced energy use in cities. *Environmental Modelling & Software, 56,* 52–62. doi:10.1016/j.envsoft.2013.12.019

Krause, L., & Petro, P. (Eds.). (2003). *Global Cities: Cinema, Architecture, and Urbanism in a Digital Age.* Rutgers University Press.

Krier, R. (1979). *Urban Space.* Academy Editions.

Krier, R., & Krier, L. (1979). *Typological and Morphological Elements of the Concept of Urban Space*. AD and Acroshaw Ltd.

Krippendorff, K. (2004). *Content Analysis: An Introduction to Its Methodology*. Sage.

Krohn, C. (2014). *Mies Van der Rohe the built work*. Birkhäuser Verlag GmbH. doi:10.1515/9783038212874

Królikowski, J., Rybak-Niedziólka, K., Skibinska, M., & Wlazlo-Malinowska, K. (2019). Indentification of the Genius Loci Category in the Process of Recognition of the City Landscape Design Issue. *Technical Transactions*, *2*, 29–40.

Kumar, P. (2015). Smart Neighbourhood to Enhance Socail Sustainability and Inclusive Planning in Smart Cities. In *National Conference*. Sonipat: Deenbandhu Chottu Ram University of Science and Technology Murthal.

Kung, J. (1977). Aristotle on essence and explanation. *Philosophical studies. An International Journal for Philosophy in the Analytic Tradition*, *31*(6), 361–383. doi:10.1007/BF01857029

Labuhn, B. (2016). Conceptualisations of Atmosphere in Text and in Image in Architectural Journalism (1991 - 2013). *Ambiances, tomorrow.* In *Proceedings of 3rd International Congress on Ambiances* (pp. 289-294). Volos, Greece: International Network Ambiances, University of Thessaly.

Labuhn, B. (2016). Conceptualisations of Atmosphere in Text and in Image in Architectural Journalism (1991 - 2013). In *Ambiances, tomorrow. Proceedings of 3rd International Congress on Ambiances* (pp. 289-294). Volos, Greece: International Network Ambiances; University of Thessaly.

Leaman, O. (2003). *Companion Encyclopedia of Middle Eastern and North African Film*. Routledge. doi:10.4324/9780203426494

LeCorbusier. (1948). *New world of space*. Reynal & Hitchcock.

Leddy, T. (2019). *Dewey's Aesthetics* (N. E. Zalta, Ed.). Retrieved January 2020, from The Stanford Encyclopedia of Philosophy (Winter Edition 2019): https://plato.stanford.edu/archives/win2019/entries/dewey-aesthetics/

Lee, S. H., Han, J. H., Leem, Y. T., & Yigitcanlar, T. (2008). Towards ubiquitous city: concept, planning, and experiences in the Republic of Korea. In *Knowledge-based urban development: Planning and applications in the information era* (pp. 148–170). IGI Global. doi:10.4018/978-1-59904-720-1.ch009

Lefebvre, H., Nicholson-Smith, & Harvey, D. (1991). The Production of Space. Oxford, UK: Blackwell Publishing.

Lefebvre, H. (2000). Plan of the Present Work. In M. Miles, T. Hall, & I. Borden (Eds.), *The City Cultures Reader*. Routledge.

Leite, J. M. V. (2009). Cultural heritage and monument, a place in memory. *City and Time*, *4*(2), 23–33.

Lenzhölzer, S. (2010). Designing Atmospheres, Research and Design for Thermal Comfort in Dutch Urban Squares. Wageningen, The Netherlands: S.n.

Lévy, P. (2011). The Semantic Sphere 1: Computation, Cognition and Information Economy 1st Edition. London: Wiley-IST.

Levy, P., & Bononno, R. (1999). *Collective Intelligence: Mankind's Emerging World in Cyberspace*. Basic Books.

Ley, D., & Murphy, P. (2001). Immigration in gateway cities: Sydney and Vancouver in comparative perspective. *Progress in Planning*, *55*(3), 119–194. doi:10.1016/S0305-9006(00)00025-8

Leydecker, S. (2016). Toolbox. In *Adaptive Reuse: Extending the Lives of Buildings* (p. 87). Birkhauser.

Lima, C. (2008). Filmic Narratives of the City. Cities in Film: Architecture, Urban Space and the Moving Image (pp. 142-148). University of Liverpool and Arts and Humanities Research Council (AHRC).

Lisdorf, A. (2020). *Demystifying Smart Cities: Practical Perspectives on How Cities Can Leverage the Potential of New Technologies*. Apress/Springer Nature.

Li, X., Hijazi, I., Reinhard Koenig, Z. L., Zhong, C., & Schmitt, G. (2016). Assessing Essential Qualities of Urban Space with Emotional and Visual Data Based on GIS Technique. *International Journal of Geo-Information, 5*(11), 1–18. doi:10.3390/ijgi5110218

López, G., Moura, P. S., Custodio, V., & Moreno, J. I. (2012). Modeling the Neighborhood Area Networks of the Smart Grid. In *Selected Areas in Communications Symposium* (pp. 3357-3361). IEEE.

Lucas, R. (2008). Acousmêtric Architecture: Filmic Sound Design and its Lessons for Architects. In Cities in Film: Architecture, Urban Space and the Moving Image (pp. 157-164). University of Liverpool and the Arts and Humanities Research Council (AHRC).

Luhrmann, B. (Director). (2013). *The Great Gatsby* [Motion Picture].

Lukermann, F. (1964). Geography as a formal intellectual discipline and the way it contributes to human knowledge. *The Canadian Geographer. Geographe Canadien, 8*(4), 167–172. doi:10.1111/j.1541-0064.1964.tb00605.x

Lynch, K. (2005). The image of the city. MIT Press.

Lynch, K. (1972). *What Time is This Place?* The MIT Press.

Lyon, D. (1999). *Postmodernity*. University of Minnesota Press.

Maeng, D. M., & Nedović-Budić, Z. (2008). Urban form and planning in the information age: Lessons from literature. *Spatium (Belgrade)*, (17-18), 1–12. doi:10.2298/SPAT0818001M

Makhail, M. N. (1992). Studies in the Short Fiction of Mahfouz and Idris. New York University Press.

Mallet-Stevens, R. (1924). *Le Cinéma et les arts: L'Architecture*. L'Herbier.

Mallgrave, H. (1996). *Otto Wagner: Reflections on the Raiment of Modernity*. Getty Publications.

Mason, R. (2007). Building women's social citizenship: A five-point framework to conceptualise the work of women-specific services in rural Australia. *Women's Studies International Forum, 30*(4), 299–312. doi:10.1016/j.wsif.2007.05.007

Massachusetts Institute of Technology (MIT). (2020). *Urban imagination and social innovation through design & science*. Retrieved from Senseable City Lab: http://senseable.mit.edu/

Massara, F., Liu, S. S., & Melara, R. D. (2010). *Adapting to a retail environment: Modeling consumer–environment interactions*. Academic Press.

Maté, A., Peral, J., Ferrández, A., Gil, D., & Trujillo, J. (2016). A hybrid integrated architecture for energy consumption prediction. *Future Generation Computer Systems, 63*, 131–147. doi:10.1016/j.future.2016.03.020

Mattern, S. (2017). A City Is Not a Computer. *Places Journal*. doi:10.22269/170207

Mayne, T. (2011). *Combinatory Urbanism A Realignment of Complex Behavior and Collective Form*. Stray Dog Cafe.

McCarthy, J. (2002). Entertainment-led regeneration: The case of Detroit. *Cities (London, England), 19*(2), 105–111. doi:10.1016/S0264-2751(02)00005-7

McClellan, S., Jimenez, J. A., & Koutitas, G. (Eds.). (2018). Smart Cities: Applications, Technologies, Standards, and Driving Factors (1st ed.). Cham: Springer International Publishing: Imprint: Springer.

McCormack, D. P. (2008). Engineering affective atmospheres on the moving geographies of the 1897 Andrée expedition. *Cultural Geographies, 15*(4), 413–430. doi:10.1177/1474474008094314

McCrum, R. (2015). *The 100 best novels: No 69 – The Heat of the Day by Elizabeth Bowen (1948).* Retrieved from https://www.theguardian.com/books/2015/jan/12/100-best-novels-elizabeth-bowen-heat-of-the-day-blitz

McGrail, L. (2018). *5 Essential Elements of Successful Mise en Scène in Film.* Retrieved from www.lightsfilmschool.com: https://www.lightsfilmschool.com/blog/mise-en-scene-in-film-afk

McGregor, K. (2016). Emotions and activism for former so-called 'comfort women' of the Japanese Occupation of the Netherlands East Indies. *Women's Studies International Forum, 54*, 67–78. doi:10.1016/j.wsif.2015.11.002

McLuhan, M. (1964). *Understanding media: The extensions of man.* McGraw-Hill.

McLuhan, M. (2004). *Understanding Me: Lectures and Interviews.* MIT Press.

McLuhan, M., & Moos, M. (1997). *Media Research: Technology, Art and Communication (Critical Voices in Art, Theory and Culture).* Routledge.

Md Sakip, S. R., & Abdullah, A. (2018). An Evaluation of Crime Prevention through Environmental Design (CPTED) Measures in a Gated Residential Area: A Pilot Survey. *Asian Journal of Environment-Behaviour Studies, 3*(6), 21. doi:10.21834/aje-bs.v3i6.232

Medak, P. (Director). (1993). *Romeo Is Bleeding (1993)* [Motion Picture].

Meeks, S., & Murphy, K. (2016). *The Past and Future City: How Historic Preservation Is Reviving America's Communities* (2nd ed.). Island Press. doi:10.5822/978-1-61091-709-4

Mehrabian, A., & Russell, J. A. (1980). *An Approach to Environmental Psychology.* Academic Press.

Mehta, V., & Bosson, J. (2018). Revisiting Lively Streets: Social Interactions in Public Space. *Journal of Planning Education and Research*, 1–13. doi:10.1177/0739456X18781453

Menezes, M., & Costa, C. People, Public Space, Digital Technology and Social Practice: an Ethnographic Approach. In *Enhancing Places Through Technology* (pp. 167–180). Edições Universitárias Lusófonas.

Merleau-Ponty, M. (2002). *Phenomenology of perception.* Psychology Press. doi:10.4324/9780203994610

Michels, C. (2015). Researching affective atmospheres. *Geographica Helvetica, 70*(70), 255–263. doi:10.5194/gh-70-255-2015

Michels, C., & Steyaert, C. (2017). By accident and by design: Composing affective atmospheres in an urban art intervention. *Organization, 24*(1), 79–104. doi:10.1177/1350508416668190

Milan, B. F., & Creutzig, F. (2017). Lifting peripheral fortunes: Upgrading transit improves spatial, income and gender equity in Medellin. *Cities (London, England), 70*(February), 122–134. doi:10.1016/j.cities.2017.07.019

Milner, A. (2004). Darker Cities: Urban Dystopia and Science Fiction Cinema. *International Journal of Cultural Studies, 7*(3), 259–279. doi:10.1177/1367877904046303

Mitchell, W. J. (1995). *City of bits: space, place, and the infobahn.* MIT Press.

Mitchell, W. J. (1999). *"e-topia:" Urban life, Jim—but not as we know it.* MIT Press. doi:10.7551/mitpress/2844.001.0001

Mizooji, K. K., Haghighat, A. T., & Forsati, R. (2012). Data clustering using bee colony optimization. In The Seventh International Multi-Conference on Computing in the Global Information Technology (pp. 189-194). IARIA.

Mossberger, K., Tolbert, C. J., & Franko, W. W. (2013). *Digital Cities: The internet and the geography of opportunity.* Oxford University Press.

Moughtin, C., Cuesta, R., Signoretta, P., & Sarris, C. (1999). *Urban Design: Method and Techniques.* Architectural Press. Retrieved from www.qualitestgroup.comhttps://www.qualitestgroup.com/white-papers/advantages-disadvantages-crowdsourcing/

Moulaert, F., Martinelli, F., Swyngedouw, E., & González, S. (2010). *Can Neighbourhoods Save the City?: Community development and social innovation.* Routledge. doi:10.4324/9780203849132

Movie-Locations.com. (2019). *Mission Impossible– Ghost Protocol.* Retrieved from movie-locations.com: http://www.movie-locations.com/movies/m/Mission-Impossible-Ghost-Protocol.php

Muminovic, M. (2015). Places as Assemblages: Paradigm Shift or Fashionable Nonsense? *Athens Journal of Architecture*, 295-309. doi:10.30958/aja.1-4-3

Munari, B. (1981). *Da Cosa Nasce Cosa.* Laterza.

Nallathiga, R. (2001, January-March). Redevelopment of Industrial Land in Urban Areas: A Case Study of Textile Mill Land Redevelopment in Mumbai. *Institute of Town Planners, India Journal*, 95-106. Retrieved November 2012, from http://itpi.org.in/pdfs/jan9_11.pdf

Narkunas, J. P. (2001). Streetwalking in the Cinema of the City: Capital Flows Through Saigon. In M. Shiel & T. Fitzmaurice (Eds.), Cinema and the City (pp. 147-157). Hoboken, NJ: Blackwell Publishing Ltd.

Nayak, S. S. (2008). *Globalization and the Indian Economy: Roadmap to Convertible Rupee.* Routledge.

Neuendrof, K. A. (2002). *The content analysis.* Sage Publications.

Ning, Z., Xia, F., Ullah, N., Kong, X., & Hu, X. (2017). Vehicular social networks: Enabling smart mobility. *IEEE Communications Magazine*, *55*(5), 16–55. doi:10.1109/MCOM.2017.1600263

Niwa, H. (2007). Article. *Development*, *134*(4), 635–646. PMID:17215298

Norberg-Schulz, C. (1971). *Existence, space and architecture.* Studio Vista Limited.

Norberg-Schulz, C. (1980). *Genius Loci: Towards a phenomenology of architecture.* Academy Editions.

Nunn, S. (2011). Police technology in cities: Changes and challenges. *Technology in Society*, *23*(1), 11–27. doi:10.1016/S0160-791X(00)00033-6

Oakley, D. (1970). *The Phenomenon of Architecture in Cultures in Change.* Pergamon.

Oliveira Neto, J. (2018). *Inclusive Smart Cities: Theory and Tools to Improve the Experience of People with Disabilities in Urban Spaces* (Unpublished doctoral dissertation). Université Paris-Saclay, Universidade de São Paulo, Brésil.

Oregon, U. o. (2015). *Sustainable Cities Institute.* Retrieved 2020, from Urbanism Next: https://sci.uoregon.edu/urbanism-next-0

Oxford Reference. (2020). *Overview: Bateau-Lavoir.* Retrieved May 2020, from https://www.oxfordreference.com/view/10.1093/oi/authority.20110803095451214#showlesscontent

Padte, R. K. (2013, June 29). Keeping Women Safe? *Economic & Political Weekly, Vol, 46*, 1–7.

Pahl, M.-O., Niedermayer, H., Kinkelin, H., & Carle, G. (2013). Enabling Sustainable Smart Neighborhoods. *Sustainable Internet and ICT for Sustainability*, 6.

Pain, R. (2000). Place, social relations and the fear of crime: A review. *Progress in Human Geography*, *24*(3), 365–387. doi:10.1191/030913200701540474

Pain, R. H. (1997). Social geographies of women's fear of crime. *Trans Inst Br Geogr NS*, *22*(2), 231–244. doi:10.1109/PRIME.2014.6872711

Pallasmaa, J. (2014). *Space, place and atmosphere. Emotion and peripheral perception in architectural experience.* Academic Press.

Pallasmaa, J. (2008). *The Architecture of Image: Existential Space in Cinema.* Rakennustieto Publishing.

Pallasmaa, J. (2011). An Architecture of the Seven Senses. In L. Weinthal (Ed.), *Theory, Toward a New Interior. An Anthology of Interior Design* (pp. 40–49). Princeton Architectural Press.

Pallasmaa, J. (2011). An architecture of the seven senses. In L. Weinthal (Ed.), *Theory, toward a New Interior: An anthology of interior design* (pp. 40–49). Princeton Architectural Press.

Pallasmaa, J. (2014). Space, Place, and Atmosphere: Peripheral Perception in Existential Experience. In C. Borch (Ed.), *Architectural Atmospheres: On the Experience and Politics of Architecture* (pp. 18–41). Birkhauser. doi:10.1515/9783038211785.18

Pandey, J. (2019). Analysis, Deductive Approach to Content. In M. Gupta, M. Shaheen, & K. P. Reddy (Eds.), *Qualitative Techniques for Workplace Data Analysis* (pp. 145–169). IGI Global., doi:10.4018/978-1-5225-5366-3.ch007

Park, J. H. (2019). Advances in future internet and the industrial internet of things. *Symmetry*, *11*(2), 244. doi:10.3390ym11020244

Park, Y., & Rogers, G. O. (2015). Neighborhood Planning Theory, Guidelines, and Research: Can area, population, and boundary guide conceptual framing? *Journal of Planning Literature*, *30*(1), 18–36. doi:10.1177/0885412214549422

Parsons, R. (2010). *New Design in historic setting*. Academic Press.

Pavlovskaya, M. (2009). Non-Quantitative GIS. In M. Cope & S. Elwood (Eds.), *Qualitative GIS: A Mixed Methods Approach* (pp. 13–38). SAGE Publications Ltd. doi:10.4135/9780857024541.n2

Paydar, M., Kamani-Fard, A., & Etminani-Ghasrodashti, R. (2017). Perceived security of women in relation to their path choice toward sustainable neighborhood in Santiago, Chile. *Cities (London, England)*, *60*, 289–300. doi:10.1016/j.cities.2016.10.002

Pelsmakers, S. (2012). *The Environmental Design Pocketbook* (J. Thompson, Ed.). Riba Publishing.

Pelton, J. N. (2019). *Preparing for the Next Cyber Revolution: How our world will be radically transformed – again!* Springer.

Pelton, J. N., & Singh, I. B. (2019). *Smart Cities of Today and Tomorrow: Better Technology, Infrastructure and Security.* Springer International Publishing. doi:10.1007/978-3-319-95822-4

Perovic, S., & Folic, N. (2012). Visual Perception of Public Open Spaces in Niksic. *Procedia: Social and Behavioral Sciences*, *68*, 921–933. doi:10.1016/j.sbspro.2012.12.277

Pezzica, C., Lopes, J., & Rebelo Paio, A. (2017). Informed Design Decision-Making: From Digital Analysis to Urban Design. *Gestão & Tecnologia de Projetos*, *12*(3), 9–26. doi:10.11606/gtp.v12i3.133936

Pierdicca, R., Malinverni, E., Khromova, A., Marcheggiani, E., Bonanno, P., Franco, F., & Martinez, A. (2017). The Integration of an Augmented Reality Module within the Way-Cyberparks App. The Case Study of Valletta City. In *Enhancing Places Through Technology* (pp. 57–68). Edições Universitárias Lusófonas.

Poe, E. A. (2017). *Great Horror Stories.* Dover Publication.

Poria, Y., Butler, R., & Airey, D. (2003). The core of heritage tourism. *Annals of Tourism Research*, *30*(1), 238–254. doi:10.1016/S0160-7383(02)00064-6

PPS. (2010). *Placemaking Meets Preservation*. https://www.pps.org/article/placemaking-meets-preservation

Prevatt, J. (1998). *Crime Prevention Through Environmental Design (CPTED) and the role of facilities planning in force protection*. Academic Press.

Project for Public Places. (2012). *The Great Movies for Placemajers*. Retrieved from pps.org: https://www.pps.org/article/ten-great-movies-for-placemakers

Rabiger, M. (2008). *Directing: Film Techniques and Aesthetics*. Focal Press.

Raymond, A. (1984). *The Great Arab Cities in the 16th-18th centuries, An Introduction*. New York University Press.

Redi, M., Aiello, L. M., Schifanella, R., & Quercia, D. (2018). The Spirit of the City: Using Social Media to Capture Neighborhood Ambiance. *Proceedings of the ACM on Human-Computer Interaction, 2*, 1–18. 10.1145/3274413

Reichler, T. (2009). Changes in the Atmospheric Circulation as Indicator of Climate Change. In T. M. Letcher (Ed.), *Climate Change: Observed* (pp. 145–165). Elsevier., doi:10.1016/B978-0-444-53301-2.00007-5

Relph, E. (2008). *Place and placelessness*. Pion Ltd.

Relph, E. (1976). *Place and placelessness*. Pion.

Relph, E. (1976). *Place and Placelessness*. Pion.

Rendell, J. (2006). *Art and architecture: A place between*. I. B. Tauris.

Rowe, J. (2020). *Urban Lab Films*. Retrieved from University College London: https://www.ucl.ac.uk/urban-lab/events/urban-lab-films

Saad, S. A. (2018). The Spatial Identity of Historic Cairo. Part two: Must Prominent Buildings be Enframed. In *Proceedings of the conference titled: Cities' Identity through Architecture and Arts 2* (pp. 183-194). Cairo: IEREK Press.

Sadler, S., & Archigram (Group). (2005). *Archigram: Architecture without architecture.* Cambridge, MA: MIT Press. Retrieved from Monoskop: https://monoskop.org/log/?p=1055

Safety, P. (n.d.). Article. *Public Safety*, 162–189.

Saitta, L., & Zucker, J.-D. (2013). *Abstraction in artificial intelligence and complex systems.* Springer Science & Business Media. doi:10.1007/978-1-4614-7052-6

Salama, A. M. (2015). *Spatial Design Education: New Directions for Pedagogy in Architecture and Beyond.* Routledge.

Salem, A. (Director). (1967). *Khan Al-Khalili* [Motion Picture].

Salingaros, N. A., & Mehaffy, M. W. (2006). *A Theory of Architecture*. UMBAU-VERLAG Harald Püschel.

Sandercock, L., & Attili, G. (2009). *Where Strangers Become Neighbours*. Integrating Immigrants in Vancouver. doi:10.1007/978-1-4020-9035-6

Sanders, A. (1988). *The Companion to A Tale of Two Cities (Dickens companions; 4)*. Unwin Hyman.

Sartre, J.-P. (1993). *The Emotions: An Outline of a Theory* (B. Frechtman, Trans.). Carol Publishing Group.

Sassen, S. (1991). *The Global City: New York, London, Tokyo*. Princeton University Press.

Sassen, S. (2006). *Cities in a World Economy* (3rd ed.). Pine Forge Press.

Sauchelli, A. (2012). On Architecture as a Spatial Art. *The Nordic Journal of Aesthetics*, (43), 53–64. doi:10.7146/nja.v23i43.7497

Sawyer, S., & Tapia, A. (2005). The sociotechnical nature of mobile computing work: Evidence from a study of policing in the United States. *International Journal of Technology and Human Interaction*, *1*(3), 1–14. doi:10.4018/jthi.2005070101

Scannell, L., & Gifford, R. (2010). Defining place attachment: A Tripartite Organizing Framework. *Journal of Environmental Psychology*, *30*(1), 1–10. doi:10.1016/j.jenvp.2009.09.006

Schaefer, D. R. (2012). Youth co-offending networks: An investigation of social and spatial effects. *Social Networks*, *34*(1), 141–149. doi:10.1016/j.socnet.2011.02.001

Schmarsow, A. (1994). The essence of architectural creation. In H. F. Mallgrave & E. Ikonomou (Eds.), *Empathy, Form and Space: Problem in German aesthetics 1873-1893* (pp. 281–297). The Getty Center for the History of Art and the Humanities.

Schmitz, H. (2014). *Atmosphären* [Atmospheres]. Freiburg: Alber.

Schmitz, H. (2016, April 27). *Atmospheric Spaces, Ambiances*. doi:10.4000/ambiances.711

Schmitz, H. (2017). The Felt Body and Embodied Communication. *Yearbook for Eastern and Western Philosophy*, 9-19. https://doi.org/ doi:10.1515/yewph-2017-0004

Schmitz, H. (2019). New Phenomenology: A Brief Introduction (Atmospheric Spaces) (T. Griffero, Ed.). Academic Press.

Schmitz, H. (1978). *System der Philosophie*. Bouvier.

Schmitz, H. (2007). *Der Leib, der Raum und die Gefühle* [Body, Space and Feeling]. Aisthesis.

Schmitz, H. (2018). Les sentiments comme atmosphères [Feelings as atmospheres]. *Communications*, *1*(102), 51–66.

Schmitz, H. (2019). *New Phenomenology: A brief introduction* (R. O. Müllan, Trans.). Milan: *Mimesis*.

Schmitz, H., Owen, R., & Slaby, M. (2011). Emotions outside the box—The new phenomenology of feeling and corporeality. *Phenomenology and the Cognitive Sciences*, *10*(2), 241–259. doi:10.100711097-011-9195-1

Schonhammer, R. (2018). Atmosphere-the life of a place, the psychology of environment and Design. In J. Weidinger (Ed.), *Designing atmospheres* (pp. 141–179). Universitätsverlag der TU Berlin.

Schwab, K. (2017). *The Fourth Industrial Revolution*. Penguin.

Schwartz, J. M., & Cook, T. (2002). Archives, records, and power: The making of modern memory. *Archival Science*, *2*(1–2), 1–19. doi:10.1007/BF02435628

Schwarzer, M. W., & Schmarsow, A. (1991). The emergence of architectural space: August Schmarsow's theory of "raumgestaltung". *Assemblage*, *15*(15), 48–61. doi:10.2307/3171125

Scott, G. (1954). *Architecture of humanism*. Doubleday Anchore Book.

Semper, G. (2004). *Style in the Technical and Tectonic Arts, Or, Practical Aesthetics.* Getty Research Institute.

Semper, G. (2011). *Four Elements of Architecture* (1851 ed.). (H. F. Mallgrave & W. Herrmann, Trans.). Cambridge University Press.

Sepasgozar, S. M., Hawken, S., Sargolzaei, S., & Foroozanfa, M. (2019). Implementing citizen centric technology in developing smart cities: A model for predicting the acceptance of urban technologies. *Technological Forecasting and Social Change*, *142*, 105–116. doi:10.1016/j.techfore.2018.09.012

Sepper, D. L. (1988). Goethe contra Newton: Polemics and the Project for a New Science of Colour. Cambridge University Press.

Sevilla-Buitrago, A. (2013). Debating contemporary urban conflicts: A survey of selected scholars. *Cities (London, England)*, *31*, 454–468. doi:10.1016/j.cities.2012.08.006

Shakibamanesh, A., & Ghorbani, M. (2017). Toward time-based design: Creating an applied time evaluation checklist for urban design research. *Frontiers of Architectural Research*, *6*(3), 290–307. doi:10.1016/j.foar.2017.05.004

Shaw, R. (2014). Beyond night-time economy: Affective atmospheres of the urban night. *Geoforum*, *51*(1), 87–95. doi:10.1016/j.geoforum.2013.10.005

Shekhar, S., & Tripathi, A. (2015). Smart Neighbourhood: A way to Sustainable Development. *Annual Conference Proceedings,* 17.

Shepard, M. (2011). Toward The Sentient City. In M. Shepard (Ed.), Sentient City: Ubiquitous Computing, Architecture, and the Future of Urban Space. Architectural League of New York, The MIT Press.

Shiel, M., & Fitzmaurice, T. (2001). *Cinema and the City: Film and Urban Societies in a Global Context.* City of Hoboken: Blackwell Publishers Ltd. doi:10.1002/9780470712948

Shoup, D. (2007). *Gone parkin'.* Retrieved from https://www.nytimes.com/2007/03/29/opinion/29shoup.html

Shoup, D. C. (2006). Cruising for parking. *Transport Policy*, *13*(6), 479–486. doi:10.1016/j.tranpol.2006.05.005

Sidanin, P. (2007). On Lynch's and Post-Lynchians Theories. *Architecture and Civil Engineering*, *5*(1), 61–69.

Silberberg, S., Lorah, K., Disbrow, R., Muessig, A., & Naparstek, A. (2013). How placemaking builds places and communities. Massachusetts Institute of Technology.

Silva, C. N. (2018). *New Approaches, Methods, and Tools in Urban E-Planning.* IGI Global. doi:10.4018/978-1-5225-5999-3

Simpson, J., & Proffitt, M. (1997). *Oxford English Dictionary Additions Series* (Vol. 3). Oxford University Press.

Sinha, U., & Shekhar, M. (2015). Comparison of Various Cloud Simulation tools available in Cloud Computing. *International Journal of Advanced Research in Computer and Communication Engineering*, *4*(3). Advance online publication. doi:10.17148/IJARCCE.2015.4342

Siramkaya, S. (2019). The Effect of Change in Urban Squares on Urban Identity: The Case of Konya. *Architecture Research*, *9*(1), 7–15.

Sitte, C. (1965). *City Planning According to Artistic Principles.* Random House.

Siva Kumar, R. (2012). *An Extraordinary Cohesiveness. Route Maps.* The Guild.

Smart City Mission Statement & Guidelines. (2015, June). Retrieved November 20, 2019, from Smart Cities Mission: shorturl.at/knIMP

Smith, P. (2010). Just Kids. London: Bloomsbury.

Sobejano, E. (2011). Combinatorial Architecture. In C. Gengnagel (Ed.), *Computational Design Modelling* (pp. 55–61). Springer. doi:10.1007/978-3-642-23435-4_7

Soeiro, D. (2011). *A cor como abrigo: a arquitectura como cuidado* [Colour as shelter: architecure as care] (PhD Thesis). Department of Philosophy. Repository of Universidade Nova de Lisboa. http://hdl.handle.net/10362/14113

Soeiro, D. (in press). Smart Cities, Well-being and Good Business: The 2030 Agenda and the role of knowledge in the era of Industry 4.0. In F. Matos, V. Vilarinhos, I. Salavisa, L. Edvinsson, & M. Massaro (Eds.), *Knowledge, People, and Digital Transformation: Approaches for a Sustainable Future* (pp. 41–53). Springer Nature.

Sönmez, M. S. (n.d.). *ISKI monitoring systems smart grid applications* [İSKİ izleme sistemleri akıllı şebeke uygulamaları]. Retrieved from https://www.tarimorman.gov.tr/SYGM/Belgeler/%C4%B0zleme%20H%C4%B0E%20SUnumlar/iski-mehmet%20sait.pdf

Stanek, L. (2012). Architecture as space, again? Notes on the 'spatial turn'. *Spéciale'Z (Paris)*, *4*, 48–53.

Steele, D., Luka, N., & Guastavino, C. (2012). Constructing ideal soundscapes: A practical study on closing the gaps between soundscape studies and urban design. In *Proceedings of the Acoustics 2012 Nantes Conference* (pp. 2163-2168). Nantes, France: Société Française D'acoustique. 10.1121/1.4709108

Steele, F. (1981). *Sense of place*. CBI Publishing Company, Inc.

Stefansdottir, H. (2017). The role of urban atmosphere for non-work activity locations. *Journal of Urban Design*, *23*(3), 319–335. doi:10.1080/13574809.2017.1383150

Steger, M. (2013). *Globalization: A Very Short Introduction*. Oxford University Press. doi:10.1093/actrade/9780199662661.001.0001

Stevenson, A. (2010). *Oxford Dictionary of English*. OUP Oxford.

Stewart, K. (2011). Atmospheric Attunements. *Environment and Planning. D, Society & Space*, *29*(3), 445–453. doi:10.1068/d9109

Stiglitz, J. (2003). *Globalization and its Discontents*. W.W. Norton.

Strydom, W., Puren, K., & Drewes, E. (2018). Exploring theoretical trends in placemaking: Towards new perspectives in spatial planning. *Journal of Place Management and Development*, *11*(2), 165–180. doi:10.1108/JPMD-11-2017-0113

Stummvoll, G. (2009). Environmental criminology and crime analysis. *Crime Prevention and Community Safety*, *11*(2), 144–146. doi:10.1057/cpcs.2008.22

Sturm, S., & Turner, S. (2012). "Built pedagogy": The University of Auckland Business School as crystal palace. *Interstices*, *12*, 23–34.

Suchocka, M., Kimic, K., Maksymiuk, G., & Kolodynska, N. (2017). Outdoor Hotspots as a Tool for Enhancing Healthy Lifestyles of ICT Users Design and Development Principles. In *Enhancing Places Through Technology* (pp. 153–165). Edições Universitárias Lusófonas.

Suchocka, M., Maksymiuk, G., Kimic, K., & Kolodynska, N. (2017). Behaviour, Expectations and Preferences of 'Digital Natives' in Regard to the Design of Urban Public Spaces. In *Enhancing Places Through Technology* (pp. 117–128). Edições Universitárias Lusófonas.

Summa, M., Koch, S. C., Fuchs, T., & Müller, C. (2012). Body memory: An integration. In S. C. Koch, T. Fuchs, M. Summa, & C. Müller (Eds.), *Body Memory, Metaphor and Movement* (pp. 417–444). John Benjamins.

Sundaresan, J. (2017). Urban planning in vernacular governance: Land use planning and violations in Bangalore, India. *Progress in Planning*, *0–1*(October). Advance online publication. doi:10.1016/j.progress.2017.10.001

Tashiro, C. (2004). Passing for the Past: Production Design and the Historical Film. *Cinéaste (New York, N.Y.)*, *29*(2), 40–44.

Tawa, M. (2017). Consilient Discrepancy: Porosity and Atmosphere in Cinema and Architecture. *Architecture_media_politics_society*, *11*(13), 1-17. doi:10.14324/111.444.amps.2017v11i3.001

Taylor, C. (1975). *Hegel*. Cambridge University Press. doi:10.1017/CBO9781139171465

Tegmark, M. (2017). *Life 3.0: Being human in the age of artificial intelligence*. Alfred A. Knopf.

Temelová, J., & Dvořáková, N. (2012). Residential satisfaction of elderly in the city centre: The case of revitalizing neighbourhoods in Prague. *Cities (London, England)*, *29*(5), 310–317. doi:10.1016/j.cities.2011.11.015

Terranova, C. N., & Tromble, M. (Eds.). (2016). *The Routledge companion to biology in art and architecture*. Routledge: Taylor & Francis. doi:10.4324/9781315687896

The United Nations. (2019). The Impact of Rapid Technological Changes. In *United Nations Conference on Trade and Development*. Geneva: United Nations Publications.

The WBDG Historic Preservation Subcommitte. (2019). *Historic Preservation*. Author.

Thibaud, J. (2002). From Situated Perception to Urban Ambiences. In *First International Workshop on Architectural and Urban Ambient Environment* (pp. 1-11). Nantes, France: Centre de recherche m'ethodologique d'architecture; Ecole d'architecture de Nantes.

Thibaud, J.-P. (2002). From Situated Perception to Urban Ambiences. In *Proceedings of the First International Workshop on Architectural and Urban Ambient Environment* (pp. 1-11). Nantes, France: Centre de recherche méthodologique Learn to pronounce d'architecture, Ecole d'architecture de Nantes.

Thibaud, J.-P. (2017). Forward. In The Aesthetics of Atmospheres (pp. 17-23). New York: Routledge.

Thibaud, J.-P. (2011). The Sensory Fabric of Urban Ambiances. *The Senses and Society*, *6*(2), 203–215. doi:10.2752/174589311X12961584845846

Thibaud, J.-P. (2015). The backstage of urban ambiances: When atmospheres pervade. *Emotion, Space and Society*, *1*(8), 39–46. doi:10.1016/j.emospa.2014.07.001

Thrift, N. (2004). Intensities of feeling: Towards a spatial politics of affect. *Geografiska Annaler. Series B, Human Geography*, *86*(1), 57–78. doi:10.1111/j.0435-3684.2004.00154.x

Tiazzoldi, C. (2008). Applied responsive devices. In M. Voyatzaki (Ed.), Emerging possibilities of testing and simulation methods and techniques (pp. 265-272). Thessaloniki: Charis ltd.

Tiazzoldi, C. (2011). *Archilovers*. Retrieved from The Socail Cave: https://www.archilovers.com/projects/107616/the-social-cave.html

Tiazzoldi, C. (2009). Prototyping the city. In E. Lupo & C. Campagnaro (Eds.), *Designing connected places* (pp. 130–150). Compositori Editori.

Tiazzoldi, C. (2016). Combinatorial Architecture Methods for the creation of ambiance in public space. In *Proceedings of 3rd International Congress on Ambiances* (pp. 865 - 872). Volos, Greece: HAL.

Tiazzoldi, C., & Tixier, N. (2008). Formal modulation for acoustic performances of a bridge. *ARCC Journal*, *5*(2), 104–115. doi:10.17831/enq:arcc.v5i2.20

Timperio, A., Veitch, J., & Carver, A. (2015). Safety in numbers: Does perceived safety mediate associations between the neighborhood social environment and physical activity among women living in disadvantaged neighborhoods? *Preventive Medicine*, *74*, 49–54. doi:10.1016/j.ypmed.2015.02.012 PMID:25735603

Tiwari, G. (2002). Urban Transport Priorities. *Cities (London, England)*, *19*(2), 95–103. doi:10.1016/S0264-2751(02)00004-5

Tiwari, P., & Gulati, M. (2013). An analysis of trends in passenger and freight transport energy consumption in India. *Research in Transportation Economics*, *38*(1), 84–90. doi:10.1016/j.retrec.2012.05.003

Toffler, A. (1981). *The Third Wave*. Bantam Books.

Townsend, A. M. (2013). *Smart cities: Big data, civic hackers, and the quest for a new utopia*. WW Norton & Company.

Transport Management Centre. (2017). *Smart Parking Systems.* Retrieved from https://uym.ibb.gov.tr/corporate/news-and-announcements/smart-parking-systems

Transport Management Centre. (2018). *Adaptive traffic management system (ATAK) in Istanbul.* Retrieved from https://uym.ibb.gov.tr/corporate/news-and-announcements/adapt%C4%B1ve-traff%C4%B1c-management-system-atak-%C4%B1n-istanbul

Transport Management Centre. (2019a). *Smart parking system of TMC.* Retrieved from https://uym.ibb.gov.tr/YHarita/Harita_EN.aspx

Transport Management Centre. (2019b). *Traffic Density Map of Istanbul.* Retrieved from https://uym.ibb.gov.tr/YHarita/Harita_EN.aspx

Transport Management Centre. (n.d.a). *Electronic violation detection systems.* Retrieved from https://uym.ibb.gov.tr/services/electronic-violation-detection-systems

Transport Management Centre. (n.d.b). *Traffic information systems.* Retrieved from https://uym.ibb.gov.tr/services/traffic-information-systems

Transport Management Centre. (n.d.c). *Traffic measuring and monitoring.* Retrieved from https://uym.ibb.gov.tr/services/traffic-measuringandmonitoring

Tripathi, S. (2007, April). Bombay's Growth Gets Shanghaied. *Far Eastern Economic Review, 170*(3).

Troiani, I., & Carless, T. (2014). Architectural Design Research through Cinematic Collage. *Architecture & Education Journal*, 255-278.

Trumbull, N. S. (2013). St. Petersburg, Russian Federation. *Cities (London, England)*, *31*, 469–490. doi:10.1016/j.cities.2011.05.003

Tuan, Y.-F. (1979). Space and Place: Humanistic perspectives. In S. Gale & G. Olsson (Eds.), *Philosophy in Geography* (pp. 387–427). Springer. doi:10.1007/978-94-009-9394-5_19

TUBITAK. (2004). *National Science and Technology Policy 2003-2023 Strategy Document* [Ulusal Bilim ve Teknoloji Politikaları 2003-2023 Strateji Belgesi]. Retrieved from https://www.tubitak.gov.tr/tubitak_content_files/vizyon2023/Vizyon2023_Strateji_Belgesi.pdf

Turkish Republic Ministry of Environment and Urbanization. (2019). *Smart cities white paper* [Akıllı şehirler beyaz bülteni]. Author.

Turkish Republic Ministry of Health. (2016). *Centralized doctor appointment system.* Retrieved from https://www.mhrs.gov.tr/Vatandas/hakkimizda.xhtml

Turkish Republic Ministry of Health. (2018). E-Pulse v.2.0 User Manual. Author.

Turley, L. W., & Milliman, R. E. (2000). *Atmospheric Effects on Shopping Behavior: A Review of the Experimental Evidence.* Academic Press.

Turner, M., & Tomer, T. (2013). Community Participation and the Tangible and Intangible Values of Urban Heritage. *Heritage & Society*, *6*(2), 185–198. doi:10.1179/2159032X13Z.00000000013

Uberoi, J. P. S. (1984). *The Other Mind of Europe – Goethe as a Scientist.* Oxford University Press.

UNESCO. (2005). *Vienna Memorandum on "World Heritage and Contemporary Architecture – Managing the Historic Urban Landscape."* Author.

UNESCO. (2011). Proposals concerning the desirability of a standard-setting instrument on historic urban landscapes 36 C/23. *Records of the General Conference, 36th Session, Paris, 25 October - 10 November 2011, v. 1: Resolutions.* UNESCO.

United Nations. (2018). *E-Government Survey 2018: Gearing E-Government to support transformation towards sustainable and resilient societies.* United Nations.

Van Vliet, H. (2018). *The measurement of atmospherics.* Plan B.

Vanden, B. G., Knapp, S., & Doe, J. (2001). *Role of reference elements in the selection of resources by psychology undergraduates.* Retrieved from http://jbr.org/articles.html

Vanolo, A. (2014). Smartmentality: The smart city as disciplinary strategy. *Urban Studies (Edinburgh, Scotland)*, *51*(5), 883–898. doi:10.1177/0042098013494427

Vidler, A. (1993). The explosion of Space: Architecture and the Filmic Imaginary. *Assemblage*, *21*(21), 44–59. doi:10.2307/3171214

von Engelhardt, D. (2009). Historical Consciousness in the German Romantic *Naturforschung.* In A. Cunningham & N. Jardine (Eds.), *Romanticism and the sciences* (pp. 13–24). Cambridge University Press.

von Goethe, J. W. (1790). The Metamorphosis of Plants. Cambridge, MA: MIT Press.

von Goethe, J. W. (1970). Theory of Colours (C.H. Eastlake, Trans.). MIT Press. (Original publication 1810)

von Goethe, J. W. (1989). Goethe's Botanical Writings (B. Mueller, Trans.). Woodbridge, Canada: Ox Bow Press.

von Goethe, J. W. (1988). *Scientific Studies* (D. Miller, Ed. & Trans.). Suhrkamp.

Wallace, C., & Abbott, P. (2009). *Social Quailty in Europe.* Academic Press.

Wang, Y., Chau, C. K., Ng, W. Y., & Leung, T. M. (2016). A review on the effects of physical built environment attributes on enhancing walking and cycling activity levels within residential neighborhoods. *Cities (London, England), 50*, 1–15. doi:10.1016/j.cities.2015.08.004

Warbis, M. (2018). *Making the Spaces of Cities Smarter: Why Cities (and Their Citizens) Should Embrace the Digital Public Realm.* Retrieved from https://medium.com/@warbismichelle/making-the-spaces-of-cities-smarter-why-cities-should-embrace-the-digital-public-realm-c12afc810aaa

Warhol, A. (Producer), Warhol, A., & Morrissey, P. (Directors). (1966). *Chelsea Girls* [Motion Picture]. The United States.

Weidinger, J. (2016). Atmosphere—an Emerging Phenomenon in Landscape Architecture. *Wolkenkuckucksheim*, 147.

Weidinger, J. (Ed.). (2018). *Designing Atmospheres*. Technical University Berlin.

Weigelt, R. (2010, Dec 16). *Samuel Taylor Coleridge, "Christabel"*. Retrieved from wiki.uiowa.edu: https://wiki.uiowa.edu/pages/viewpage.action?pageId=41878932

Weisberg, G. (2001). *Montmartre and the Making of Mass Culture*. Rutgers University Press.

Werth, P. T. (1999). *Text Worlds: Representing Conceptual Space in Discourse*. Longman.

Wessner, G. (2011). Preface. In M. Shepard (Ed.), *Sentient City: Ubiquitous Computing, Architecture, and the Future of Urban Space* (pp. 8–9). Architectural League of New York, The MIT Press.

Western Australian Planning Commission. (2007). *Liveable Neighbourhoods Community Design Code: A Western Australian Government Sustainable Cities Initiative*. Western Australian Planning Commission.

WHC-UNESCO. (1985). Inscription: Historic areas of Istanbul: Turkey. In *9th Session of the Committee*. Retrieved from http://whc.unesco.org/en/decisions/3864

Whitman, C. T., Reid, C., von Klemperer, J., Radoff, J., & Roy, A. (2008, March). New Songdo city–The making of a new green city. *Proceedings of the 8th World Congress Council on Tall Buildings and Urban Habitat (CTBUH'08).*

Whyte, W. (1968). *The Last Landscape*. University of Pennsylvania Press.

Whyte, W. (1980). *The Social Life of Small Urban Spaces*. Conservation Foundation.

Whyte, W. H. (1989). *City: Rediscovering the Center*. University of Pennsylvania Press.

Wigley, M. (1998). Architecture of Atmosphere. *Daidalos, 68*, 18–27.

Wilkins, J. (2018). *The Discovery of a World in the Moone*. CreateSpace Independent Publishing Platform.

Williams, D. E. (1988). Ideology as Dystopia: An Interpretation of "Blade Runner". *International Political Science Review / Revue internationale de science politique, 9*(4), 381-394.

Wilson, E. J. (2004). *The information revolution and developing countries*. MIT Press.

Winner, L. (1997). Technology today: Utopia or dystopia? *Social Research, 66*(3), 998–1017.

Wittgenstein, L. (1979). Bemerkungen über die Farben/Remarks on Colour (G.E.M. Anscombe, Ed.; L.L. McAlister & M. Schättle, Trans.). Hoboken, NJ: Blackwell.

Wong, L. (2016). Toolbox. In Adaptive Reuse: Extending the Lives of Buildings (p. 87). Birkhauser.

World Heritage Center. (2013). *New life for historic cities: The historic urban landscape approach explained.* https://unesdoc.unesco.org/ark:/48223/pf0000220957_chi

Wyckoff, M. A. (2017). *Placemaking as an Economic Development Tool: A Placemaking Guidebook*. Academic Press.

Wyndham-Davies, J. (Producer), & Morahan, C. (Director). (1997). *The Heat of the Day* [Motion Picture]. Retrieved from https://www.amazon.com/Heat-Day-Masterpiece-Theatre-VHS/dp/156442667X

Yamawaki, Y., & Filho, F. M. (2020). Do cultural issues matter on the adoption of technologies on public spaces? The case of Brazil through Black Mirror series. *Cities (London, England)*, 98.

Yeomans, R. (2006). *The Art And Architecture of Islamic Cairo*. Garnet Publishing Limited.

Yigitcanlar, T., Kamruzzaman, M., Foth, M., Sabatini-Marques, J., da Costa, E., & Ioppolo, G. (2019). Can cities become smart without being sustainable? A systematic review of the literature. *Sustainable Cities and Society*, *45*, 348–365. doi:10.1016/j.scs.2018.11.033

Yuen, B. (2004). Safety and dwelling in Singapore. *Cities (London, England)*, *21*(1), 19–28. doi:10.1016/j.cities.2003.10.005

Zalta, E. N. (2006). Dewey's Aesthetics. In *The Stanford Encyclopedia of Philosophy*. Metaphysics Research Lab, Stanford University. https://plato.stanford.edu/entries/dewey-aesthetics/

Zeng, H., & Zhao, Y. (2011). Sensing Movement: Microsensors for Body Motion Measurement. *Sensors (Basel)*, *11*(1), 638–660. doi:10.3390110100638 PMID:22346595

Zhan, D., Kwan, M. P., Zhang, W., Fan, J., Yu, J., & Dang, Y. (2018). Assessment and determinants of satisfaction with urban livability in China. *Cities (London, England)*, *79*(March), 92–101. doi:10.1016/j.cities.2018.02.025

Zhang, Q., Yung, E. H., & Chan, E. H. (2018). Towards Sustainable Neighborhoods: Challenges and Opportunities for Neighborhood Planning in Transitional Urban China. *Sustainability*, *10*(406), 23. doi:10.3390u10020406

Zhang, X., Cha, K. L., & Constable, M. (2014). Atmospheric Perspective Effect Enhancement of Landscape Photographs Through Depth-Aware Contrast Manipulation. *IEEE Transactions on Multimedia*, *16*(3), 653–667. doi:10.1109/TMM.2014.2299511

Zhong, J. (2012). Reconnecting lost place: Comprehensive green space design based on historic preservation principles in compact Changsha City. *China*, 53.

Zou, C., Deng, H., & Qiu, Q. (2013). Design and Implementation of Hybrid Cloud Computing Architecture Based on Cloud Bus. In *9th International Conference on Mobile Ad-hoc and Sensor Networks* (p. 5). IEEE. 10.1109/MSN.2013.72

Zucker, P. (1970). *Town and Square: From the Agora to the Village Green*. MIT Press.

Zumthor, P. (1998). *Thinking architecture*. Lars Müller.

Zumthor, P. (1998). *Thinking Architecture*. Lars Müller.

Zumthor, P. (2006). *Atmospheres: Architectural Environments - Surrounding Objects*. Birkhauser.

Zumthor, P. (2006). *Atmospheres: Architectural environments-Surrounding objects*. Birkhauser.

Zumtor, P. (1998). Questioning Images. Interview with Peter Zumtor. *Daidalos*, *68*, 90–101.

About the Contributors

Hisham Abusaada is a Professor of Architecture and Urban Design at the Housing and Building National Research Centre (HBRC), Egypt. During 1995-2015, he taught urban design course at universities in Saudi Arabia and Egypt. He is the author and co-authors of several scientific manuscripts and books in both Arabic and English published in scientific journals and conference proceedings. His area of interest is in urban design between theory and practice.

Abeer Elshater, PhD, is Professor of Urban Morphology at Ain Shams University (ASU), Cairo, Egypt, teaching and supervising multidisciplinary topics in the ideology of urban design. Her area of interest is in urban design disciplines. She has worked on some international research projects with international universities. Elshater is an author and co-author of seven books as well as a number of scientific manuscripts in international journals and conference proceedings. Recently, she acts as an ambassador of the Regional Studies Association, United Kingdom.

* * *

Nada Majid Alhakkak has received her PhD in Computer Science, Information Technology. Nada has published more than 25 papers. Her area of interest in all topics related to IT. More information about her list of publications could be tracked from her ResearchGate account (https://www.researchgate.net/profile/Nada_Al_Hakkak).

Burak Belli is an urban planner/designer and a research assistant from Istanbul, Turkey. Belli received his Bachelor's Degree in Urban and Regional Planning from Istanbul Technical University in 2016 and later attended the Urban Design Master's Degree Program of ITU where he was graduated in July 2019. Belli has been appointed by ITU in 2018, as a "Research Assistant in priority areas" which is smart cities and sustainable development. In his master's thesis titled "Dimensional Evaluation of Retrofitting Smart Cities: The Historical Peninsula Case", he brought upon a critical point of view to the smart city concept, while assessing the Historical Peninsula's progress in the deployment of smart applications. His work experiences include a sustainable urban redevelopment project and Turkey's first national spatial strategic plan. Belli is currently a Ph.D. student in Istanbul Technical University Urban and Regional Planning Ph.D. Program.

Vrushali Dhage, PhD in Art History and Aesthetics, with a specialization in Indian Contemporary Art with respect to economic globalization. She has taught at the Department of Art History and Aesthetics, M. S. University of Baroda, and as a visiting faculty at architecture, design and Fine Arts colleges and institutions in India. Her work has been published in various Indian and international art magazines, portals, and in research journals. Further, she has played a role of a curator for select exhibitions; and has also been a participant and facilitator for art residencies. Currently she lives and works in San Jose.

Saurav Dixit is a creative person with a passion for research and technology up-gradation and works dedicatedly to achieve his goals, led to a good number of publications in peer-reviewed Scopus indexed journals. He is having an excellent publication record (15 publications and 168 number of citations on the Scopus database, and having a ResearchGate (total research interest of 200, with more than 16,500 publication reads). Saurav is contributing to the knowledge of research by serving as an active reviewer for several scientific journals that include: Journal of Cleaner Production (Elsevier), International Journal of Construction Management (Taylor & Francis), Smart and sustainable built environment (Emerald), and Ain Shams Engineering Journal (Elsevier). He received the award for Outstanding Contribution in Reviewing in 2018 from Ain Shams Engineering Journal (Elsevier). He also served as the member of Editorial Advisory Board for IGI Global, Engineering Science Reference. His area of expertise is construction project management, Building Economics, Research Methodology, Construction Productivity, Construction management, Construction delays, Built environment studies, Sustainability studies, Lean construction, Sustainable construction, Economic development, and Project management studies.

Aya Elgobashi is an Egyptian architect, who holds a B.Sc. of Architectural Engineering and Environmental Design from October University for Modern Sciences and Arts is located in Cairo, Egypt and she received a British B.Sc. degree in architecture from University of Greenwich. Currently she is a teaching assistant with a demonstrated history of working in the education management industry and M.Sc. candidate at the Department of Architecture Engineering and Environmental Design in Arab Academy for Science, Technology & Maritime Transport in smart village. Her research concerned on public spaces attached to heritage site in the Carine context and its impact on the social quality of life. Moreover, she published paper in AR-UP 2019 third international conference of architecture and urban planning, Faculty of Engineering, Ain Shams University and in Improving Sustainability Concept in Developing Countries (ISCDC) – 4th Edition Conference by IEREK press.

Yasmeen Elsemary, architect, researcher and assistant professor, currently works at the Department of Architectural Engineering, Arab Academy for Science, Technology and Maritime Transport, smart village campus, Egypt. She graduated from Cairo University, Faculty of Engineering. Her PhD is in Architecture and Urban Design awarded by Cairo University in 2014, under the title of "Thinking Processes for Contextual Design". Her MSc in Architecture was awarded in 2009 under the title of "Rhetoric of Architecture: Metaphor and Icon on the conceptual and practical levels". Her published papers appear in journals such as International Journal of Science and Engineering Applications (IJSEA).

Rafee Hakky holds a Bachelor of Architectural Engineering, Damascus University, 1981, Master of Landscape Architecture, Ball State University, Muncie, IN., USA, 1985, PhD in Environmental Design and Planning, Virginia Tech, Blacksburg, VA, USA. 1992. He has taught in Ball State University, Virginia Tech, King Faisal University (Currently Dammam University), Bahrain University, International

University for Science and Technology, Damascus, Syria. Currently, he is a professor of architecture at Al Rasheed University, Damascus, Syria. His research interests are in the areas of architectural education, history and theory of architecture and urban design, and landscape architecture.

Turgay Kerem Koramaz received PhD degree in Urban and Regional Planning with the dissertation about the perception of urban historic sites through computer-aided three-dimensional visualization techniques. He was a visiting scholar at the Royal Institute of Technology, Graduate School of Built Environment and Architecture – Stockholm (Sweden) in 2008. He has been working in ITU, the Department of Urban and Regional Planning since 2000. He has publications about spatial analysis, internal migration in Turkey, urban conservation and renewal. He is teaching site conservation, information and communication technologies, computer systems in urban studies and research methods for urban design. He is involved in academic and professional activities about spatial analysis in urban planning, urban conservation, and regeneration.

Diana Soeiro (Ph.D. Philosophy, Universidade Nova de Lisboa/UNL, 2011). Her research interests are Philosophical Anthropology, Architecture and Urbanism, and Political Science. She conducted her Postdoctoral studies in Philosophy and Urban Studies (UNL, 2012-2018) funded by Fundação para a Ciência e a Tecnologia (FCT). Recently she was Postdoctoral researcher at the project ROCK – Regeneration and Optimization of Cultural heritage in creative and Knowledge cities (2017-2020), funded by the European Union under Horizon 2020. She concluded a Post Grad in Economics and Public Policy at ISCTE - Instituto Universitário de Lisboa (ISCTE-IUL, 2018). In 2017 she was appointed Ambassador for United Nations' 2030 Sustainable Development Goals Agenda (Portugal).

Caterina Tiazzoldi, PhD, MSAAD, Columbia University, B.Arch. Politecnico di Torino, is an architect and educator. She focuses her research on the idea of performative modulation in multiscale projects. Caterina Tiazzoldi won of the Architecture Prize Architetture Rivelate, was a finalist at the Renzo Piano Award for a Young Italian Talent with the Project Toolbox and was invited as the keynote speaker in the symposium MoMoWo - Women's Creativity since the Modern Movement. Tiazzoldi works with several institutions including the UNESCO, Interior Department Piedmont Region Italy, National Park Service USA and several foundations. Her work as an architect has been published in several national and internationals national newspapers, scientific magazine and books. She teaches at Rhode Island School of Design, Politecnico di Torino, and she is the former founder and director of the research lab of Non-Linear Solutions Unit at Columbia University.

Index

4IR 23-25, 32, 35-37

A

ABC 114, 156
Affective Atmospheres 2, 11, 13, 15-18, 41-47, 50-54, 221-230, 232-239
Al-Hara 226, 229-230, 235, 237
Ambiance 2, 12, 126-130, 132-133, 135, 145-146, 151-155, 164-167, 170-171, 176, 199, 201, 205, 208, 210, 216, 222
Architecture 1-2, 5-7, 9, 14-18, 24, 30, 32-37, 41-45, 47-54, 62, 84-85, 119, 128, 130, 139, 151, 166, 180, 199-200, 202, 205-206, 208, 214, 216, 221-230, 232-233, 236-239
artworks 2-4, 7, 17, 175-176, 179, 183, 188, 196, 221-222, 236-238
Atmospheres 1-18, 23, 29-33, 35, 37, 41-47, 50-54, 62, 69, 126, 165, 200, 203-205, 208, 214, 221-230, 232-239
Atmospheric Architecture 15
Attached space 157

B

Base Milano 201, 211-212
Benjamin H. Bratton 24
Built Environment 16-17, 23-24, 46, 51, 53, 126, 128-130, 133-135, 137, 140, 146-149, 213, 222, 225, 229, 233

C

Citizen Engagement 107-108
City Design 199
Cluster Analysis 126, 133, 137-138
Clustering 114-115, 118, 120-121, 203
Colour 8, 24, 27-29, 32, 35, 37, 224-225, 230
Combinatory Urbanism 204-205
Coworking 201, 208-211, 216

E

Economic Globalization 176
Egyptian alley 226, 231, 237

F

Feeling 1-3, 9, 11-12, 14-17, 24, 29-30, 32, 37, 46-47, 49-50, 61, 64, 156, 163, 214, 225
Flâneur 175-176, 179, 196

G

Gastronomy of the Eye 175
Genius Loci 13, 16-17, 42, 46-47, 51
Gentrification 86, 182-183, 190
Gernot Böhme 4, 14, 24, 29

H

Hermann Schmitz 4, 14, 24, 29-30
Historic preservation 146-148, 158, 165
Historic urban landscape 149-151
Historical setting 151

I

Information and Communication Technology (ICT) 65, 118, 176
Internet of Things (IoT) 36, 60, 65, 69

J

Jan Gehl 60
Johann W. von Goethe 24-25

L

Landscape Architecture 1-2, 222, 224-225, 227, 236
Ludwig Wittgenstein 24, 28

M

mise en scène 224, 226-227, 232
Mood 3, 7-9, 14, 31, 36, 42, 48-50, 61-62, 76, 222, 224, 227-228

N

Naguib Mahfouz 221, 223, 226

O

Open Spaces 41, 60-61, 63-69, 73, 77, 114, 146, 148-149, 154, 166

P

Perceptual Dimension 2, 8, 14, 16, 41-43, 46, 49-50, 53
Perceptual Qualities 1, 18, 43-44, 47, 52, 54, 222, 236
Phenomenology 3, 8-9, 11, 14, 24, 28-29, 31-32, 37, 43, 62
Placemaking 145-146, 151-152, 157-161, 163-164, 169, 171, 221-222
Productivity 93, 126, 132-135, 137, 150

S

Seating 63-64, 163
Sense of Place 2-5, 13, 16-17, 42-43, 46-47, 51, 127, 146-147, 163
Servicescape 155
Smart Applications 86, 106-107
Smart Cities 23-24, 32, 35, 37, 67, 86, 103, 107-108, 116, 121, 128, 175, 177, 191, 196, 200
Smart Technologies 82-84, 95, 103, 106-108, 178, 199, 201-202, 208, 216
Sociability 60-61, 63-64, 73, 161, 163, 213-214
Spirit of Place 13, 17, 41-44, 51-52, 54
Spiritual Dimension 42-43, 50, 52, 54
State of Mind 62, 68, 75

T

Technocapitalism 177, 180
Technology 23-25, 29, 32-33, 35, 37, 60, 65-69, 76-77, 83-85, 88, 96, 100, 103-104, 106-108, 115-116, 118, 126, 139-140, 176-178, 187, 199-204, 208, 211, 213-214, 224, 227

U

Urban Ambiance 2, 12, 126, 128, 132-133, 135, 176, 199, 201, 205, 216, 222
Urban Atmospheres 2, 9, 12-13, 15-18, 50, 200, 203, 222, 237-238
Urban Environments 1, 17, 23, 25, 29-30, 32-33, 35-37, 41-44, 46-47, 50, 53-54, 82, 84, 127, 130, 222
Urban Governance 95
Urban Planning 1, 15, 17-18, 42-44, 47, 50, 53-54, 82-84, 86, 103, 106, 108, 128, 199, 216, 222, 238
Urban Studies 1-2, 41-44, 47, 52, 128, 200
Usability 68, 73, 77

V

visitor experience 152-153, 167

W

William Whyte 60, 65, 158

www.ingramcontent.com/pod-product-compliance
Lightning Source LLC
LaVergne TN
LVHW062152300926
849161LV00010B/2652

* 9 7 8 1 7 9 9 8 3 8 5 6 2 *